Eel Physiology

Eel Physiology

Editors

Francesca Trischitta
Department of Biological and Environmental Sciences
University of Messina
Italy

Yoshio Takei
Atmosphere and Ocean Research Institute
The University of Tokyo
Japan

Philippe Sébert
Laboratoire ORPHY
Université de Brest
France

CRC Press
Taylor & Francis Group
Boca Raton London New York

CRC Press is an imprint of the
Taylor & Francis Group, an **informa** business

A SCIENCE PUBLISHERS BOOK

CRC Press
Taylor & Francis Group
6000 Broken Sound Parkway NW, Suite 300
Boca Raton, FL 33487-2742

First issued in paperback 2019

Cover Illustrations: Reproduced by kind courtesy of Philippe Sébert

ISBN-13: 978-1-4665-9827-0 (hbk)
ISBN-13: 978-0-367-37963-6 (pbk)

Library of Congress Cataloging-in-Publication Data

Eel physiology / editors, Francesca Trischitta, Yoshio Takei.
       pages cm
   "A CRC title, part of the Taylor & Francis imprint, a member of
the Taylor & Francis Group, the academic division of T&F Informa
plc."
   Includes bibliographical references and index.
   ISBN 978-1-4665-9827-0 (hardcover : acid-free paper) 1. Eels-
-Physiology. I. Trischitta, Francesca. II. Takei, Y. (Yoshio),
1951-
   QL637.9.A5E456 2013
   597'.43--dc23
                                                      2013010319

Visit the Taylor & Francis Web site at
http://www.taylorandfrancis.com

CRC Press Web site at
http://www.crcpress.com

Science Publishers Web site at
http://www.scipub.net

Preface

Eels are fascinating creatures with an incredible life cycle; preleptocephalus, leptocephalus, glass eel, elver, yellow eel and silver eel are distinctly different stages of life separated by dramatic changes in morphology and probably also in physiology. Eels are catadromous semelparous species that are born, spawn and die in the ocean but spend most of their life in inland fresh waters, thus facing two long journeys in opposite directions during their life. Such a mysterious life history of eels has aroused our curiosity for more than a century, and although efforts in Europe and Japan have revealed the ecology of their migration to some extent, little is yet known about their physiology during this migratory period. Indeed, since the pioneering works of Homer H. Schmidt in the 1930's, eels have attracted the interest of many physiologists, especially those studying various aspect of adaptive physiology.

In recent years, dramatic declines of eel stocks have been reported for European eels (*Anguilla anguilla*), Japanese eels (*A. japonica*), and American eels (*A. rostrata*), and all of these are categorized as endangered species. To sustain these biologically and economically important resources, the governments in Europe, Japan and United States encourage researchers all over the world to elucidate the ecology and physiology of eels so as to serve as a basis for species conservation. Therefore, now is the most appropriate time to publish a book summarizing the current knowledge of eel physiology.

This book provides coverage of many areas of eel physiology and it is the result of the enthusiastic contribution of specialists in various physiological fields. Each chapter, besides providing a close examination of the related topics, provides perspectives and future directions of research. The book contains numerous figures and tables and an extensive up-to-date bibliography. It will be of interest not only to researchers, teachers and students but also to professionals engaged in fishery works including aquaculture who face in practical terms collapsing populations and hence the necessity to protect and conserve these species.

At last, but not least, as editors we would like to express our sincere thanks to all authors of this book that should serve as a landmark for fish physiology. We recognize their expert contributions. We also thank our colleagues who carefully read the manuscripts and provided helpful and constructive comments.

February, 2013,

Francesca Trischitta
Yoshio Takei
Philippe Sébert

Contents

Advances in Eel Reproductive Physiology and Endocrinology

Karine Rousseau,[1,a] Anne-Gaëlle Lafont,[1] Jérémy Pasquier,[1] Gersende Maugars,[1] Cécile Jolly,[1] Marie-Emilie Sébert,[1] Salima Aroua,[1] Catherine Pasqualini[2] and Sylvie Dufour[1,*]

Introduction: Eel Life and Reproductive Cycle

Eels have a complex migratory life cycle (Fig. 1) with the occurrence of two metamorphoses (for reviews: Sinha and Jones 1975, Tesch 1977, Haro 2003, Dufour and Rousseau 2007, Rousseau and Dufour 2008, Rousseau et al. 2009, 2012). They present a typical larval (first) metamorphosis, leptocephali larva being transformed into glass eels. After this drastic transformation, the growth phase starts in the continental waters and glass eels become elvers then "yellow" eels. After many years in freshwater, the yellow eels transform into "silver" eels which stop growing, cease feeding and start their downstream migration towards the ocean and the area of reproduction.

[1]Research Unit BOREA "Biology of Aquatic Organisms and Ecosystems" Muséum National d'Histoire Naturelle, CNRS 7208, IRD 207, UPMC, 7 rue Cuvier, CP32, 75231 Paris Cedex 05, France.
[a]E-mail: karine.rousseau@mnhn.fr
[2]CNRS, UPR 3294, Neurobiologie et Développement, Av. de la Terrasse, Gif-sur-Yvette, 91198 Cedex.
E-mail: sylvie.dufour@mnhn.fr
*Corresponding author

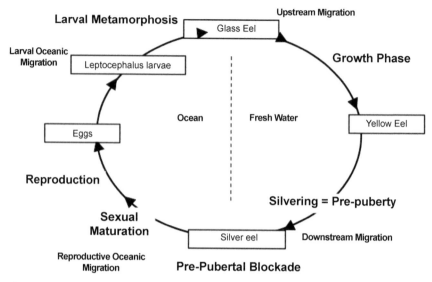

Figure 1. Eel biological life cycle.

To allow this transition from sedentary life in freshwater to migrant life in seawater, eels undergo their second metamorphosis, known as silvering. Silvering not only preadapts the eel to deep-sea conditions (Sébert 2003), but also prepares the sexual maturation, which will only be completed during their reproductive migration towards the tropics: the Sargasso sea for European eel, *Anguilla anguilla* (Tesch 1982, 1989, Dufour and Fontaine 1985, Fontaine 1985, Dufour 1994), possibly in the South Fiji basin for New Zealand longfin eel, *Anguilla dieffenbachii* (Jellyman and Tsukamoto 2010) and towards the waters around the southern part of the West Mariana Ridge for Japanese eel, *Anguilla japonica* (Tsukamoto et al. 2003, 2011, Tsukamoto 2006, Chow et al. 2009). The eel, which have ceased feeding at the silver stage, will not resume feeding during the reproductive migration as shown by Chow et al. (2010) in Japanese eel using stable isotope analysis.

Eels remain blocked at the silver prepubertal stage, as long as the reproductive migration is prevented (Dufour et al. 2003). Until recently, silver prepubertal stage was the last stage known in natural conditions since eels had never been caught during their reproductive migration nor observed at their spawning site. In 2008, Tsukamoto and collaborators found mature eels in the open ocean at the southern part of the West Mariana Ridge (Chow et al. 2009). The three male eels, one giant mottled eel (*Anguilla marmorata*) and two Japanese eels presented dark brown or blackish gray body color, larger eye index than silver eels and bigger gonadosomatic index (GSI) compared to silver eels or even compared to fully matured male eels induced by hormonal injection (Chow et al. 2009). In 2011, collections of eggs, larvae and spawning-condition adults of Japanese and giant mottled eels in their shared spawning area in the Pacific allowed to get further information on reproductive characteristics of matured eels and spawning conditions (Tsukamoto et al. 2011). All eels were silvery black in colour with greatly enlarged eyes. Tsukamoto et al. (2011) could also observe the highly modified bodies and degenerated condition of spawning adults, which confirm that they have only one spawning season in their life and die afterwards.

Eel fisheries and farming activities in Europe and Asia totally rely on wild stocks, which dramatically declined in recent decades (Dekker et al. 2003, Stone 2003). The European eel population has collapsed to the point where it is now listed as endangered species. Recent advances in experimental reproduction in the Japanese eel led to complete the reproductive life and even obtain second generation in captivity, but success rate remains very low. For other eel species, eggs and early larval stages have been also artificially produced, but further development have not yet been obtained (see "Experimental sexual maturation" section). Raising new knowledge on the neuroendocrine control of eel reproduction is still required to further decipher the mechanisms of blockade or activation of eel sexual maturation, and to develop innovative and more physiological protocoles for inducing eel reproduction.

Most of the studies described in this chapter were performed on females. Several reasons could be drawn for this discrepancy. Prepubertal blockade in males is less drastic than in females and experimental maturation is easier to obtain (Olivereau et al. 1986, Palstra et al. 2008b). Induction of sexual maturation and of production of sperm are thus easily obtained in males, after a few weekly injections of ready-to-use commercial human chorionic gonadotropin (hCG). In contrast, sexual maturation in female eels requires several month-treatments with fish pituitary extracts. At the end of the pituitary extract treatment, administration of progestagen is also required for inducing final oocyte maturation and ovulation (see "Experimental sexual maturation" section). In addition to this long and complex treatment, many difficulties are encountered in the females, such as the lack of standardized pituitary extracts, eel mortality during the treatment, large individual variations in the response to the hormonal treatment, with no, slow and fast responders, and general poor quality eggs. Despite most research efforts are therefore put on the females, more data on male experimental maturation and sperm production are still necessary, and some current studies concern sperm quality and storage (Peñaranda et al. 2010a, b).

The Brain-Pituitary Gonadotropic Axis in the Eel

Gonadotropin-Releasing Hormone (GnRH)

Up to now, 25 forms of GnRH variants have been characterized, 14 in vertebrates (12 in gnathostomes and 2 in lampreys) and 11 in invertebrates (for review: Kah et al. 2007, Tostivint 2011). In the European eel, two GnRH molecular forms were demonstrated by High Performance Liquid Chromatography (HPLC) and specific radioimmunoassay (RIA); they are similar to mammalian GnRH (mGnRH) and to chicken GnRH-II (cGnRH-II), respectively (King et al. 1990). The occurrence of two GnRH forms in the eel brain was confirmed by the isolation of their cDNAs and genes in the Japanese eel (Okubo et al. 1999a, b). In addition, this study demonstrated the occurrence of three splicing variants of the messenger RNA coding for mGnRH, revealing further diversity of mGnRH potential roles and regulation (Okubo et al. 1999a, b, Okubo et al. 2002). Using immunocytochemistry, mGnRH and cGnRH-II were localized in different neurons (European eel: Montero et al. 1994, Japanese eel: Chiba et al. 1999). In the European eel, Montero et al. (1994) showed that the mGnRH immunoreactive neurones were located in the anterior brain, along a continuum from the olfactory bulbs to the telencephalic and preoptic area, whereas cGnRH-II immunoreactive neurones were limited to the midbrain tegmentum. This study also revealed that eel pituitary is mostly innervated by mGnRH immunoreactive axonal endings, which represent the major hypophysiotropic GnRH form, while only a few cGnRH-II axons could be detected (Montero et al. 1994). Using qPCR assays, Sébert et al. (2008a) confirmed the differential brain localisation of the two GnRH.

GnRH stimulates the expression, synthesis and release of pituitary gonadotropins, luteinizing hormone (LH), and follicle-stimulating hormone (FSH), in all vertebrate investigated (Fig. 2). GnRHs exert their actions through interactions with specific receptors (GnRH-Rs) that belong to the rhodopsin-like G protein-coupled receptor (GPCR). In contrast to mammals, which have one or two GnRH-Rs, up to five GnRH-R genes have been identified in teleosts (for review: Kah et al. 2007). In the Japanese eel, two receptors have been cloned for now (Okubo et al. 2000, Okubo et al. 2001).

Kisspeptin

Recent studies have shown that a peptide called kisspeptin (or metastin), product of the kiss-1 gene, and its receptor (Kissr, previoulsy named GPR54) play a major role in the onset of puberty in mammals, by activating GnRH neurons (for reviews: Kauffman et al. 2007, Roa et al. 2008). When Kissr

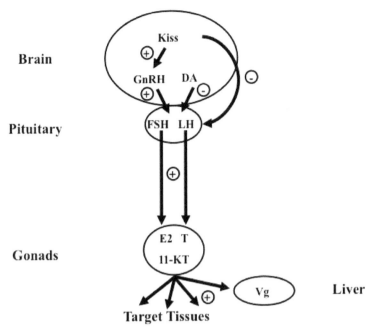

Figure 2. Brain-Pituitary gonadotropic axis in the eel. Kiss=kisspeptin; DA=dopamine; GnRH=gonadotropin-releasing hormone; LH=luteinizing hormone; FSH=follicle-stimulating hormone; E2=estradiol; T=testosterone; 11KT=11-ketotestosterone; Vg=vitellogenin.

is absent or muted, mice or humans are not able to undergo puberty, due to a blockade of GnRH release (Messager et al. 2005). The hypothalamic expressions of kisspeptin and its receptor are activated around the onset of puberty and modulated by sex steroids. Injections of kisspeptin in mammals are able to induce GnRH release, as well as FSH and LH release. Kisspeptin system may also be able to directly stimulate gonadotropin release. A few studies have characterized kisspeptins and their receptors in teleosts and suggested that kisspeptin may be involved in the onset of puberty as it does in mammals (for review: Tena-Sempere et al. 2012). Besides kisspeptin system, another RF-amide peptide, GnIH (gonadotropin inhibitory hormone), may also be important in the inhibitory control of reproduction in vertebrates, specially in birds (for review: Tsutsui et al. 2010).

In the European eel, we recently cloned the full-length coding sequence of a kisspeptin receptor (Kissr) in the eel (Pasquier et al. 2011). Comparison of Kissr sequences assigned the eel Kissr to a basal position in a clade including most of the known teleost Kissr, in agreement with the eel phylogenetical position, as a representative species of the Elopomorphs, one of the most ancient teleost groups. Eel Kissr tissue distribution was analyzed by quantitative real-time PCR. Eel Kissr was highly expressed in the brain, especially in the telencephalon and di-/mes-encephalon, while a very low or undetectable expression was observed in various peripheral organs. A high expression of Kissr was also found in the pituitary indicating a possible direct pituitary role of kisspeptin. Primary cultures of eel pituitary cells were performed to investigate the direct effects of kisspeptin on pituitary hormone expression (Pasquier et al. 2011). Human/lamprey kisspeptin exerted a time- and dose-dependent inhibitory effect on LHβ expression. All other tested kisspeptins had a similar inhibitory effect on LHβ expression.

Thanks to the recent sequencing of the European eel genome (Henkel et al. 2012a), we further characterized eel kisspeptin system (Pasquier et al. 2012 and in press). We reported the cloning of two other Kissr (Pasquier et al. 2012), giving a total of three Kissr in the eel, which is the maximal number of Kissr ever found in a teleost. As measured by quantitative RT-PCR, we showed that the three eel Kissr were mainly, but differentially, expressed in the brain-pituitary-gonadal axis (Pasquier et al. 2012). In the pituitary, Kissr-1 and Kissr-2 expressions are observed, while Kissr-3 expression is undetectable. Finally, they are, all three, expressed in the gonads, especially Kissr-1. We also observed a differential regulation of these receptors after experimental maturation: in the brain, Kissr-1 expression was up-regulated in experimentally matured eels, as compared to prepubertal controls; in the pituitary both Kissr-1 and Kissr-2 expressions were down-regulated; in the ovary, Kissr-3 expression was down-regulated. Subfunctionalisation, as shown by these differences in tissue distribution and regulation, may have represented significant evolutionary constraints for the conservation of multiple Kissr paralogs in this species. Concerning kisspeptin (Fig. 2), we predicted the sequence of two Kiss genes in the eel genome, as in various other teleosts. We are currently investigating their tissue distribution and their regulation after experimental maturation. Our preliminary data indicate an increase in Kiss2 brain expression in experimentally matured female eels (Pasquier 2012 and unpublished data). Thus, the parallel increases in mGnRH (Dufour et al. 1991), Kiss2 and Kissr1 (Pasquier et al. 2012) in matured eels, suggest that the Kiss2/Kissr1 brain system may be involved in the stimulatory control of GnRH neurons and reproduction in the eel (Pasquier 2012).

Dopamine (DA)

Our previous works demonstrated that dopamine (DA) played a key role in the eel prepubertal blockade by exerting a strong inhibitory action on LH synthesis and secretion (Vidal et al. 2004) (Fig. 2). Since 1990s, DA has been investigated in the eel brain using different methods such as HPLC, immunohistochemistry, as well as *in situ* hybridization and qPCR for tyrosine hydroxylase (TH) (for review: Sébert et al. 2008b). TH was found to be mostly expressed in the olfactory bulbs, whereas lower mRNA levels were found in the telencephalon/POA and in the diencephalon/mesencephalon (Boularand et al. 1998, Weltzien et al. 2005, 2006, Sébert et al. 2008a). In particular, TH-immunoreactive cell bodies were observed in the preoptic anteroventral nucleus (NPOav), with a dense tract of immunoreactive fibers reaching the pituitary proximal pars distalis, where the gonadotroph cells are located (Vidal et al. 2004). These NPOav hypophysiotropic neurons are responsible for the dopaminergic inhibitory control of pituitary gonadotropic function in the silver eel. Some of our recent studies showed an androgen-dependent stimulation of brain dopaminergic systems (Weltzien et al. 2006).

Concerning DA receptors, we previously characterized DA D1 receptors and demonstrated that more subtypes (A, B and C) exist in eel than in mammals (Kapsimali et al. 2000). None of them are expressed in the proximal pars distalis of the eel pituitary (Kapsimali et al. 2000). As antagonists of mammalian D2 receptors could release the inhibition on puberty onset (Dufour et al. 1988, Vidal et al. 2004), we then focused our research on DA D2 receptors. Dopamine D2 receptors are membrane receptors belonging to the seven transmembrane domain G protein-coupled family. We isolated and cloned two different cDNA sequences coding for DA D2 type receptors in the eel. They potentially encode two distinct receptor proteins exhibiting only 79% identity and exhibit closer phylogenetical relationships to vertebrate D2 receptors compared to D3 or D4 DA receptors. They were thus designated as D2A- and D2B-R (Pasqualini et al. 2009). Thus, in contrast to mammals and birds in which only one gene coding for D2-R is present, at least two are found in the European eel, as in a few other teleosts. They likely result from the major gene duplication event that occurred specifically in the teleost lineage. From the deduced amino acid sequences, it appeared that the predicted transmembrane domains TM1 to TM7 are identical in the two D2-Rs, suggesting that the two receptors may have similar DA binding capacity. By contrast, the large sequence divergence within the third intracellular loop of the two receptors, strongly suggests a difference in their coupling to G-proteins and intracellular signalling pathways.

We recently developed real-time RT-PCR and *in situ* hybridization for D2-R, in order to study the specific brain distribution of the two receptors and to determine which one is expressed in gonadotropic cells.

We expressed eel recombinant D2A- and D2B-R in heterologous cells (HEK 293 cell lines) to study their individual pharmacological profiles and pinpoint potential differences (Jolly et al. unpublished data). Both recombinant receptors were functional, as measured by activation of G protein and induction of inositol phosphate accumulation. Only specific agonists of mammalian D2 receptors were able to activate both eel receptors. Similarly, only antagonists of mammalian D2 receptors could block both eel receptors. We showed that while the affinity of D2A and D2B was identical for their natural ligand, DA, they exhibited different affinities for synthetic agonists. Similarly, D2A and D2B displayed a differential affinity for various antagonists of the mammalian D2 receptor.

As measured by specific qPCR, we could detect both DA D2A- and DA D2B-R expressions in the European eel whole pituitary, as well as in primary cultures of eel pituitary cells (Jolly et al. unpublished data). In order to further investigate the precise location of these receptors in the pituitary, we also performed *in situ* hybridization using highly sensitive RNA probes developed for each receptor, combined with immunocytochemistry using rabbit antibodies against LH beta or FSH beta (Jolly in preparation). Data revealed that DA D2B-R was expressed by LH and by FSH cells, as well as by some other pituitary cells. In contrast, DA D2A-R was not expressed by LH nor by FSH cells, but only by some other pituitary cells. This study revealed a specific expression of DA D2B receptors by eel gonadotropic cells, indicating that the DA inhibitory control of reproduction in the eel is mediated by the pituitary DA D2B receptor (Jolly et al. unpublished data).

Eel pituitary cells were incubated with DA and DA antagonists to test the ability of DA antagonists to counteract DA inhibitory effect on LH expression. The comparison between the results in HEK cells expressing eel recombinant DA receptors, and in eel pituitary cells, showed some similar but also some differential effects of DA D2 receptor antagonists. This highlighted the importance of testing antagonist effects on native DA receptors in eel pituitary cells, and not only on recombinant eel DA receptors in heterologous cell systems. These data will allow selecting the most adequate antagonist to be used for the development of innovative protocols able to induce eel sexual maturation.

Luteinizing Hormone and Follicle-Stimulating Hormone

Studies in the eel have focused on the two pituitary hormones involved in the control of reproduction, the gonadotropins (luteinizing hormone, LH and follicle-stimulating hormone, FSH) (Fig. 2). In *A. japonica* (Han et al. 2003a), as well as in *A. anguilla* (Aroua et al. 2005), variation in mRNA levels of the alpha and the beta subunits of the gonadotropins were observed throughout silvering. In *A. japonica*, authors observed a concomitant increase in mRNA of the different subunits, LHβ, FSHβ and the glycoprotein alpha (GPα) (Han et al. 2003a). In *A. anguilla*, LH and FSH were shown to be differentially expressed during the silvering process, with an early increase in FSHβ expression and a late increase of LHβ expression (Aroua et al. 2005). These data suggest that FSH could play an early role in the activation of gonads, while LH may have an important role later in the silvering process. Indeed, a concomitance exists between the increase in FSH expression and the start of lipid incorporation in oocytes (also called "endogenous vitellogenesis"), which suggests that FSH could be responsible for the initiation of lipidic vitellogenesis. The early increase in FSH may also be responsible for the first increase in steroid (estradiol) production. In contrast, the later increase in vitellogenin (Vg) plasma levels, concomitant with the later increase in LH expression, suggests that LH may participate in the induction of Vg production and initiation of the "exogenous vitellogenesis". Similarly, LH may also participate in the second increase in sex steroid levels, in silver eels (Aroua et al. 2005).

Gonadotropin receptors (LH-R and FSH-R) were cloned in Japanese eel (partial: Ohta et al. 2007, Jeng et al. 2007, full: Kazeto et al. 2012). The ovary was the only major source for LH-R expression, while FSH-R was higly expressed in both ovary and forebrain (Kazeto et al. 2012). LH-R and FSH-R were differentially expressed during artificially induced ovarian development (Kazeto et al. 2012). The ovarian transcript level of FSH-R rose significantly at the early vitellogenic stage and remained highly expressed during the vitellogenic stage, while ovarian LH-R expression was acutely induced at the late vitellogenic stage. In contrast, FSH-R expression in the forebrain was not regulated in experimentally matured eels. Studies are currently being performed in the European eel in our laboratory. Recently, the team of G. Van den Thillart successfully developed functional bioassays using cell lines that stably express eel recombinant LH or FSH receptors and carry a firefly luciferase reporter gene driven by a cAMP-responsive element (Minegishi et al. 2012). These assays allow to quantify LH or FSH levels in pituitary extracts and blood samples (Minegishi et al. 2012).

Steroid Feedback and Aromatase

Low levels of androgens, testosterone and teleost specific androgen, 11-keto-testosterone (11-KT) are detected in the plasma of male silver eels (European eel: Khan et al. 1987, Japanese eel: Miura et al. 1991). Androgen production by eel testis is largely stimulated during experimental maturation induced by human chorionic gonadotropin (hCG) (European eel: Khan et al. 1987, Japanese eel: Ohta and Tanaka 1997, Miura et al. 1991).

In the female eel, plasma levels of T, 11-KT and estradiol (E2), significantly increase between the pre-vitellogenic (yellow) and early vitellogenic (silver) stages as shown in *A. anguilla* (Sbaihi et al. 2001, Aroua et al. 2005) as well as in other eel species (*A. australis* and *A. dieffenbachii*: Lokman et al. 1998, *A. rostrata*: Cottrill et al. 2001, *A. japonica*: Han et al. 2003b). Throughout the silvering process, we observed a first increase in E2 levels at the early stages, then a further increase of E2, accompanied by an increase in 11-KT and T at the final steps of silvering (Aroua et al. 2005). Further increases in androgens and in E2 levels are observed in female eels during experimental maturation induced by gonadotropic (fish pituitary extract) treatments (Leloup-Hatey et al. 1988, Peyon et al. 1997). The similarity in plasma levels of androgens and estradiol is a remarkable feature in the female eel, likely related to various androgen-specific regulated processes.

Our early investigations showed that castration abolished the mGnRH and LH increases in experimentally matured eels, indicating a feedback by gonadal hormones (Dufour et al. 1991). Experiments using exogenous sex steroids confirmed these results and demonstrated a differential regulation of gonadotropins and GnRHs by estrogens and androgens (Dufour et al. 1983a, b, Montero et al. 1995).

The cloning of P-450 aromatase (CYP19) was obtained from ovaries in both Japanese (Ijiri et al. 2003) and European eels (Tzchori et al. 2004). The same transcript was found in eel ovary and brain (Ijiri et al. 2003, Tzchori et al. 2004). This contrasts with the situation in other teleosts, which possess two genes for aromatase, likely resulting from the teleost specific genome duplication (Mayer and Schartl 1999), one expressed in the gonads and the other one in the brain. Analysis of the recently sequenced genomes of the European eel (Henkel et al. 2012a) and of the Japanese eel (Henkel et al. 2012b) allowed us to confirm the presence of a single aromatase gene in the eel (Jeng et al. 2012a). This single copy is expressed in both brain and gonad and we previously showed, in Japanese eel, that eel brain aromatase had an activity much lower than in other teleosts but similar to that in mammals (Jeng et al. 2005). This low aromatase activity allows, in the eel, androgen specific actions to be exerted, not only by non-aromatizable androgens such as 11-KT but also by aromatizable androgens, such as testosterone. Accordingly, testosterone-specific and estradiol-specific feedbacks were evidenced in the eel (Montero et al. 1995, Aroua et al. 2007). There was no significant sex difference in aromatase activity in the forebrain, midbrain, hindbrain, or pituitary. In contrast, there was in the gonads, where aromatase could be detected in the ovaries but not in testes, in accordance with the role of this enzyme in ovarian differentiation (Jeng et al. 2005). Variations of aromatase expression/activity during gonadal development, after experimental maturation or after steroid treatments were investigated in the Japanese eel. Aromatase activity as well as expression (Jeng et al. 2012b) significantly increased in the pituitary, brain and ovary of experimentally matured female eels compared to controls (Jeng et al. 2005, 2012b). Treatments with sex steroids (E2 or T) revealed that the increase in aromatase expression in the brain may result from E2-specific induction, as only E2, and not T, could increase aromatase expression both in the forebrain and midbrain (Jeng et al. 2012b). In contrast, the increase in aromatase expression in the ovary of experimentally matured eels is a result of steroid-independent control, likely due to a direct effect of gonadotropins contained in the pituitary extract, as no effects of E2 nor of T were observed (Jeng et al. 2012b). Recent immunohistochemistry and *in situ* hybridization studies showed that aromatase expression in the brain was localised in the radial glial cells and that experimental maturation strongly stimulated aromatase messenger and protein expression in radial glial cells and pituitary cells (Jeng et al. 2012a).

Two androgen nuclear receptor subtypes (ARα and ARβ) have been evidenced in the eel (Ikeuchi et al. 1999, Todo et al. 1999), as in various other teleosts, likely resulting from the teleost-specific whole genome duplication. Two major subtypes of estrogen nuclear receptors (Er-α and ER-β) were also previously reported in European (Palstra et al. 2010a) and Japanese (Todo et al. 1996) eels, a situation similar to that observed in mammals. However, analysis of the recently sequenced genomes of the European and Japanese eels (Henkel et al. 2012a, b) allowed us to characterize the presence of three estradiol nuclear receptors (Er-α, ER-β1 and ER-β2), the two ER-β subtypes likely resulting from the teleost-specific genome duplication (Lafont et al. unpublished data). We are currently investigating the tissue distribution and regulation of the expression of ER and AR subtypes in the European eel (Lafont et al. unpublished data), in order to decipher which subtypes may be involved in the sex steroid feedbacks on brain GnRHs and pituitary gonadotropins.

Other Hormones Involved in Eel Reproductive Physiology

Activin/Follistatin System

Besides the sex steroids, gonadal peptides, activins and inhibins, also play an important role in the positive and negative regulations of gonadotropins in mammals (for review: Ying 1988). In addition, another gonadal peptide, follistatin was found to neutralize activin bioactivity (Nakamura et al. 1990). The presence of activin and follistatin not only in gonads but also in a wide range of tissues including pituitary (for review: Bilezikjian and Vale 2011) suggested that their actions could be exerted locally as autocrine/paracrine effects on gonadotropins.

In teleosts, actions of activins and inhibins in the control of FSH and/or LH were reported in goldfish (Yam et al. 1999), zebrafish (Lin and Ge 2009), tilapia (Yaron et al. 2001) and coho salmon (Davies et al. 1999). In the European eel, we recently reported the presence and tissue distribution of activin and follistatin (Aroua et al. 2012). We showed that both activin and follistatin were expressed in the brain, pituitary and gonads. Using primary cultures of European eel pituitary cells, we demonstrated that pituitary

FSH was stimulated while LH was inhibited by activin, and that follistatin antagonized both effects. Our results evidenced activin as the first major stimulator of FSH expression in the eel and suggested auto-/para-crine role for activin in addition to potential endocrine function (Aroua et al. 2012).

Somatotropic Axis

Studies focusing on growth hormone (GH) cell regulation have shown that basal release and synthesis of GH persist *in vitro*, in the absence of secretagogues or serum, using organ-cultured pituitaries (European eel: Baker and Ingleton 1975, Japanese eel: Suzuki et al. 1990) or primary cultures of pituitary cells (European eel: Rousseau et al. 1998, 1999, rainbow trout: Yada et al. 1991, turbot: Rousseau et al. 2001). All these observations lead to the suggestion that the major control of GH cells *in vivo* in teleosts is an inhibitory control (Rousseau and Dufour 2004). Our *in vitro* study in the European eel demonstrated that the brain inhibitory control of GH is exerted by somatostatin (SRIH), as in other vertebrates. We also showed that insulin-like growth factor 1 (IGF-1), a growth factor produced by the liver under GH stimulatory effect, exerts a negative feedback on GH, in the eel, as in other vertebrates.

Comparing maturing European eels from 38 different locations throughout Europe and South Africa, Vollestad (1992) reported geographic variation in age at metamorphosis, which could be explained by variation in growth rate according to both latitude and longitude. Various studies showed that favourable growth conditions cause eels to silver rapidly (Vollestad 1988, 1992, De Leo and Gatto 1995), such as growth in brackish water and at low latitudes (Lee 1979, Fernandez-Delgado et al. 1989), as well as growth in aquaculture farms and under experimental conditions (Tesch 1991, Fernandez-Delgado et al. 1989).

Silvering seems thus to be triggered by a period of high growth. Indeed, an important increase in GH was observed during summer in pre-migrant European eels (stage III = beginning of silvering) (Durif et al. 2005). This suggests that while GH might not be involved in the control of the silvering process itself (Aroua et al. 2005), the somatotropic axis may participate earlier in the initiation of the silvering. A similar growth surge is observed at pre-puberty in mammals.

In vivo experiments indicated a potentiating role of GH on E2-induced vitellogenesis in the female silver European eel (Burzawa-Gérard and Delevallée-Fortier 1992, Peyon et al. 1997). Using primary cultures of eel hepatocytes, Kwon and Mugiya (1994) in the Japanese eel, and Peyon et al. (1996, 1997) in European eel, demonstrated that GH potentiates the stimulatory effect of E2 on both vitellogenin synthesis and secretion, by a direct effect on hepatocytes.

Using primary cultures of eel pituitary cells, we showed that insulin-like growth factor IGF-1 was able to inhibit GH (Rousseau et al. 1998) and stimulate LH release (Huang et al. 1998), leading to the hypothesis that this growth factor could be the link between body growth and induction of puberty in the eel. The high GH levels observed during spring in the eel could stimulate liver secretion of IGF-1, which could trigger the initial increase in LH synthesis.

Other *in vitro* studies, on testis or ovarian cultures, demonstrated a potential role of GH and IGF-I on eel gametogenesis. In the Japanese eel, Nader et al. (1999) reported that IGF-I, in the presence of 11-KT, played an essential role in the onset, progress, and regulation of spermatogenesis in the testis. Treatment of previtellogenic ovarian fragments from eel, *A. australis*, with recombinant human IGF-I resulted in increased oocyte diameters (Lokman et al. 2007). Recently, GH was shown to be produced by testis of Japanese eel and able to stimulate the proliferation of spermatogonia (Miura et al. 2011).

All these data suggest that the somatotropic axis may play a critical role in eel reproduction, by contributing to the activation of the gonadotropic axis at silvering, and by potentiating the effects of reproductive hormones at the levels of the liver and gonads.

Corticotropic Axis

Only a few studies have focused on the corticotropic axis during the transition from yellow to silver stage. It is probably because of the difficulty of sampling blood in order to measure plasma cortisol levels without stressing animals. Van Ginneken et al. (2007a) demonstrated elevated cortisol levels in silver eels prior to migration. This is in agreement with the fact that during the downstream migration, eels are fasting and it is well known that the production of cortisol is induced in response to starvation. A role of cortisol may be to permit the mobilization of energy stores needed by the fish at this critical period.

Forrest et al. (1973) found that Na_+, K_+-ATPase activity rose very slowly after transfer of yellow American eels (*A. rostrata*) to seawater (SW). If the eels were treated with cortisol before transfer, the enzyme activity increased to levels found in SW-acclimated eels and did not increase further following transfer to SW. Injection of yellow American eels with cortisol, in addition to increasing Na_+, K_+-ATPase of gill and intestine, caused their ventral surface to turn silver (Epstein et al. 1971).

In addition to the effect of cortisol on energy mobilization and seawater adaptation, we previously demonstrated in the European eel that cortisol had also a strong positive effect on LH production *in vivo* as well as *in vitro* (Huang et al. 1999). This stimulation was stronger when eels were treated by a combination of cortisol and androgens, indicating synergistic actions of these hormones on LH synthesis (Huang 1998, Sbaihi 2001). It is interesting to note that while in amphibians, cortisol has a synergistic effect with thyroid hormones on metamorphosis, a synergy between cortisol and sex steroids is observed in the control of eel silvering. In the eel, cortisol also induces a regression of the digestive tract, and can mobilize mineral stores from vertebral skeleton (Sbaihi et al. 2009).

The various effects of cortisol demonstrated in the eel indicate that the corticotropic axis may play an important role throughout sexual maturation by permitting energy and metabolite mobilization. Cortisol may therefore control the metabolic challenge occurring during both metamorphosis and puberty/reproduction in teleosts.

Fat Stores

In general, a minimum, critical amount of fat is needed, as a trigger for metamorphosis or sexual maturation, in order to withstand the metabolic needs due to metamorphosis, migration and/or reproduction. In the eel, as mentioned above, there is a wide variability in the age at silvering (Rossi and Colombo 1976, Vollestad and Jonsson 1986, Poole and Reynolds 1996, Svedäng et al. 1996). In 2006, Durif et al. showed, in a study collecting silver European eels from 5 different locations in France, that ages ranged from 5 to 24 years. Many maturing eels may delay their migration to increase their fat stores (Durif et al. 2009, 2011). The idea of a 'critical fat mass' is discussed in Larsson et al. (1990), based on the comparison of the percentage of lipid in tail muscular and skin tissues, between yellow and silver female eels. They found that a fat % of 28 (in the tail) may be a prerequisite for the change from yellow to silver stage and under a critical fat mass, silvering may not even be initiated (Larsson et al. 1990). However, a number of large yellow eels had fat above 28 and a number of silver eels had fat below 28, indicating that a certain fat % may not be sufficient as triggering factor for silvering, and may need to be associated with other internal and environmental triggering factors. Svedäng and Wickström (1997) suggested that silver eels with low fat concentrations may temporarily halt migration, revert to a feeding stage, and "bulk up" until fat reserves are sufficient to carry out successful migration to the spawning area. Combining both Boëtius and Boëtius (1985) and Larsson et al. (1990) studies, Larsen and Dufour (1993) found that silver eels (regardless of sex) rarely have a fat % lower than 20. Analysis of the energy budget of migratory eels have been made by Boëtius and Boëtius (1980, 1985), and more recently by the group of Van den Thillart, who developed swimtunnels to mimic the challenge of the eel trans-oceanic (Van Ginneken and Van den Thillart 2000, Van den Thillart et al. 2004). They showed that silver eels contain energy reserves and organic material sufficient for both the long migration and the extensive gonadal growth. Van Ginneken and Van den Thillart (2000) found that the energetic cost of the 6,000-km migration was actually low, with 60% of the fat store remaining available for the gonadal growth. By artificially inducing sexual maturation with carp pituitary extracts (CPE) and monitoring individual characteristics of female European eels, Durif et al. (2006) demonstrated that the initial state of the eel in terms of energy stores partly explained the variability of subsequent sexual maturation. Indeed, the best response to CPE treatment came from eels with the highest initial condition factor and largest initial body diameter. A recent study on *A. anguilla* from Baltic Sea reported that a large proportion of female silver eels would have inadequate or suboptimal reserves for successful migration and reproduction, as calculations found that 20.4% of eels would have completely exhausted all initial fat reserves and that 45% would be within 90% of complete energy depletion after migration and reproduction (Clevestam et al. 2011). This phenomenom may be one of the causes of eel population decline.

Leptin and Ghrelin

Metabolic hormones, such as leptin and ghrelin, which are primarly involved in the control of energy balance, were also shown to be implicated in the cross-talk between metabolism and reproduction and the neuroendocrine control of reproduction in mammals (for reviews: leptin: Hausman et al. 2012, ghrelin: Muccioli et al. 2011).

First investigations in some teleost species indicated that leptin mRNA is expressed primarily in the liver, differently from the situation in mammals in which leptin is mainly produced by the adipose tissue (for review: Copeland et al. 2011). Its role in regulating energy balance by modulating food intake and consequently body weight, as well as by stimulating lipolytic pathways, is well known in mammals and is also observed in fish (Copeland et al. 2011). *In vivo* and *in vitro* treatments with heterologous leptin are able to increase gonadotropin in sea bass (Peyon et al. 2001) and trout (Weil et al. 2003). So far in teleosts, up to two leptin genes have been identified but only one receptor (for review: Copeland et al. 2011). The analysis of the European eel genome led us to characterize two leptin genes and, for the first time, two leptin receptor genes (Lafont et al. 2012). Based on this new discovery, we performed phylogenetic analyses of both leptin and its receptor families. We developed specific qPCRs to investigate the differential tissue distribution of the two leptins and the two leptin receptors. We also analyzed their differential regulation in the organs of the brain-pituitary-gonad (BPG) axis during experimental maturation. This provides the first evidence of a conserved duplicated leptin/leptin receptor system in teleosts and of a potential differential involvement of these duplicated peptides and receptors in the eel sexual maturation (Lafont et al. 2012).

The primary structure of ghrelin has been determined in various teleost fish species (for review: Kang et al. 2011). The role of ghrelin on energy balance may differ among teleosts. In tilapia, while long-term treatment with implanted n-decanoic ghrelin increased food consumption (Riley et al. 2005), fasting for 4 and 8 days did not affect ghrelin levels in the plasma or stomach (Fox et al. 2009), suggesting an absence of role in meal initiation or short-term energy homeostasis in this species. In zebrafish, preproghrelin expression in the brain and gut is increased in fasted fish and decreased after refeeding, showing an orexigenic role of ghrelin in zebrafish (Amole and Unniappan 2009). In goldfish also, ghrelin has an orexigenic action and intraperitoneal (ip), as well as intracerebroventricular (icv) injections stimulates food intake (Miura et al. 2009). In contrast, an anorexigenic action of ghrelin is reported in rainbow trout, in which ip injection of ghrelin did not affect food intake, while central injection decreased it through a corticotropin-releasing hormone (CRH)-mediated pathway (Jönsson et al. 2010). Considering these data, it will be of interest to investigate the potential roles of leptin and ghrelin in the control of silvering, fasting and sexual maturation of the eel.

Silvering and Pre-puberty

Reproductive Changes During Eel Silvering

Silvering consists of various morphological, physiological and behavioral changes. Among the modified organs, some are related to sensory functions, others to hydrostatic pressure or seawater adaptation, similarly to changes observed during smoltification in salmonids, which traditionally led to eel silvering being defined as a metamorphosis. However, unlike smoltification, silvering also includes some changes related to an onset of sexual maturation such as some gonad development, which led to the hypothesis that silvering corresponds to a pubertal event (Aroua et al. 2005, Rousseau et al. 2012). Puberty, the major post-embryonic developmental event in the life cycle of all vertebrates, encompasses various morpho-physiological and behavioral changes, which unlike metamorphic changes are induced by sexual steroids (for review, see Romeo 2003).

Gonadal changes

In females, the gonadosomatic index (GSI) raises progressively in yellow eels from 0.3 to ≥1.5 in silver eels, with an increase of follicular diameter, thickening of follicular wall and appearance of many lipidic vesicles in the ooplasma (Fontaine et al. 1976, Lopez and Fontaine 1990). This increase in gonad size was shown to be a good criterion to estimate the state of advancement of the silvering process in the eel (Marchelidon et al. 1999, Durif et al. 2005). Durif et al. (2005) described five stages with physiological and morphological validation. In this study, a growth phase (stages I and II), a pre-migrating stage (stage III) and a migrating phase (stages IV and V) were defined. Stages I and II correspond to the previous "yellow" stage with a GSI<0.4%; the gonads show small primary, non-vitellogenic oocytes, with a dense ooplasma and a dense nucleus with a large nucleolus (Aroua et al. 2005). Stage III corresponds to the previous "intermediate" or "yellow/silver" stage with $0.4\% \leq$ GSI $< 1.2\%$; oocytes are larger and a few lipidic vesicles are observed in the ooplasma, which indicates the initiation of the incorporation of lipidic stores in the oocytes, also referred as "endogenous vitellogenesis" (Aroua et al. 2005). Stages IV and V correspond to the previous "silver" stage with a GSI$\geq$1.2%; oocytes are further enlarged with a large nucleus and small nucleoli at a peripheral position and numerous lipidic vesicles in the ooplasma, which is the oil-droplet stage of early vitellogenesis (Aroua et al. 2005). In the most advanced stage of silvering, vitellogenin can be observed in the ooplasm, as well as in the plasma, which corresponds to the start of exogenous vitellogenesis. Similarly, in the Japanese eel, the gonadosomatic index (GSI) of females increased significantly from 0.27% for yellow phase, to 0.55% for pre-silver and 1.32% for silver phase (Han et al. 2003c). Oocyte development progressed from the chromatin nucleolus stage in the yellow phase eel, through the peri-nucleolus stage in the pre-silver phase eel and to the oil-drop stage in the silver phase eel (Han et al. 2003c).

There have been less studies on silvering in male eels. In male Japanese eels, GSI was shown to increase from 0.07% in yellow males to 0.15% in silver males (Han et al. 2003c). In European eel, Durif et al. (2005) also reported an increase in GSI from 0.04% in yellow males to 0.16% in silver males.

Hormonal changes

The study of the expression profiles of TSH showed a non-significant or a weak increase in TSHβ mRNA between yellow and silver eels (*A. anguilla*: Aroua et al. 2005, *A. japonica*: Han et al. 2004). Moreover, measurement of plasma levels of thyroid hormones in yellow and silver eels showed a moderate increase in thyroxine (T4) and no significant variations in triiodothyronine (T3) during silvering (*A. anguilla*: Marchelidon et al. 1999, Aroua et al. 2005, *A. japonica*: Han et al. 2004). These results suggest that the thyrotropic axis is poorly implicated in the neuroendocrine control of the silvering process. The weak variations observed on TSHβ mRNA and

T4 plasma level could be involved in the increased activity of eels related to their migratory behavior.

In contrast, measurement of sexual steroids, estrogens (E2) and androgens (T and 11-KT) in the plasma of female eels showed an increase between yellow and silver stage (*A. australis and A. dieffenbachii*: Lokman et al. 1998, *A. anguilla*: Sbaihi et al. 2001, Aroua et al. 2005, *A. rostrata*: Cottrill et al. 2001, *A. japonica*: Han et al. 2003b, Sudo et al. 2011a). In males, the few studies available demonstrated an increase in androgen levels during silvering (*A. australis* and *A. dieffenbachii*: Lokman and Young 1998, *A. japonica*: Han et al. 2003b, Sudo et al. 2012). All these results suggest that the gonadotropic axis plays a critical role in the silvering process.

Studies on LH and FSH showed in *A. japonica* (Han et al. 2003a), as well as in *A. anguilla* (Aroua et al. 2005), variations in mRNA levels of the alpha and the beta subunits of the gonadotropins throughout silvering. In *A. japonica*, authors observed a concomitant increase in mRNA of the different subunits, LHβ, FSHβ and the glycoprotein alpha (GPα) (Han et al. 2003a). In *A. anguilla*, LH and FSH were shown to be differentially expressed during the silvering process, with an early increase in FSHβ expression and a late increase of LHβ expression (Aroua et al. 2005). These data suggest that FSH could play an early role in the activation of gonads, while LH may have an important role later in the silvering process.

Considering the start of gonadal maturation, silvering should be considered as an initiation of puberty. As the development of gonads and sexual maturity are blocked at this stage until the occurrence of oceanic reproductive migration, our group defined eel silvering as a prepuberty. These data concerning gonadal maturation show that eel silvering is quite different from salmon smoltification, which occurs before a growth phase and is not associated with changes related to reproduction.

Experimental Induction of Silvering

Silvering is a crucial step in eel life cycle, as this second "metamorphosis" prepares the eel not only to changes of environments during migration but also to sexual maturation. Experimental approaches were conduced in order to further assess the potential role of the gonadotropic axis in the control of silvering, differently from the classical role of the thyrotropic axis in typical metamorphic events.

Experimental data using exogenous sex steroid treatments are in agreement with the involvement of gonadotropic axis in the induction of silvering. Early studies showed that injections of male silver European eels with 17α-methyltestosterone resulted in enlarged eye diameter, increased skin thickness and darkened head and fins (*A. anguilla*: Olivereau and Olivereau 1985). Similarly, implants of testosterone induced an increase of eye size in male silver eels (*A. anguilla*: Boëtius and Larsen 1991). Moreover, immature female *A. australis* which received implants of 11-KT, a non-aromatizable androgen, for 6 weeks presented the external morphological changes observed during silvering, such as increased eye diameter, larger gonads and thicker dermis, compared to controls (Rohr et al. 2001). Our studies in *A. anguilla* showed that treatment with testosterone induced a decrease in the digestive tract-somatic index (Vidal et al. 2004, Aroua et al. 2005) and an increase in ocular index (Aroua et al. 2005), while E2 had no effect (Aroua et al. 2005). Finally, recent data in *A. japonica* reported that 11-KT administration similarly induced the early stage of oocyte growth, eye enlargement, degeneration of digestive tract and development of the swimbladder (Sudo et al. 2012). All these data suggest that the silvering changes are androgen-dependent.

In contrast, we showed that a 3-month-treatment of yellow eels with thyroid hormone (T3) did not induce any changes in ocular index nor digestive tract-somatic index (Aroua et al. 2005). This further supported the conclusion that eel silvering is not triggered by the thyrotropic axis, differently from a true metamorphic event (for review: Rousseau et al. 2012). However, cortisol may have a synergistic action with steroids in this complex process of eel silvering, as we demonstrated that concomitant administration of sex steroids and cortisol was most efficient in inducing the silvering of the skin in eels (Sbaihi 2001).

All these data demonstrate the involvement of androgens as crucial actors in the morpho-physiological changes of eel silvering. However, despite the possible experimental induction of silvering by androgens, eels still remain blocked at a pre-pubertal stage as long as the oceanic reproductive migration does not occur.

Experimental Sexual Maturation

Due to the decline of eel populations, there is an urgent need for success in artificial reproduction. Current studies aim at developing new strategies to induce eel sexual maturation and reproduction, by deciphering regulatory mechanisms of the dual neuroendocrine control of puberty by GnRH and DA, by investigating new key-factors such as kisspeptin, and by using combinations and/or variations of environmental factors.

Protocols for Artificial Maturation and Recent Advances

Pioneer work by M. Fontaine and co-workers, at the National Museum of Natural History, demonstrated that injection of urinary extract from pregnant women (known later to contain human chorionic gonadotropin, hCG) to male European eels (Fontaine 1936), or of carp pituitary extract to females (Fontaine et al. 1964), induced full spermatogenesis and ovarian development respectively. These experiments led to the first observation of mature gonads in the eel, suggesting that the arrest in gonadal development was due to a deficit in the pituitary gonadotropic function.

Based on the pioneer experiments in the European eel, similar treatments (hCG in males and fish pituitary extract in females) have been since currently employed to induce sexual maturation (gametogenesis and steroidogenesis) in various eel species (*A. japonica*: Yamamoto and Yamauchi 1974, Yamauchi et al. 1976, Ohta et al. 1996, 1997, Tanaka et al. 2001, *A. rostrata*: Edel 1975, Sorensen and Winn 1984, *A. dieffenbachii*: Todd 1979, Lokman and Young 2000). In addition, the application of 17α, 20β-dihydroxy-4-pregnen-3-one (DHP) for final maturation and ovulation resulted in increasing fertility and hatching rates (*A. japonica*: Ohta et al. 1996) and is commonly used in current protocols. Thus, since the first attempts to artificial reproduction of European eel by Pr. M. Fontaine in 1936, the improvement of protocols and techniques have yielded to the production of fertilized eggs and larvae in the Japanese eel in 1974 (Yamamoto and Yamauchi 1974), then rearing of preleptocephalus larvae was obtained in 1976 (Yamauchi et al. 1976). In 2003, the first production of glass eel of Japanese eel opened new hopes for aquaculture (Tanaka et al. 2003, Kagawa et al. 2005) and recently, artificial breeding has been accomplished (Masuda et al. 2011). In other eel species, pre-pubertal blockade seems stronger as more injections of pituitary extracts are needed to induce maturation, making progress in obtaining full life cycle slower. In the European eel, Palstra et al. (2005) succeeded to fertilise eggs and produce embryos, and Tomkiewicz and Soerensen (2008) were able to raise yolk-sac larvae up to 18 days. For *A. rostrata*, attempts to induce maturation and reproduction have been few and have produced limited results until the maintenance of larvae for up to 6 days reported by Oliveira and Hable (2010). The authors underlined that the crucial factor for successful fertilization was the stage of the oocyte at the time of induced ovulation.

Hormonal Changes During Experimental Maturation

In the female European eel, long-term treatment with carp pituitary extract stimulated ovarian vitellogenesis, leading to a gradual increase in gonado-somatic index, which reached up to 30–40% after several months, an index much higher than in control silver eels (1.5–2%) (Fontaine et al. 1964, Dufour et al. 1989, 1993, Schmitz et al. 2005, Durif et al. 2005).

Our first studies by radioimmunoassay, employing antibodies recognizing all forms of GnRH, indicated a positive effect of experimental sexual maturation on total GnRH level in the brain of female or male silver eels (males treated with hCG and females with estradiol: Dufour et al. 1985; females treated with pituitary extract: Dufour et al. 1989). This effect was even more marked in the pituitary, reflecting the accumulation of GnRH in the axonal endings, which are directly innervating the adenohypophysis in the eel as in other teleosts. These data were confirmed by immunocytochemical observation (Kah et al. 1989), which indicated a strong accumulation of GnRH in the pituitary and, in particular, in the axonal endings of the hypophysiotropic neurons. Moreover, castration was able to abolish the increase in brain and pituitary GnRH content, which indicates that gonadal hormones are responsible for this positive effect (Dufour et al. 1989). Later on, using specific RIAs for each native form of GnRH in the eel, we could perform more specific analyses of the effect of experimental maturation on mGnRH and cGnRH-II. We were able to demonstrate an opposite regulation of the two forms with an increase in mGnRH levels in the brain and pituitary, whereas a decrease in cGnRH-II levels in the brain was found, cGnRH-II levels being not detectable by RIA in the pituitary (Dufour et al. 1993). This opposite regulation suggests that mGnRH and cGnRH-II play drastically different roles during eel sexual maturation, and that mGnRH would play a major role in the neuroendocrine control of pituitary gonadotropins.

In the European eel, our early studies, using heterologous radioimmunoassay for carp LH β subunit, showed a large increase in pituitary LH content in artificially matured eels, namely in females treated with carp pituitary extract or in males treated with human chorionic gonadotropin (Dufour 1985). The effect of carp pituitary extract on pituitary LH content was prevented by ovariectomy (Dufour et al. 1989). These data suggested the involvement of gonadal hormones in the stimulation of endogenous pituitary LH during experimental maturation. In contrast, no change in plasma LH level was found in experimentally matured male eels, in spite of largely elevated pituitary LH content (Dufour 1985). This indicates that endogenous LH synthesis but not release is stimulated during experimental maturation. The situation is likely the same in experimentally matured female eels, but the recognition of exogenous carp LH by RIA prevented the determination of endogenous LH plasma levels. Recently, the cloning of eel FSHβ and LHβ subunits allowed us to demonstrate that during experimental maturation induced by carp pituitary extract in females, LH and FSH undergo an opposite regulation with a large increase in LHβ mRNA levels but a decrease in FSHβ mRNA levels (Schmitz et al. 2005).

In the Japanese eel, repeated treatment with salmon gonadotropin (Sato et al. 1996) is required for the artificial induction of ovarian maturation, and stimulation of endogenous LH synthesis is observed after treatment with salmon pituitary homogenate (Nagae et al. 1996a, b, 1997), salmon GtH (Yoshiura et al. 1999) or sex steroids (Lin et al. 1998). An increase in both

LHβ and GPα mRNA levels was observed during the induction of ovarian development (Nagae et al. 1996a, b, 1997, Yoshiura et al. 1999, Suetake et al. 2002, Saito et al. 2003). A dramatic decrease in FSHβ mRNA levels was reported in experimentally matured female Japanese eels, and FSHβ mRNA levels were undetectable after 14 weeks of gonadotropic treatment (Yoshiura et al. 1999). Saito and colleagues found profound differences in FSHβ and LHβ mRNA profiles between artificially maturing Japanese eels and naturally maturing New Zealand longfinned eels, *A. dieffenbachii* (Saito et al. 2003). Indeed, FSHβ mRNA level was high at the previtellogenic stage in Japanese eels, but low in New Zealand longfinned eels, and then increased at the mid-vitellogenic stage in New Zealand longfinned eels, but decreased in Japanese eels; LHβ mRNA level increased remarkably at the mid-vitellogenic stage in Japanese eels, but only slightly in New Zealand longfinned eels. This reveals that the opposite variations in FSHβ and LHβ mRNA pituitary levels, observed in artificially maturing Japanese as well as European eels, may largely differ from their natural profiles.

Concerning LH and FSH receptors, quantitative real time RT-PCR indicate an increase in the ovarian expression of both receptors during induced maturation (Jeng et al. 2007).

A 'from brain to testis' study was recently performed by Peñaranda et al. (2010a) investigating the regulatory mechanisms of the artificial maturation of European eel males. They found a differential regulation of GnRHs by repeated injections of hCG, cGnRH II expression being decreased after the 4th week of treatment and mGnRH increased after 3 week of treatment. One injection of hCG was enough to dramatically decrease FSHβ expression, while increasing LHβ synthesis. These differential regulations of gonadotropin expression in males were already reported in Japanese eels (Yoshiura et al. 1999). Concerning sex steroids, plasma levels of 11-KT were highest when later stages of spermatogenesis (late meiosis and spermiogenesis) were obtained (before 5 weeks of treatment), whereas the highest 17, 20β-P levels were observed in the 5th week of treatment and were maintained throughout the treatment, supporting an essential role of this progestin for sperm motility acquisition (Miura et al. 1991).

Treatments based on Brain Neuro-hormones

Current protocols for inducing eel gametogenesis are still based on direct activation of the gonads by gonadotropic treatments (hCG in males; carp, salmon or catfish pituitary extracts in females). However, these treatments lead to extra-physiological activation of steroidogenesis, inadequate kinetic of vitellogenesis, inappropriate oocyte stores and poor quality eggs, even in *A. japonica*. Deciphering the neuroendocrine control of eel reproduction would lead to the development of new strategies by triggering the release of eel endogenous gonadotropins via activation of brain-pituitary axis.

In the European eel, we proved that the deficit in the pituitary gonadotropic function at the silver stage resulted from both a lack of stimulatory input from gonadotropin-releasing hormone (GnRH) and low pituitary sensitivity to GnRH, as well as from a strong inhibition by dopamine (DA; Dufour et al. 2003, 2005, 2010, Vidal et al. 2004). Indeed, synthesis of LH can be stimulated by administration of sexual steroids (Dufour et al. 1983a), but despite a strong accumulation of LH in the pituitary, no stimulation of LH release and, consequently, no significant ovarian development could be observed (Dufour et al. 1983a). In addition even in steroid-pre-treated females, a GnRH treatment alone is unsuccessful in inducing LH release (Dufour et al. 1988). It is only the triple treatment with steroid, GnRH agonist and DA-D2 receptor antagonist (pimozide) which could induce dramatic increases in LH synthesis and release, as well as in vitellogenin (Vg) plasma levels and a stimulation of ovarian vitellogenesis (Dufour et al. 1988, Vidal et al. 2004). This revealed that a dual brain control was responsible for the arrest of eel sexual maturation at a prepubertal stage.

In experimentally matured female eels, injection of progesterone (i.e.; dihydroxy progesterone, DHP) is now used at the end of the gonadotropic treatment, in order to induce final oocyte maturation and ovulation (Japanese eel: Ohta et al. 1997, Kagawa et al. 2012, American eel: Oliveira and Hable 2010). However, Kagawa et al. (2003) showed the importance of the developmental stage of the oocyte at the time of induction. DHP-induced ovulation, *in vitro*, occurred earlier in large oocytes (>850 μm) compared to smaller oocytes (800 μm). Current protocols are developed to aim at recruiting into maturation only those oocytes that are capable of responding to DHP, using lower doses of steroid. Another approach would consist in inducing the release of endogenous LH, which is synthezised and accumulated in the pituitary in experimentally matured eels, under the effect of steroid positive feedbacks (Dufour et al. 1989, Aroua et al. 2007). Based on the dual brain control of LH (Dufour et al. 1988, Vidal et al. 2004), our current investigations aim at developing protocols for acute injections of both GnRH agonist and DA-D2 receptor antagonist, in experimentally matured eels, at the end of the chronic gonadotropic treatment. The objective is to trigger the release of LH accumulated in the pituitary, leading to an endogenous ovulatory peak of LH, in order to induce final oocyte maturation and ovulation *via* physiological plasma levels of endogenous LH and DHP.

Studies of Migration-related Factors

As eel species remain blocked at pre-puberty as long as their reproductive migration does not occur, various studies have tried to investigate the effects on sexual maturation of swimming and of the different environmental factors that these fishes encounter during this journey.

Swimming

The swimming physiology of European silver eels and the effects on sexual maturation and reproduction, using swimmtunnels, has been recently reviewed (Palstra and Van den Thillart 2010). Swimming induced an increase of eye diameter in different swim trials (Palstra et al. 2007, 2008a). Swimming also induced changes in gonad histology of both females and males (Palstra et al. 2007, 2008b). For instance, swimming female eels (age of 13 to 21 years) from Lake Balaton had higher GSI and oocyte diameter than resting eels (Palstra et al. 2007). This increase in GSI was not found with younger (3 years) silver farmed eels (Van Ginneken et al. 2007b), suggesting that older eels are more responsive and sensitive for maturation. After long term-swimming (5, 500-km, 173 days) in freshwater, young farmed female eels showed increased plasma levels of estradiol and 11-ketotestosterone, as well as of pituitary levels of LH (Van Ginneken et al. 2007b). In contrast, female silver eels that swam for 1.5 or 3 months had reduced ERα, vitellogenin 1 and 2 liver expression and lower plasma calcium levels in their blood, indicative of suppressed hepatic vitellogenesis (Palstra et al. 2010b).

Male silver eels showed a significant increase in pituitary LHβ expression after three month-swimming, as well as males injected with GnRH agonist, compared to controls (Palstra et al. 2008b). Both swimming and GnRH treatment also caused a three- to five-fold increase in GSI and some induction of the first steps of spermatogenesis (Palstra et al. 2008b). However, in this study, females were not stimulated by swimming nor by GnRH agonist, and even showed regression of maturation over time as demonstrated by lower LHβ expression, GSI and oocyte diameters in all groups after 3 months (Palstra et al. 2008b). This unresponsiveness of females may be due to the fact that, in contrast to males, their pituitaries were not sensitized and still under dopaminergic control.

Different mechanisms have been suggested by Van den Thillart and collaborators (Palstra et al. 2009a). Swimming may regulate GnRH-LH axis, stimulate lipid metabolism and/or inhibit vitellogenesis. A possible role of cortisol in this phenomenon has been hypothetized as higher cortisol levels have been measured in swimming eels (Palstra et al. 2009b).

In summary, swimming trials performed by Van den Thillart's group showed different effects in males and females. In silver males, swimming trials induced some sexual maturation. In contrast, suppressed exogenous vitellogenesis, associated with stimulated lipid deposition in oocytes is observed after swimming trials in silver females. These latter observations in females suggest that in nature, a different sequence of events may occur compared to artificial maturation and eels may undergo exogenous vitellogenesis and final maturation near or at the spawning grounds.

Photoperiod and Melatonin

Melatonin is a hormone principally released by the pineal organ during the night, and known to act as a transmitter of environmental cues, notably photic information, and to play a role in synchronising various behaviors and physiological processes (for review: Zachmann et al. 1992, Pandi-Perumal et al. 2006).

In the eel, a first experiment conducted by Dufour and Fontaine (1985) reported that a three-month stay at obscurity had no effect on GSI, while LH pituitary content displayed a decreasing tendency. More recently, a five-month melatonin treatment was shown to increase brain tyrosine hydroxylase (TH, the rate limiting enzyme of DA synthesis) mRNA expression in a region dependent way (Sébert et al. 2008a). Melatonin stimulated the dopaminergic system of the preoptic area, which is involved in the inhibitory control of gonadotrophin (LH and FSH) synthesis and release. Moreover, the increased TH expression was consistent with melatonin binding site distribution as shown by $2[^{125}I]$-melatonin labelling studies. On the other hand, melatonin had no significant effects on the two eel native forms of GnRH (mGnRH and cGnRH-II) mRNA expression. Concerning the pituitary-gonad axis, melatonin treatment decreased both gonadotrophin β-subunit (LHβ, FSHβ) mRNA expression and reduced sexual steroid (11-ketotestosterone, oestradiol) plasma levels. This indicates that melatonin treatment had a negative effect on eel reproductive function. By this mechanism melatonin could represent one pathway by which environmental factors could modulate reproductive function in the eel (Sébert et al. 2008a).

As several studies demonstrated that melatonin secretion can be modulated by various photic factors and nonphotic environmental cues (such as temperature, blue wavelengths, oxygen pressure, changes in water current direction or water pressure), it is conceivable that various factors encountered by eels during their reproductive migration could interact to reinforce or suppress melatonin-DA inhibitory tone on gonadotropic axis (Sébert et al. 2008a).

Hydrostatic Pressure

As a first clue, spawning areas of the different eel species have been approximately localized by the presence of the smallest leptocephalus larvae. These areas were situated over deep oceanic regions (*A. anguilla, A. rostrata*: Sargasso Sea: Schmidt 1923, Boëtius and Harding 1985, *A. celebensis*: Sumatra island: Delsman 1929, Jespersen 1942, *A. australis; A. dieffenbacchii*: East of Fidji islands: Jellyman, 1987, 2003, *A. japonica*: Marianas trench: Tsukamoto 1992, 2006). Moreover, tracking data are in agreement with a presumed migration at great depth. For example, silver eels and some artificially mature individuals have been released in the Atlantic and tracked at more than 1000 m depth (for review: Tesch and Rohlf 2003). Another experiment conducted near Gibraltar demonstrated that eels would swim between 400 m and 600 m depth. This study also reported that eels would display daily vertical migration with swimming in shallow waters at night and in deeper waters during the day (for review: Tesch and Rohlf 2003). Release of silver eels near their supposed spawning areas also showed that eels swam between 250 m and 270 m depth (Fricke and Kaese 1995) and up to 700 m depth (Tesch 1989).

Fishing reports gave additional arguments for a migration at great depth. Indeed, in the north-east of Faroe islands, one silver eel was taken back up from 325 m depth (Ernst 1975). An apparently mature female eel was also observed near Bahamas at 2000 m depth by the submersible ALVIN (Robins et al. 1979). Another specimen with a relatively high gonadosomatic index (GSI of 10) was caught at 500 m depth in the Azores region (Bast and Klinkhardt 1988). Finally, few silver eels were also found in deep fish's stomach, captured for some of them deeper than 700 m (Reinsch 1968). Based on experimental facts, Sébert (2008) recently calculated that the maximal depth at which eel could have normal muscle activity was 2,000 m.

In addition to these observations, some of the anatomical and physiological changes occuring at silvering would prepare eels for a life at depth. Among these changes, the contrast between black back and silver belly (Scott and Crossman 1973, Pankhurst and Lythgoe 1982), the enlargement of ocular diameter (Pankhurst 1982, Pankhurst and Lythgoe 1983) and the change of retinal sensitivity from green to blue wave lengths (Archer et al. 1995, Hope et al. 1998, Zhang et al. 2000) are specific features of mesopelagic teleost fishes. Moreover the capillaries network associated with the swimbladder expands to enhance gas excretion efficiency and the swimbladder wall gets thicker allowing a constant volume although eels swim at great depth (Kleckner 1980, Yamada et al. 2000).

Recently, observations of eel migration, using cutting edge technology of pop-up satellite archival transmitting tags, were performed on European eel and indicate diel vertical migrations in Atlantic ocean, predominantly between depths of 200 and 1000 m (Aarestrup et al. 2009). The use of pop-up tags on three New Zealand longfin eels (*A. dieffenbachii*) also showed that they undergo daytime dives to depths of 600 to 900 m, followed by nighttime ascents to depths of 200 to 300 m (Jellyman and Tsukamoto 2002, 2005, 2010). In Japanese eel, pop-up archival transmitting tag informations also reported diel vertical migrations, eels predominantly swimming between depths of about 100 to 500 m at night and 500 to 800 m in the daytime (Manabe et al. 2011).

Furthermore, experimental studies have highlighted the incredible hydrostatic pressure resistance of eels as compared to other teleost fishes (for review: Sébert 2003). Such a resistance was also shown to be higher in silver than in yellow eels, notably due to an increase in cell membrane fluidity (Vettier et al. 2006).

In 1980s, preliminary experiments were performed using female silver eels kept in a cage 450 m in depth in the Mediterranean sea during 3 months (Dufour and Fontaine 1985, Fontaine et al. 1985). A slight but significant increase in GSI was observed, and the pituitary LH content was 27 times higher in immersed eels. Serum LH levels, undetectable in control eels, could be measured by RIA in immersed eels. Since laboratory experiments showed that neither salinity nor obscurity succeeded in increasing pituitary LH levels, it was hypothesized that the stimulatory effect of immersion was related to hydrostatic pressure (Dufour and Fontaine 1985, Leloup-Hatey et al. 1985). In a more recent study, female silver eels were submitted to 101 ATA, in a hyperbaric chamber equipped with a freshwater recirculation system, for 3 weeks (Sébert et al. 2007). Significant increases in oocyte diameter, and in plasma levels of 11-KT, E2 and vitellogenin were measured in females submitted to high hydrostatic pressure (HP), as compared to control eels. At the pituitary level, the exposure to HP had specific effects on the expression of both gonadotropins: an increase in LH-ß and a decrease in FSH-ß mRNA levels. A similar opposite regulation of LH and FSH has been reported to occur at the end of silvering (Aroua et al. 2005) and during artificial maturation (Schmitz et al. 2005), due to a differential feedback by sexual steroids, positive on LH and negative on FSH (Schmitz et al. 2005). These data suggest that hydrostatic pressure may be one the environmental factor of the oceanic migration that may be involved in the regulation of the eel gonadotropic axis.

At the end of the oceanic migration, spawning would occur in shallower layers of 150–200 m and not at great depths, as indicated by catches of *A. japonica* and *A. marmorata* on their spawning ground (Tsukamoto et al. 2011). This is agreement with previous hypothesis from Chow et al. (2009).

An increase in temperature associated with continuous swimming in the shallower water could be a trigger to final maturation (Palstra et al. 2009a), and would also increase the likelihood of eels encountering olfactory cues that they might use to find conspecifics at the spawning area (Tsukamoto 1992).

Temperature

The rate of maturation is affected by temperature (Boëtius and Boëtius 1980, Sato et al. 2006): while warm temperatures are necessary to achieve full maturity, continued exposure to cool temperatures inhibits maturation. In Japanese eel, water temperatures of 18 to 22°C appear optimal for maturation, as at 10°C, final maturation could not be induced and at 14°C (Sato et al. 2006), spawning was not achieved (Dou et al. 2008). All observations of natural silver eel swimming behavior reported daytime diving (600 to 900 m) and nighttime ascents (200 to 300 m). Jellyman and Tsukamoto (2010) and others (Van Ginneken and Maes 2005, Aarestrup et al. 2009) recently proposed that eels are using regular exposure to cool temperatures to regulate the rate of maturity and avoid over-maturation

In Japanese eel, gradual water temperature decrease (from 28°C to 15°C over 39 days) induced an early stage of ovarian development (Sudo et al. 2011b), suggesting that other environmental factors are needed for further maturation. As water temperature in the hypothetical spawning area for European eels has been considered to be 20°C, experimental maturation had been performed around 20°C (Boëtius and Boëtius 1967, 1980). However, a study in European eel showed that during experimental maturation by CPE injections, a variable thermal regime, which progressively increased water temperature from 10 to 14 then 17°C, gave more suitable hormonal profiles (FSH β expression and E2 plasma levels) than those observed at constant high temperature regime (Perez et al. 2011). These data suggested that lower temperatures could promote the first vitellogenic steps, while higher temperatures would be necessary for the final steps of oocyte maturation (Perez et al. 2011).

Salinity

Salinity acclimation seems to potentiate eel responses to gonadotropic treatments, so that transfer to SW is currently used before starting hormonal treatments for experimental maturation (*A. japonica*: Ohta et al. 1996, Kagawa et al. 2003, *A. anguilla*: Pedersen et al. 2003, Palstra et al. 2005). Transfer to SW was shown to increase E2 production (Dufour and Fontaine 1985, Quérat et al. 1987). Additional mechanisms could be involved, such as the increase in cortisol, which may exert a stimulatory role on the gonadotropic axis (Huang et al. 1999).

Natural Sexual Maturation

The recent studies from Tsukamoto et al. (Chow et al. 2009, Tsukamoto et al. 2011), using collection of matured and spawning-condition Japanese eels, allowed getting important informations on spawning ecology, reproductive characteristics of spawners (Chow et al. 2009, Tsukamoto et al. 2011) as well as for the first time on natural hormonal levels (Tsukamoto et al. 2011) of such eels. In Tsukamoto et al. (2011), all females had gonads in a post-spawning condition with degenerated follicles after ovulation, but showed relatively high GSI (9–13.4%) compared with migrating eels (1–4%), suggesting several possible spawnings; to compare, GSI of artificially matured eels can reach 40–70% at their spawning. All males had well-developed testis with GSI values of 13.4–40.3%, compared to migrating (<1%) and artificially matured (10–15%) males. The testes contained a large amount of sperm, whereas cysts composed of spermatids and late type B spermatogonia were occasionally observed, which is similar to artificially matured testes with active spermiation. Plasma levels of DHP (oocyte maturation/ovulation-inducing steroid hormone in Japanese eel; Adachi et al. 2003) were high (0.7–1.5 ng/ml) in females collected by Tsukamoto et al. (2011), compared to levels found in artificially matured eels. Estradiol (in females) and testosterone (in males) levels were also high but at levels observed in artificially matured eels. These first data on natural gonadal and hormonal parameters of eel spawners provide unvaluable basic information for future improvement of protocols for inducing eel experimental maturation.

Conclusions

During eel silvering, there is an overall activation of the gonadotropic axis, with increases of gonadotropin and sex steroid levels. Moreover, exogenous sex steroids are able to induce peripheral morphological changes observed during this process. This let us regard eel silvering as a pubertal rather than a metamorphic event. The term "prepuberty" was first used by our group, as puberty is blocked at an early stage in silver eels, and further sexual maturation only occurs during the reproductive migration. Other endocrine axes may participate in the control of eel silvering. This is the case of somatotropic and corticotropic axes acting in synergy with the gonadotropic axis.

The pre-pubertal blockade of the eels is due to still a low production of pituitary gonadotropins at the silver stage, which results from both a lack of stimulatory input from gonadotropin-releasing hormone (GnRH) and low pituitary sensitivity to GnRH, as well as from a strong inhibition by dopamine. Current protocols of experimental maturation are based on gonadotropic treatment (hCG in males and carp/salmon pituitary extracts in females). A final injection of DHP is then used to induce final oocyte maturation and ovulation. However, the quality of eggs and larvae are still not sufficient and current studies aim at developing new strategies to trigger the release of eel endogenous gonadotropins. Sébert et al. (2009) argued that the unsuccessful experimental reproduction could be due to the fact that all the current experiments were performed at atmospheric pressure, neglecting the pressure effect on egg buoyancy as well as on larval development and viability.

Eel sexual maturation has not yet been obtained when only variations of environmental and migration-related factors such as swimming, hydrostatic pressure or temperature are used. In the future, investigations should consider using combinations and/or variations of internal/hormonal and environmental factors. Great advances have been made in recent years thanks to campaigns and collections of eggs, larvae and matured adults (*A. japonica* and *A. marmorata*) at spawning site, as well as the use of pop-up tags to follow eels during their migration. However, they are scarce because time-consuming and costly. In addition, recent report points out the dramatic effect of pop-up satellite tags on eel swimming (Burgerhout et al. 2011).

Thus, investigations to decipher regulatory mechanisms involved in the control of eel reproduction and to improve protocols to induce sexual maturation are still needed. Further studies should aim at investigating the neuroendocrine interactions between internal/hormonal and environmental factors in the control of silvering and sexual maturation, leading to the activation of the gonadotropic axis and reproduction.

References

Aarestrup, K., F. Okland, M.M. Hansen, D. Righton, P. Gargan, M. Castonguay, L. Bernatchez, P. Howey, H. Sparholt, M.I. Pedersen and R.S. McKinley. 2009. Oceanic spawning migration of the European eel (*Anguilla anguilla*). Science 325: 1660.

Adachi, S., S. Ijiri, Y. Kazeto and K. Yamauchi. 2003. Oogenesis in the Japanese eel, *Anguilla japonica*. In: K. Aida, K. Tsukamoto and K. Yamauchi [eds]. Eel Biology. Springer, Tokyo, Japan. pp. 502–518.

Amole, N. and S. Unniappan. 2009. Fasting induces preproghrelin mRNA expression in the brain and gut of zebrafish, *Danio rerio*. Gen. Comp. Endocrinol. 161: 133–137.

Archer, S., A. Hope and J.C. Partridge. 1995. The molecular basis for the green-blue sensitivity shift in the rod visual pigments of the European eel. Proc. R. Soc. Series B. 262: 289–295.

Aroua, S., M. Schmitz, S. Baloche, B. Vidal, K. Rousseau and S. Dufour. 2005. Endocrine evidence that silvering, a secondary metamorphosis in the eel, is a pubertal rather a metamorphic event. Neuroendocrinology 82: 221–232.

Aroua, S., F.A. Weltzien, N. Le Belle and S. Dufour. 2007. Development of real-time RT-PCR assays for eel gonadotropins and their application to the comparison of *in vivo* and *in vitro* effects of sex steroids. Gen. Comp. Endocrinol. 153: 333–343.

Aroua, S., G. Maugars, S.-R. Jeng, C.-F. Chang, F.-A. Weltzien, K. Rousseau and S. Dufour. 2012. Pituitary gonadotropins FSH and LH are oppositely regulated by the activin/follistatin system in a basal teleost, the eel. Gen. Comp. Endocrinol. 175: 82–91.

Baker, B.I. and P.M. Ingleton. 1975. Secretion of prolactin and growth hormone by teleost pituitaries *in vitro*. II. Effect of salt concentration during long-term organs culture. J. Comp. Physiol. 100: 269–282.

Bast, H.D. and M.B. Klinkhardt. 1988. Fang eines Silberaales (*Anguilla Anguilla* (L., 1758)) im Iberischen Becken (Nordostatlantik) (Teleostei: Anguillidae). Zool. Anz. 221: 386–398.

Bilezikjian, L.M. and W. Vale. 2011. The local control of the pituitary by activin signaling and modulation. Open Neuroendocrinol. J. 4: 90–101.

Boëtius, J. and I. Boëtius. 1967. Studies in the European eel, *Anguilla Anguilla* (L.) experimental induction of the male sexual cycle, its relation to temperature and other factors. Medd. Da. Fisk. Havunders 4: 339–405.

Boëtius, J. and I. Boëtius. 1980. Experimental maturation of female silver eels, *Anguilla anguilla*. Estimates of fecundity and energy reserves for migration and spawning. Dana 1: 1–28.

Boëtius, I. and J. Boëtius. 1985. Lipid and protein content in *Anguilla anguilla* during growth and starvation. Dana 4: 1–17.

Boëtius, J. and E.F. Harding. 1985. A re-examination of Johannes Schmidt's Atlantic eel investigations. Dana 4: 129–162.

Boëtius, I. and L.O. Larsen. 1991. Effects of testosterone on eye size and spermiation in silver eels, *Anguilla anguilla*. Gen. Comp. Endocrinol. 82: 238.

Boularand, S., N.F. Biguet, B. Vidal, M. Veron, J. Mallet, J.D. Vincent, S. Dufour and P. Vernier. 1998. Tyrosine hydroxylase in the European eel (*Anguilla anguilla*): cDNA cloning, brain distribution, and phylogenetic analysis. J. Neurochem. 71: 460–470.

Burgerhout, E., E. Manabe, S.A. Brittijn, J. Aoyama, K. Tsukamoto and G.E.E.J.M. Van den Thillart. 2011. Dramatic Effect of Pop-up Satellite Tags on Eel Swimming. Naturwissenschaften 98: 631–634.

Burzawa-Gérard, E. and B. Delevallée-Fortier. 1992. Implication of growth hormone in experimental induction of vitellogenesis by estradiol-17 beta in female silver eel (*Anguilla anguilla* L.). C. R. Acad. Sci. III. 314: 411–6.

Chiba, H., M. Nakamura, M. Iwata, Y. Sakuma, K. Yamauchi and I.S. Parhar. 1999. Development and differentiation of gonadotropin hormone-releasing hormone neuronal systems and testes in the Japanese eel (*Anguilla japonica*). Gen. Comp. Endocrinol. 114: 449–459.

Chow, S., H. Kurogi, N. Mochioka, S. Kaji, M. Okazaki and K. Tsukamoto. 2009. Discovery of mature freshwater eels in the open ocean. Fish Science 75: 257–259.

Chow, S., H. Kurogi, S. Katayama, D. Ambe, M. Okazaki, T. Watanabe, T. Ichikawa, M. Kodama, J. Aoyama, A. Shinoda, S. Watanabe, K. Tsukamoto, S. Miyazaki, S. Kimura, Y. Yamada, K. Nomura, H. Tanaka, Y. Kazeto, K. Hata, T. Handa, A. Tada and N. Mochioka. 2010. Japanese eel *Anguilla japonica* do not assimilate nutrition during oceanic spawning migration: evidence from stable isotope analysis. Mar. Ecol. Prog. Ser. 402: 233–238.

Clevestam, P.D., M. Ogonowski, N.B. Sjöberg and H. Wickström. 2011. Too short to spawn? Implications of small body size and swimming distance on successful migration and maturation of the European eel *Anguilla anguilla*. J. Fish Biol. 78: 1073–1089.

Copeland, D.L., R.J. Duff, Q. Liu, J. Prokop and R.L. Londraville. 2011. Leptin in teleost fishes: an argument for comparative study. Front. Physiol. 2: 26.

Cottrill, R.A., R.S. McKinley, G. van der Kraak, J.-D. Dutil, K.B. Reid and K.J. McGrath. 2001. Plasma non-esterified fatty acid profiles and 17b-oestradiol levels of juvenile immature and maturing adult American eels in the St. Lawrence River. J. Fish Biol. 59: 364–379.

Davies, B., J. Dickey and P. Swanson. 1999. Regulation of FSH (GTH I) and LH (GTH II) in coho salmon (*Oncorhynchus kisutch*): the action of recombinant human activin and a salmon testis extract. *In*: B. Norberg, O.S. Kjesbu, G.L. Taranger, E. Andersson and S.O. Stefansson [eds]. Proceedings of the 6th International Symposium on Reproductive Physiology of Fish, Bergen, 1999. University of Bergen, Bergen, 2000. pp. 486.

De Leo, G.A. and M. Gatto. 1995. A size and age-structured model of the European eel (*Anguilla anguilla* L.). Can. J. Fish. Aquat. Sci. 52: 1351–1367.

Dekker, W., J.M. Casselman, D.K. Cairns, K. Tsukamoto, D. Jellyman and H. Lickers. 2003. Worldwide decline of eel resources necessites immediate action: Quebec declaration of concern. Fisheries 28: 28–30.

Delsman, H.C. 1929. The distribution of freshwater eels in Sumatra and Borneo. Treubia 11: 287–292.

Dou, S.-Z., Y. Yamada, A. Okamura, A. Shinoda, S. Tanaka and K. Tsukamoto. 2008. Temperature influence on the spawning performance of artificially-matured Japanese eel, *Anguilla japonica*, in captivity. Environ. Biol. Fish. 82: 151–164.

Dufour, S. 1985. La fonction gonadotrope de l'anguille européenne, *Anguilla anguilla* L., au stade argenté (au moment du depart pour la migration de reproduction): les mécanismes de son blocage et sa stimulation expérimentale. PhD thesis University Paris VI, Paris, France.

Dufour, S. 1994. Neuroendocrinologie de la reproduction de l'anguille: de la recherche fondamentale aux problèmes appliqués. Bull. Fr. Peche Piscicult. 335: 187–211.

Dufour, S. and Y.A. Fontaine. 1985. La migration de reproduction de l'anguille européenne (*Anguilla anguilla* L.): un rôle probable de la pression hydrostatique dans la stimulation de la fonction gonadotrope. Bull. Soc. Zool. Fr. 110: 291–299.

Dufour, S. and K. Rousseau. 2007. Neuroendocrinology of fish metamorphosis and puberty: evolutionary and ecophysiological perspectives. J. Mar. Sci. Tech. Special Issue 55–68.

Dufour, S., N. Delerue-Le Belle, and Y.A. Fontaine. 1983a. Effects of steroid hormones on pituitary immunoreactive gonadotropin in European freshwater eel, *Anguilla anguilla* L. Gen. Comp. Endocrinol. 52: 190–197.

Dufour, S., N. Delerue-Le Belle and Y.A. Fontaine. 1983b. Development of a heterologous radioimmunoassay for eel (*Anguilla anguilla*) gonadotropin. Gen. Comp. Endocrinol. 49: 404–13.

Dufour, S., Y.A. Fontaine and B. Kerdelhué. 1985. Increase in brain and pituitary radioimmunoassayable gonadotropin releasing hormone (GnRH) in the European silver eel treated with sexual steroid or human chorionic gonadotropin. Neuropeptides 6: 495–502.

Dufour, S., E. Lopez, F. Le Menn, N. Le Belle, S. Baloche and Y.A. Fontaine. 1988. Stimulation of gonadotropin release and ovarian development, by the administration of a gonadoliberin agonist and of dopamine antagonists, in female silver eel pretreated with estradiol. Gen. Comp. Endocrinol. 70: 20–30.

Dufour, S., N. Le Belle, S. Baloche and Y.A. Fontaine. 1989. Postive feedback control by the gonads on gonadotropin (GTH) and gonadoliberin (GnRH) levels in experimentally matured female silver eels, *Anguilla anguilla*. Fish Physiol. Biochem. 7: 157–162.

Dufour, S., M. Bassompierre, M. Montero, N. Le Belle, S. Baloche and Y.A. Fontaine. 1991. Stimulation of pituitary gonadotropic function in female silver eels treated by a gonadoliberin agonist and dopamine antagonist. *In*: A.P. Scott, J.P. Sumpter, D.E. Kime and M.S. Rolfe [eds]. Proc IVth Int Symp Reproductive Physiology of Fish. Edited by FishSymp 91, Sheffield. UK. pp. 54–56.

Dufour, S., M. Montero, N. Le Belle, M. Bassompierre, J.A. King, R.P. Millar, R.E. Peter and Y.A. Fontaine. 1993. Differential distribution and response to experimental sexual maturation of two forms of brain gonadotropin-releasing hormone (GnRH) in the European eel, *Anguilla anguilla*. Fish Physiol. Biochem. 11: 99–106.

Dufour, S., E. Burzawa-Gérard, N. Le Belle, M. Sbaihi and B. Vidal. 2003. Reproductive endocrinology of the European eel, *Anguilla anguilla*. *In*: K. Aida, K. Tsukamoto and K. Yamauchi [eds]. Eel Biology. Springer, Tokyo, Japan. pp. 107–117.

Dufour, S., F.-A. Weltzien, M.E. Sébert, N. Le Belle, B. Vidal, P. Vernier and C. Pasqualini. 2005. Dopaminergic inhibition of reproduction in teleost fishes: ecophysiological and evolutionary implications. Ann. New York Acad. Sci. 1040: 9–21.

Dufour, S., M.E. Sébert, F.A. Weltzien, K. Rousseau and C. Pasqualini. 2010. Neuroendocrine control by dopamine of teleost reproduction. J. Fish Biol. 76: 129–60.

Durif, C., S. Dufour and P. Elie. 2005. The silvering process of *Anguilla anguilla*: a new classification from the yellow resident to the silver migrating stage. J. Fish Biol. 66: 1025–1043.

Durif, C.M.F., S. Dufour and P. Elie. 2006. Impact of silvering stage, age, body size and condition on the reproductive potential of the European eel. Mar. Ecol. Prog. Ser. 327: 171–181.

Durif, C.M.F., V. Van Ginneken, S. Dufour, T. Müller and P. Elie. 2009. Seasonal evolution and individual differences in silvering eels from different locations. *In*: G. Van den Thillart, S. Dufour and J.C. Rankin [eds]. Spawning Migration of the European Eel. Springer, Netherland. pp. 13–38.

Durif, C.M.F., J. GjØsaeter and A. Vollestad. 2011. Influence of oceanic factors on *Anguilla anguilla* (L.) over the twentieth century in coastal habitats of the Skagerrak, southern Norway. Proc. R. Soc. Lond. B. 278: 464–473.

Edel, R.K. 1975. The induction of maturation of female American eels through hormone injections. Helgol. Mar. Res. 27: 131–138.

Epstein, F.H., M. Cynamon and W. McKay. 1971. Endocrine control of Na-K-ATPase and seawater adaptation in *Anguilla rostrata*. Gen. Comp. Endocrinol. 16: 323–328.

Ernst, P. 1975. Catch of an eel (*A. Anguilla*) northeast of the Faroe Islands. Ann. Biologiques Cons. Perm. Explor. Mer. 32: 175.

Fernandez-Delgado, C., J.A. Hernando, Y. Ferrera and M. Bellido. 1989. Age and growth of yellow eels, *Anguilla anguilla*, in the estuary of the Guadalquivir river (south-west Spain). J. Fish Biol. 34: 561–570.

Fontaine, M. 1936. Sur la maturation complète des organes génitaux de l'anguille mâle et l'émission spontanée des produits sexuels. C. R. Acad. Sci. Paris 202: 1312–1315.

Fontaine, M. 1985. Action de facteurs anormaux du milieu sur l'écophysiologie d'anticipation des poissons migrateurs amphihalins. Ichtyophysiologica Acta 9: 11–25.

Fontaine, M., E. Bertrand, E. Lopez and O. Callamand. 1964. Sur la maturation des organes génitaux de l'anguille femelle (*Anguilla anguilla* L.) et l'émission spontanée des oeufs en aquarium. C. R. Hebd. Seances Acad. Sci. 259: 2907–2910.

Fontaine, Y.-A., E. Lopez, N. Delerue-Le Belle, E. Fontaine-Bertrand, F. Lallier and C. Salmon. 1976. Stimulation gonadotrope de l'ovaire chez l'anguille (*Anguilla anguilla* L.) hypophysectomisée. J. Physiol. Paris 72: 871–892.

Fontaine, Y.-A., S. Dufour, J. Alinat and M. Fontaine. 1985. A long immersion in deep-sea stimulates the pituitary gonatropic function of the female European eel (*Anguilla anguilla* L.). C. R. Séances Acad. Sci. 300: 83–87.

Forrest, J.N. 1973. Na transport and Na-K-ATPase in gills during adaptation to seawater: effects of cortisol. Am. J. Physiol. 224: 709–713.

Fox, B.K., J.P. Breves, T. Hirano and E.G. Grau. 2009. Effects of short- and long-term fasting on plasma and stomach ghrelin, and the growth hormone/insulin-like growth factor I axis in the tilapia, *Oreochromis mossambicus*. Domest. Anim. Endocrinol. 37: 1–11.

Fricke, H. and R. Kaese. 1995. Tracking artificially matured eels (*Anguilla anguilla*) in the Sargasso Sea and the problem of the eel's spawning site. Naturwissenschaften 82: 32–36.

Han, Y.S., I.C. Liao, Y.S. Huang, W.N. Tzeng and J.Y. Yu. 2003a. Profiles of PGH-alpha, GTH I-beta, and GTH II-beta mRNA transcript levels at different ovarian stages in the wild female Japanese eel *Anguilla japonica*. Gen. Comp. Endocrinol. 133: 8–16.

Han, Y.-S., I.-C. Liao, W.-N. Tzeng, Y.-S. Huang and J.Y.-L. Yu. 2003b. Serum estradiol-17β and testosterone levels during silvering in wild Japanese eel *Anguilla japonica*. Comp. Biochem. Physiol. 136B: 913–920.

Han, Y.-S., I.-C. Liao, Y.-S. Huang, J.-T. He, C.-W. Chang and W.-N. Tzeng. 2003c. Synchronous changes of morphology and gonadal development of silvering Japanese eel *Anguilla japonica*. Aquaculture 219: 783–796.

Han, Y.S., I.C. Liao, W.N. Tzeng and J.Y.L. Yu. 2004. Cloning of the cDNA for thyroid stimulating hormone ß subunit and changes in activity of the pituitary-thyroid axis during silvering of the Japanese eel, *Anguilla japonica*. J. Mol. Endocrinol. 32: 179–194.

Haro, A. 2003. Downstream migration of silver-phase anguillid eels. *In*: K. Aida, K. Tsukamoto and K. Yamauchi [eds]. Eel Biology. Springer, Tokyo, Japan. pp. 215–222.

Hausman, G.J., C.R. Barb and C.A. Lents. 2012. Leptin and reproductive function. Biochimie 94: 2075–81.

Henkel, C.V., E. Burgerhout, D.L. de Wijze, R.P. Dirks, Y. Minegishi, H.J. Jansen, H.P. Spaink, S. Dufour, F.A. Weltzien, K. Tsukamoto and G.E. Van den Thillart. 2012a. Primitive duplicate Hox clusters in the European eel's genome. PLoS One 7:e32231.

Henkel, C.V., R.P. Dirks, D.L. de Wijze, Y. Minegishi, J. Aoyama, H.J. Jansen, B. Turner, B. Knudsen, M. Bundgaard, K.L. Hvam, M. Boetzer, W. Pirovano, F.A. Weltzien, S. Dufour, K. Tsukamoto, H.P. Spaink and G.E. Van den Thillart. 2012b. First draft genome sequence of the Japanese eel, *Anguilla japonica*. Gene 511: 195–201.

Hope, A.J., J.C. Partridge and P.K. Hayes. 1998. Switch in rod opsin gene expression in the European eel, *Anguilla anguilla* (L.). Proc. Biol. Sci. 265: 869–874.

Huang, Y.S. 1998. Rôle des steroïdes sexuels et des hormones métaboliques dans le contrôle direct hypophysaire de l'hormone gonadotrope (GtH-II) chez l'anguille européenne, *Anguilla anguilla*. PhD thesis University Paris VI, Paris, France.

Huang, Y.S., K. Rousseau, N. Le Belle, B. Vidal, E. Burzawa-Gerard, J. Marchelidon and S. Dufour. 1998. Insulin-like growth factor-I stimulates gonadotrophin production from eel pituitary cells: a possible metabolic signal for induction of puberty. J. Endocrinol. 159: 43–52.

Huang, Y.S., K. Rousseau, M. Sbaihi, N. Le Belle, M. Schmitz and S. Dufour. 1999. Cortisol selectively stimulates pituitary gonadotropin beta-subunit in primitive teleost, *Anguilla anguilla*. Endocrinology 140: 1228–1235.

Ijiri, S., Y. Kazeto, P.M. Lokman, S. Adachi and K. Yamauchi. 2003. Characterization of a cDNA encoding P-450 aromatase (CYP19) from Japanese eel ovary and its expression in ovarian follicles during induced ovarian development. Gen. Comp. Endocrinol. 130: 193–203.

Ikeuchi, T., T. Todo, T. Kobayashi and Y. Nagahama. 1999. cDNA cloning of a novel androgen receptor subtype. J. Biol. Chem. 274: 25205–25209.

Jellyman, D.J. 1987. Review of the marine life history of Australian temperate species of *Anguilla*. Am. Fish. Soc. Symp. 1: 276–285.

Jellyman, D.J. 2003. The distribution and biology of the south Pacific species of *Anguilla*. *In*: Eel Biology. Springer, Tokyo, Japan. pp. 275–292.

Jellyman, D.J. and K. Tsukamoto. 2002. First use of archival transmitters to track migrating freshwater eels *Anguilla dieffenbachii* at sea. Mar. Ecol. Prog. Ser. 233: 207–215.

Jellyman, D.J. and K. Tsukamoto. 2005. Swimming depths of offshore migrating longfin eels *Anguilla dieffenbachii*. Mar. Ecol. Prog. Ser. 286: 261–267.

Jellyman, D. and K. Tsukamoto. 2010. Vertical migrations may control maturation in migrating female *Anguilla dieffenbachii*. Mar. Ecol. Prog. Ser. 404: 241–247.

Jeng, S.-R., S. Dufour and C.-F. Chang. 2005. Differential expression of neural and gonadal aromatase enzymatic activities in relation to gonadal development in Japanese eel, *Anguilla japonica*. J. Exp. Zool. 303A: 802–812.

Jeng, S.-R., W.-S. Yueh, G.-R. Chen, Y.-H. Lee, S. Dufour and C.-F. Chang. 2007. Differential expression and regulation of gonadotropins and their receptors in the Japanese eel, *Anguilla japonica*. Gen. Comp. Endocrinol. 154: 161–173.

Jeng, S.-R., W.-S. Yueh, Y.-T. Pen, M.-M. Gueguen, J. Pasquier, S. Dufour, C-F. Chang and O. Kah. 2012a. Expression of aromatase in radial glial cells in the brain of the Japanese eel provides insight into the evolution of the cyp191a gene in actinopterygians. PLoS ONE 7(9): e44750.

Jeng, S.-R., J. Pasquier, W.-S. Yueh, G.-R. Chen, Y.-H. Lee, S. Dufour and C.-F. Chang. 2012b. Differential regulation of the expression, of cytochrome P450 aromatase, estrogen and androgen receptor subtypes in the brain-pituitary-ovarian axis of the Japanese eel (*Anguilla japonica*) reveals steroid dependent and independent mechanisms. Gen. Comp. Endocrinol. 175: 163–172.

Jespersen, P. 1942. Indo-Pacific leptocephalids of the genus *Anguilla*: systematic and biological studies. Dana Rep. 22: 1–128.

Jönsson, E., H. Kaiya and B.T. Björnsson. 2010. Ghrelin decreases food intake in juvenile rainbow trout (*Oncorhynchus mykiss*) through the central anorexigenic corticotropin-releasing factor system. General and Comparative Endocrinology 166: 39–46.

Kagawa, H., H. Tanaka, T. Unuma, H. Ohta, K. Gen and K. Okuzawa. 2003. Role of prostaglandin in the control of ovulation in the Japanese eel *Anguilla japonica*. Fisheries Sci. 69: 234–241.

Kagawa, H., H. Tanaka, H. Ohta, T. Unuma and K. Nomura. 2005. The first success of glass eel production in the world : basic biology on fish reproduction advances new apllied technology in aquaculture. Fish Physiol. Biochem. 31: 193–199.

Kagawa, H., Y. Sakurai, R. Horiuchi, Y. Kazeto, K. Gen, H. Imaizumi and Y. Masuda. 2012. Mechanism of oocyte maturation and ovulation and its application to seed production in the Japanese eel. Fish Physiol Biochem. Published online: 26 January 2012.

Kah, O., S. Dufour, S. Baloche and B. Breton. 1989. The GnRH systems in the brain and pituiatry of normal and hCG-treated European silver eels. Fish Physiol. Biochem. 6: 279–284.

Kah, O., C. Lethimonier, G. Somoza, L.G. Guilgur, C. Vaillant and J.J. Lareyre. 2007. GnRH and GnRH receptors in metazoa: a historical, comparative, and evolutive perspective. Gen. Comp. Endocrinol. 153: 346–64.

Kang, K.S., S. Yahashi and K. Matsuda. 2011. Central and peripheral effects of ghrelin on energy balance, food intake and lipid metabolism in teleost fish. Peptides 32: 2242–7.

Kapsimali, M., B. Vidal, A. Gonzalez, S. Dufour and P. Vernier. 2000. Distribution of the mRNA encoding the four dopamine D(1) receptor subtypes in the brain of the european eel (*Anguilla anguilla*): comparative approach to the function of D(1) receptors in vertebrates. J. Comp. Neurol. 419: 320–43.

Kauffman, A.S., D.K. Clifton and R.A. Steiner. 2007. Emerging ideas about kisspeptin-GPR54 signaling in the neuroendocrine regulation of reproduction. Trends in Neuroscience 30: 504–11.

Kazeto, Y., M. Kohara, T. Tosaka, K. Gen, M. Yokoyama, C. Miura, T. Miura, S. Adachi and K. Yamauchi. 2012. Molecular characterization and gene expression of Japanese eel (*Anguilla japonica*) gonadotropin receptors. Zool. Sci. 29: 204–211.

Khan, I.A., E. Lopez and J. Leloup-Hatey. 1987. Induction of spermatogenesis and spermiation by a single injection of human chorionic gonadotropin in intact and hypophysectomized immature European eel (*Anguilla anguilla* L.). Gen. Comp. Endocrinol. 68: 91–103.

King, J.A., S. Dufour, Y.A. Fontaine and R.P. Millar. 1990. Chromatographic and immunological evidence for mammalian GnRH and chicken II GnRH in eel (*Anguilla anguilla*) brain and pituitary. Peptides 11: 507–514.

Kleckner, R.C. 1980. Swimbladder volume maintenance related to initial oceanic migratory depth in silver-phase *Anguilla rostrata*. Science 208: 1481–1482.

Kwon, H.C. and Y. Mugiya. 1994. Involvement of growth hormone and prolactin in the induction of vitellogenin synthesis in primary hepatocyte culture in the eel, *Anguilla japonica*. Gen. Comp. Endocrinol. 93: 51–60.

Lafont, A.G., M. Morini, J. Pasquier, K. Rousseau and S. Dufour. 2012. Duplicated leptin/ leptin receptor system in a basal teleost, the European eel. 7th International Symposium on Fish Endocrinology, September 3-6 2012, Buenos Aires, Argentina.

Larsen, L.O. and S. Dufour. 1993. Growth, reproduction and death in lampreys and eels. *In*: J.C. Rankin and F.B. Jensen [eds]. Fish Ecophysiology. Chapman and Hall, London, UK. pp. 72–104.

Larsson, P., S. Hamrin and L. Okla. 1990. Fat content as a factor inducing migratory behavior in the eel (*Anguilla anguilla* L.) to the Sargasso sea. Naturwissenschaften 77: 488–490.

Lee, T.W. 1979. Dynamique des populations d'anguilles *Anguilla anguilla* (L.) des lagunes du basin d'Arcachon. PhD thesis, Université des Sciences et Techniques du Languedoc, Montpellier, France.

Leloup-Hatey, J., J.P. Oudinet and E. Lopez. 1985. Testicular steroidogenesis during gonadotropin-induced spermatogenesis in male European eel (*Anguilla anguilla* L.). *In*: B. Lofts and W.N. Holmes [eds]. Current Trends in Comparative Endocrinology. Hong Kong University Press, Hong Kong. pp. 229–232.

Leloup-Hatey, J., A. Hardy, K. Nahoul, B. Quérat and Y. Zohar. 1988. Influence of gonadotrophic treatment upon the ovarian steroidogenesis in European silver eel (*Anguilla anguilla* L.). INRA Paris, Les colloques de l'INRA, N°44, Reproduction chez les poissons, Bases fondamentales et appliquées en endocrinologie et génétique. Tel-Aviv, Nov. 10–12. pp. 127–130.

Lin, H.R., G. Xie, L.H. Zhang, X.D. Wang and L.X. Chen. 1998. Artificial induction of gonadal maturation and ovulation in the Japanese eel (*Anguilla japonica* T & S). Bull. Fr. Peche Piscicult. 349: 163–176.

Lin, S.-W. and W. Ge. 2009. Differential regulation of gonadotropins (FSH and LH) and growth hormone (GH) by neuroendocrine, endocrine, and paracrine factors in the zebrafish—An *in vitro* approach. Gen. Comp. Endocrinol. 160: 183–193.

Lokman, P.M. and G. Young. 1998. Gonad histology and plasma steroid profiles in wild New Zealand freshwater eels (*Anguilla dieffenbachii* and *A. australis*) before and at the onset of the natural spawning migration. II. Males. Fish Physiol. Biochem. 19: 339–347.

Lokman, P.M. and G. Young. 2000. Induced spawning and early ontogeny of New Zealand freshwater eels (*Anguilla dieffenbachii* and *A. australis*). N. Z. J. Mar. Freshwat. Res. 34: 135–145.

Lokman, P.M., G.J. Vermeulen, J.G.D. Lambert and G. Young. 1998. Gonad histology and plasma steroid profiles in wild New Zealand freshwater eels (*Anguilla dieffenbachii* and *A. australis*) before and at the onset of the natural spawning migration. I. Females. Fish Physiol. Biochem. 19: 325–338.

Lokman, P.M., K.A. George, S.L. Divers, M. Algie and G. Young. 2007. 11-Ketotestosterone and IGF-I increase the size of previtellogenic oocytes from shortfinned eel, *Anguilla australis*, *in vitro*. Reproduction 133: 955–67.

Lopez, E. and Y.A. Fontaine. 1990. Stimulation hormonale *in vivo* de l'ovaire d'anguille européenne au stade jaune. Reproduction Nutrition Development 30: 577–582.

Manabe, R., J. Aoyama, K. Watanabe, M. Kawai, M.J. Miller and K. Tsukamoto. 2011. First observations of the oceanic migration of Japanese eel, from pop-up archival transmitting tags. Mar. Ecol. Prog. Ser. 437: 229–240.

Marchelidon, J., N. Le Belle, A. Hardy, B. Vidal, M. Sbaihi, E. Burzawa-Gérard, M. Schmitz and S. Dufour. 1999. Etude des variations de paramètres anatomiques et endocriniens chez l'anguille européenne (*Anguilla anguilla*) femelle, sédentaire et d'avalaison: application à la caractérisation du stade argenté. Bull. Fr. Peche Piscicult. 355: 349–368.

Masuda, Y., H. Imaizumi, K. Oda, K. Hashimoto, K. Teruya and H. Usuki. 2011. Japanese eel *Anguilla japonica* larvae can metamorphose into glass eel within 131 days after hatching in captivity. Nippon Suisan Gakk. 77: 416–418.

Mayer, A. and M. Schartl. 1999. Gene and genome duplications in vertebrates: the one-to-four (-to-eight in fish) rule and the evolution of novel gene functions. Curr. Opin. Cell Biol. 11: 699–704.

Messager, S., E.E. Chatzidaki, D. Ma, A.G. Hendrick, D. Zahn, J. Dixon, R.R. Thresher, I. Malinge, D. Lomet, M.B. Carlton, W.H. Colledge, A. Caraty and S.A. Aparicio. 2005. Kisspeptin directly stimulates gonadotropin-releasing hormone release *via* G protein-coupled receptor 54. Proc. Natl. Acad. Sci. USA 102: 1761–6.

Minegishi, Y., R.P. Dirks, D.L. de Wijze, S.A. Brittijn, E. Burgerhout, H.P. Spaink and G.E.E.J.M. Van den Thillart. 2012. Quantitative bioassays for measuring biologically functional gonadotropins based on eel gonadotropic receptors. Gen. Comp. Endocrinol. 178: 145–152.

Miura, T., K. Yamauchi, Y. Nagahama and H. Takahashi. 1991. Induction of spermatogenesis in male Japanese eel, *Anguilla japonica*, by a single injection of human chorionic gonadotropin. Zool. Sci. 8: 63–73.

Miura, T., K. Maruyama, H. Kaiya, M. Miyazato, K. Kangawa, M. Uchiyama, S. Shioda and K. Matsuda. 2009. Purification and properties of ghrelin from the intestine of the goldfish, *Carassius auratus*. Peptides 30: 758–65.

Miura, C., Y. Shimizu, M. Uehara, Y. Ozaki, G. Young and T. Miura. 2011. Gh is produced by the testis of Japanese eel and stimulates proliferation of spermatogonia. Reproduction 142: 869–77.

Montero, M., B. Vidal, J.A. King, G. Tramu, F. Vandesande, S. Dufour and O. Kah. 1994. Immunocytochemical localization of mammalian GnRH (gonadotropin-releasing hormone) and chicken GnRH-II in the brain of the European silver eel (*Anguilla anguilla* L.). J. Chem. Neuroanat. 7: 227–241.

Montero, M., N. Le Belle, J.A. King, R.P. Millar and S. Dufour. 1995. Differential regulation of the two forms of gonadotropin-releasing hormone (mGnRH and cGnRH-II) by sex steroids in the European female silver eel (*Anguilla anguilla*). Neuroendocrinology 61: 525–35.

Muccioli, G., T. Lorenzi, M. Lorenzi, C. Ghè, E. Arnoletti, G.M. Raso, M. Castellucci, O. Gualillo and R. Meli. 2011. Beyond the metabolic role of ghrelin: a new player in the regulation of reproductive function. Peptides 32: 2514–21.

Nader, M.R., T. Miura, N. Ando, C. Miura and K. Yamauchi. 1999. Recombinant human insulin-like growth factor I stimulates all stages of 11-ketotestosterone-induced spermatogenesis in the Japanese eel, *Anguilla japonica*, in vitro. Biol. Reprod. 61: 944–7.

Nagae, M., T. Todo, K. Gen, Y. Kato, G. Young, S. Adachi and K. Yamauchi. 1996a. Molecular cloning of the cDNAs encoding pituitary glycoprotein hormone α- and gonadotropin IIβ-subunits of the Japanese eel, *Anguilla japonica*, and increase in their mRNAs during ovarian development induced by injection of chum salmon pituitary homogenate. J. Mol. Endocrinol. 16: 171–181.

Nagae, M., S. Adachi and K. Yamauchi. 1996b. Changes in transcription of pituitary glycoprotein hormone α and gonadotropin IIβ subunits during ovarian development induced by repeated injections of salmon pituitary homogenate in the Japanese eel, *Anguilla japonica*. Fish Physiol. Biochem. 17: 179–186.

Nagae, M., S. Adachi and K. Yamauchi. 1997. Changes in transcription of pituitary glycoprotein hormones α and gonadotropin IIβ subunits during ovarian development induced by repeated injections of salmon pituitary homogenate in the Japanese eel, *Anguilla japonica*. Fish Physiol. Biochem. 17: 179–186.

Nakamura, T., K. Takio, Y. Eto, H. Shibai, K. Titani and H. Sugino. 1990. Activin-binding protein from rat ovary is follistatin. Science 247: 836–838.

Ohta, H. and H. Tanaka. 1997. Relationship between serum levels of human chorionic gonadotropin (hCG) and 11-ketotestosterone after a single injection of hCG and induced maturity in the male Japanese eel, *Anguilla japonica*. Aquaculture 153: 123–134.

Ohta, H., H. Kagawa, H. Tanaka, K. Okuzawa and K. Hirose. 1996. Changes in fertilization and hatching rates with time after ovulation induced by 17α, 20β-dihydroxy-4-pregnen-3-one in the Japanese eel, *Anguilla japonica*. Aquaculture 139: 291–301.

Ohta, H., H. Kagawa, H. Tanaka, K. Okuzawa and K. Hirose. 1997. Artificial induction of maturation and fertilization in the Japanese eel, *Anguilla japonica*. Fish Physiol. Biochem. 17: 163–169.

Ohta, T., H. Miyake, C. Miura, H. Kamei, K. Aida and T. Miura. 2007. Follicle-stimulating hormone induces spermatogenesis mediated by androgen production in Japanese eel, *Anguilla japonica*. Biol. Reprod. 77: 970–7.

Okubo, K., H. Suetake and K. Aida. 1999a. Expression of two gonadotropin-releasing hormone (GnRH) precursor genes in various tissues of the Japanese eel and evolution of GnRH. Zool. Sci. 16: 471–478.

Okubo, K., H. Suetake and K. Aida. 1999b. A splicing variant for the prepro-mammalian gonadotropin-releasing hormone (prepro-mGnRH) mRNA is present in the brain and various peripheral tissues of the Japanese eel. Zool. Sci. 16: 645–651.

Okubo, K., H. Suetake and K. Aida. 2002. Three mRNA species for mammalian-type gonadotropin-releasing hormone in the brain of the eel *Anguilla japonica*. Mol. Cell. Endocrinol. 192: 17–25.

Okubo, K., H. Suetake, T. Usami and K. Aida. 2000. Molecular cloning and tissue-specific expression of a gonadotropin-releasing hormone receptor in the Japanese eel. Gen. Comp. Endocrinol. 119: 181–192.

Okubo, K., S. Nagata, R. Ko, H. Kataoka, Y. Yoshiura, H. Mitani, M. Kondo, K. Naruse, A. Shima and K. Aida. 2001. Identification and characterization of two distinct gonadotropin-releasing hormone receptor subtypes in a teleost, the medaka *Oryzias latipes*. Endocrinology 142: 4729–4739.

Oliveira, K. and W.E. Hable. 2010. Artificial maturation, fertilization and early development of the American eel (*Anguilla rostrata*). Can. J. Zool. 88: 1121–1128.

Olivereau, M. and J. Olivereau. 1985. Effects of 17 α-methyltestosterone on the skin and gonads of freshwater male silver eels. Gen. Comp. Endocrinol. 57: 64–71.

Olivereau, M., P. Dubourg, P. Chambolle and J. Olivereau. 1986. Effects of estradiol and mammalian LHRH on the ultrastructure of the pars distalis of the eel. Cell Tissue Res. 246: 425–37.

Palstra, A.P. and G.E.E.J.M. Van den Thillart. 2010. Swimming physiology of European silver eels (*Anguilla anguilla* L.): energetic costs and effects on sexual maturation and reproduction. Fish Physiol. Biochem. 36: 297–322.

Palstra, A.P., E.G.H. Cohen, P.R.W. Niemantsverdriet, V.J.T Van Ginneken and G.E.E.J.M. Van den Thillart. 2005. Artificial maturation and reproduction of European silver eel: Development of oocytes during final maturation. Aquaculture 249: 533–547.

Palstra, A.P., D. Curiel, M. Fekkes, M. de Bakker, C. Székely, V. Van Ginneken and G. Van den Thillart. 2007. Swimming stimulates oocyte development in European eel (*Anguilla anguilla* L.). Aquaculture 270: 321–332.

Palstra, A.P., V. Van Ginneken and G. Van den Thillart. 2008a. Cost of transport and optimal swimming speeds in farmed and wild European eels (*Anguilla anguilla*). Comp. Biochem. Physiol. 151A: 37–44.

Palstra, A.P., D. Schnabel, M.C. Nieveen, H.P. Spaink and G.E.E.J.M. Van den Thillart. 2008b. Male silver eels mature by swimming. BMC Physiol. 8: 14.

Palstra, A.P., V. Van Ginneken and G. Van den Thillart. 2009a. Effects of swimming on silvering and maturation. *In*: G. Van den Thillart, S. Dufour and C. Rankin [eds]. Spawning Migration of the European Eel. Springer, Netherlands. pp. 309–332.

Palstra, A.P., D. Schnabel, M.C. Nieveen, H.P. Spaink and G. Van den Thillart. 2009b. Temporal expression of hepatic *estrogen receptor 1*, *vitellogenin1* and *vitellogenin2* in European silver eels. Gen. Comp. Endocrinol. 166: 1–11.

Palstra, A.P., D. Schnabel, M.C. Nieveen, H.P. Spaink and G. Van den Thillart. 2010a. Temporal expression of hepatic estrogen receptor 1, vitellogenin1 and vitellogenin2 in European silver eels. Gen. Comp. Endocrinol. 166: 1–11.

Palstra, A.P., D. Schnabel, M.C. Nieveen, H.P. Spaink and G. Van den Thillart. 2010b. Swimming suppresses hepatic vitellogenesis in European female silver eels as shown by expression of the estrogen receptor 1, vitellogenin1 and vitellogenin2 in the liver. Reprod. Biol. Endocrinol. 8: 27–36.

Pandi-Perumal, S.R., V. Srinivasan, G.J.M. Maestroni, D.P. Cardinali, B. Poeggeler and R. Hardeland. 2006. Melatonin. Nature's most versatile biological signal? FEBS J. 273: 2813–2838.

Pankhurst, N.W. 1982. Relation of visual changes to the onset of sexual maturation in the European eel, *Anguilla anguilla* (L). J. Fish Biol. 21: 127–140.

Pankhurst, N.W. and J.N. Lythgoe. 1982. Structure and color of the tegument of the European eel *Anguilla anguilla* (L.). J. Fish Biol. 21: 279–296.

Pankhurst, N.W. and J.N. Lythgoe. 1983. Changes in vision and olfaction during sexual maturation in the European eel *Anguilla anguilla* (L.). J. Fish Biol. 23: 229–240.

Pasqualini, C., F.-A. Weltzien, B. Vidal, S. Baloche, C. Rouget, N. Gilles, D. Servent, P. Vernier and S. Dufour. 2009. Two distinct dopamoine D2 receptor genes in the European eel: molecular characterization, tissue-specific transcription, and regulation by sex steroids. Endocrinology 150: 1377–1392.

Pasquier, J. 2012. Evolution du contrôle neuroendocrinien de la reproduction: origine et rôle du système kisspeptine? PhD Thesis University Paris VI, Paris, France.

Pasquier, J., A.G. Lafont, J. Leprince, H. Vaudry, K. Rousseau and S. Dufour. 2011. First evidence for a direct inhibitory effect of kisspeptins on LH expression in the eel, *Anguilla anguilla*. Gen. Comp. Endocrinol. 173: 216–25.

Pasquier, J., A.G. Lafont, S.R. Jeng, M. Morini, R. Dirks, G. Van den Thillart, J. Tomkiewicz, H. Tostivint, C.F. Chang, K. Rousseau and S. Dufour. 2012. Multiple kisspeptin receptors in early osteichthyans provide new insights into the evolution of this receptor family. PLoS One 7: e48931.

Pasquier, J., A.-G. Lafont, H. Tostivint, H. Vaudry, K. Rousseau and S. Dufour. 2012. Comparative evolutionary histories of kisspeptins and kisspeptin receptors in vertebrates reveal both parallel and divergent features. Frontiers in Endocrinology, Front. Endocrinol. 3: 173.

Pedersen, B.H. 2003. Induced sexual maturation of the European eel *Anguilla anguilla* and fertilisation of the eggs. Aquaculture 224: 323–338.

Peñaranda, D.S., L. Pérez, V. Gallego, M. Jover, H. Tveiten, S. Baloche, S. Dufour and J.F. Asturiano. 2010a. Molecular and physiological study of the artificial maturation process in European eel males: from brain to testis. Gen. Comp. Endocrinol. 166: 160–71.

Peñaranda, D.S., L. Pérez, V. Gallego, R. Barrera, M. Jover and J.F. Asturiano. 2010b. European eel sperm diluent for short-term storage. Reprod. Dom. Anim. 45: 407–415.

Perez, L., D.S. Pearanda, S. Dufour, S. Baloche, A.P. Palstra, G.E.E.J.M. Van den Thillart and J.F. Asturiano. 2011. Influence of temperature regime on endocrine parameters and vitellogenis during experimental maturation of European eel (*Anguilla anguilla*) females. Gen. Comp. Endocrinol. 174: 51–59.

Peyon, P., S. Baloche and E. Burzawa-Gerard. 1996. Potentiating effect of growth hormone on vitellogenin synthesis induced by 17 beta-estradiol in primary culture of female silver eel (*Anguilla anguilla* L.) hepatocytes. Gen. Comp. Endocrinol. 102: 263–73.

Peyon, P., S. Baloche and E. Burzawa-Gérard. 1997. Investigation into the possible role of androgens in the induction of hepatic vitellogenesis in the European eel: *in vivo* and *in vitro* studies. Fish Physiol. Biochem. 16: 107–118.

Peyon, P., S. Zanuy and M. Carrillo. 2001. Action of leptin on *in vitro* luteinizing hormone release in the European sea bass (*Dicentrarchus labrax*). Biol. Reprod 2001 65: 1573–8.

Poole, W.R. and J.D. Reynolds. 1996. Growth rate and age at migration of *Anguilla anguilla*. J. Fish Biol. 48: 633–642.

Quérat, B., K. Nahoul, A. Hardy, Y.A. Fontaine and J. Leloup-Hâtey. 1987. Plasma concentrations of ovarian steroids in the freshwater European silver eel (*Anguilla anguilla* L.): effects of hypophysectomy and transfer to sea water. J. Endocrinol. 114: 289–94.

Reinsch, H.H. 1968. Fund von Fluss-Aalen, *Anguilla anguilla* (L.) im Nordatlantik. Archiv für Fischereiwissenschaft 19: 62–63.

Riley, L.G., B.K. Fox, H. Kaiya, T. Hirano and E.G. Grau. 2005. Long-term treatment of ghrelin stimulates feeding, fat deposition, and alters the GH/IGF-I axis in the tilapia, *Oreochromis mossambicus*. Gen. Comp. Endocrinol. 142: 234–40.

Roa, J., E. Aguilar, C. Dieguez, L. Pinilla and M. Tena-Sempere. 2008. New frontiers in kisspeptin/GPR54 physiology as fundamental gatekeepers of reproductive function. Front. Neuroendocrinol. 29: 48–69.

Robins, C.R., D.M. Cohen and C.H. Robins. 1979. The eels, Anguilla and Histiobranchus, photographed on the floor of the deep Atlantic in the Bahamas. Bull. Mar. Sci. 29: 401–405.

Rohr, D.H., P.M. Lokman, P.S. Davie and G. Young. 2001. 11-ketotestosterone induces silvering-related changes in immature female short-finned eels, *Anguilla australis*. Comp. Biochem. Physiol. 130A: 701–714.

Romeo, R.D. 2003. Puberty: a period of both organizational and activational effects of steroid hormones on neurobehavioral development. J. Neuroendocrinol. 15: 1185–1192.

Rossi, R. and G. Colombo. 1976. Sex ratio, age and growth of silver eels in two brackish lagoons in the northern Adriatic Valli of Commacchio and Valli Nuova. Archivio di Oceanografia e Limnologia 18: 327–341.

Rousseau, K. and S. Dufour. 2004. Phylogenetic evolution of the neuroendocrine control of growth hormone: contribution from teleosts. Cybium 28: 181–198.

Rousseau, K. and S. Dufour. 2008. Endocrinology of migratory fish life cycle in special environments: the role of metamorphoses. *In*: P. Sébert, D.W. Onyango and B.G. Kapoor [eds]. Fish Life in Special Environments. Science Publishers, Enfield (NH), USA. pp. 193–231.

Rousseau, K., Y.S. Huang, N. Le Belle, B. Vidal, J. Marchelidon, J. Epelbaum and S. Dufour. 1998. Long-term inhibitory effects of somatostatin and insulin-like growth factor 1 on growth hormone release by serum-free primary culture of pituitary cells from European eel (*Anguilla anguilla*). Neuroendocrinology 67: 301–309.

Rousseau, K., N. Le Belle, J. Marchelidon and S. Dufour. 1999. Evidence that corticotropin-releasing hormone acts as a growth hormone-releasing factor in a primitive teleost, the European eel (*Anguilla anguilla*). J. Neuroendocrinol. 11: 385–392.

Rousseau, K., N. Le Belle, K. Pichavant, J. Marchelidon, B.K.C. Chow, G. Boeuf and S. Dufour. 2001. Pituitary growth hormone secretion in turbot, a phylogenetically recent teleost, is regulated by a species-specific pattern neuropeptides. Neuroendocrinology 74: 375–385.

Rousseau, K., S. Aroua, M. Schmitz, P. Elie and S. Dufour. 2009. Silvering: metamorphosis or puberty ? *In*: G. Van den Thillart, S. Dufour and C. Rankin [eds]. Spawning Migration of the European Eel. Springer, Netherlands. pp. 39–64.

Rousseau, K., S. Aroua and S. Dufour. 2012. Eel secondary metamorphosis: silvering. *I   n   :* S. Dufour, K. Rousseau and B.G. Kapoor [eds]. Fish Metamorphosis. Science Publishers, Enfield (NH), USA. pp. 216–249.

Saito, K., P.M. Lokman, G. Young, Y. Ozaki, H. Matsubara, H. Okumura, Y. Kazeto, Y. Yoshiura, K. Aida, S. Adachi and K. Yamauchi. 2003. Follicle-stimulating hormone β, luteinizing hormone β, and glycoprotein hormone α subunit mRNA levels in artificially maturing Japanese eel *Anguilla japonica* and naturally maturing New Zealand long-finned eel *Anguilla dieffenbachii*. Fisheries Sci. 69: 146–153.

Sato, N., I. Kawazoe, Y. Suzuki and K. Aida. 1996. Use of an emulsion prepared with lipophilized gelatin for the induction of ovarian maturation in the Japanese eel *Anguilla japonica*. Fisheries Sci. 62: 806–814.

Sato, N., I. Kawazoe, Y. Suzuki and K. Aida. 2006. Effects of temperature on vitellogenesis in Japanese eel *Anguilla japonica*. Fisheries Sci. 72: 961–966.

Sbaihi, M. 2001. Interaction des stéroïdes sexuels et du cortisol dans le contrôle de la reproduction et du métabolisme calcique chez un téléostéen migrateur, l'anguille (*Anguilla anguilla*). PhD thesis University Paris VI, Paris, France.

Sbaihi, M., M. Fouchereau-Peron, F. Meunier, P. Elie, I. Mayer, E. Burzawa-Gérard, B. Vidal and S. Dufour. 2001. Reproductive biology of the conger eel from the south coast of Brittany, France and comparison with the European eel. J. Fish Biol. 59: 302–318.

Sbaihi, M., K. Rousseau, S. Baloche, F. Meunier, M. Fouchereau-Peron and S. Dufour. 2009. Cortisol mobilizes mineral stores from vertebral skeleton in the European eel: an ancestral origin for glucocorticoid-induced osteoporosis? J. Endocrinol. 201: 241–252.

Schmidt, J. 1923. Breeding places and migration of the eel. Nature 111: 51–54.

Schmitz, M., S. Aroua, B. Vidal, N. Le Belle, P. Elie and S. Dufour. 2005. Differential regulation of luteinizing hormone and follicle-stimulating hormone expression during ovarian development and under sexual steroid feedback in the European eel. Neuroendocrinology 81: 107–119.

Scott, W.B. and E.J. Crossman. 1973. Freshwater fishes of Canada. Fisheries Research Board of Canada Bulletin 184: 624.

Sébert, P. 2003. Fish adaptations to pressure. *In*: A.L. Val and B.G. Kapoor [eds]. Fish Adaptation. Science Publishers, Enfield (NH), USA. pp. 73–95.

Sébert, P. 2008. Fish muscle function and pressure. *In*: P. Sébert, D. Onyango and B.G. Kapoor [eds]. Fish Life in Special Environments. Science Publishers, Enfield (NH), USA. pp. 233–255.

Sébert, M.-E., A. Amérand, A. Vettier, F.-A. Weltzien, C. Pasqualini, P. Sébert and S. Dufour. 2007. Effects of high hydrostatic pressure on the pituitary-gonad axis in the European eel, *Anguilla anguilla*. Gen. Comp. Endocrinol. 153: 289–298.

Sébert, M.-E., C. Legros, F.-A. Weltzien, B. Malpaux, P. Chemineau and S. Dufour. 2008a. Melatonin activates brain dopaminergic systems in the eel with an inhibitory impact on reproductive function. J. Neuroendocrinol. 20: 917–929.

Sébert, M.E., F.-A. Weltzien, C. Moisan, C. Pasqualini and S. Dufour. 2008b. Dopaminergic systems in the European eel : characterization, brain distribution, and potential role in migration and reproduction. Hydrobiologia 602: 27–46.

Sébert, P., A. Vettier, A. Amérand and C. Moisan. 2009. High pressure resistance and adaptation of European eels. *In*: G. Van den Thillart, S. Dufour and C. Rankin [eds]. Spawning Migration of the European Eel. Springer, Netherlands. pp. 99–127.

Sinha, V.P.R. and J.W. Jones. 1975. The European Freshwater Eel. Liverpool University Press, Liverpool, UK.

Sorensen, P.W. and H.E. Winn. 1984. The induction of maturation and ovulation in American eels, *Anguilla rostrata* (LeSueur), and the relevance of chemical and visual cues to male spawning behavior. J. Fish Biol. 25: 261–268.

Stone, R. 2003. Freshwater eels are slip-sliding away. Science 302: 221–222.

Sudo, R., H. Suetake, Y. Suzuki, T. Utoh, S. Tanaka, J. Aoyama and K. Tsukamoto. 2011a. Dynamics of reproductive hormones during downstream migration in Japanese eels, *Anguilla japonica*. Zool. Sci. 28: 180–188.

Sudo, R., R. Tosaka, S. Ijiri, S. Adachi, H. Suetake, Y. Suzuki, N. Horie, S. Tanaka, J. Aoyama and K. Tsukamoto. 2011b. Effect of temperature decrease on oocyte development, sex steroids, and gonadotropin β-subunit mRNA expression levels in female Japanese eel *Anguilla japonica*. Fisheries Sci. 77: 575–582.

Sudo, R., R. Tosaka, S. Ijiri, S. Adachi, J. Aoyama and K. Tsukamoto. 2012. 11-ketotestosterone synchronously induces oocyte development and silvering-related changes in the Japanese eel, *Anguilla japonica*. Zool. Sci. 29: 254–259.

Suetake, H., K. Okubo, N. Sato, Y. Yoshiura, Y. Suzuki and K. Aida. 2002. Differential expression of two gonadotropin (GTH)β subunit genes during ovarian maturation induced by repeated injection of salmon GTH in the Japanese eel *Anguilla japonica*. Fisheries Sci. 68: 290–298.

Suzuki, R., M. Kishida and T. Hirano. 1990. Growth hormone secretion during longterm incubation of the pituitary of the Japanese eel, *Anguilla japonica*. Fish Physiol. Biochem. 8: 159–165.

Svedäng, H. and H. Wickström. 1997. Low fat contents in female silver eels: indications of insufficient energetic stores for migration and gonadal development. J. Fish. Biol. 50: 575–586.

Svedäng, H., E. Neuman and H. Wickström. 1996. Maturation patterns in female European eel: age and size at the silver eel age. J. Fish Biol. 48: 342–351.

Tanaka, H., H. Kagawa and H. Ohta. 2001. Production of leptocephali of Japanese eel (*Anguilla japonica*) in captivity. Aquaculture 201: 51–60.

Tanaka, H., H. Kagawa, H. Ohta, T. Unuma and K. Nomura. 2003. The first production of glass eel in captivity: fish reproductive physiology facilitates great progress in aquaculture. Fish Physiol. Biochem. 28: 493–497.

Tena-Sempere, M., A. Felip, A. Gómez, S. Zanuy and M. Carrillo. 2012. Comparative insights of the kisspeptin/kisspeptin receptor system: lessons from non-mammalian vertebrates. Gen. Comp. Endocrinol. 175: 234–243.

Tesch, F.W. 1977. The Eel: Biology and Management of Anguillid Eels. Chapman and Hall, London, UK.

Tesch, F.W. 1982. The sargasso Sea Eel Expedition 1979. Helgoländer Meeresunters 35: 263–277.

Tesch, F.W. 1989. Changes in swimming depth and direction of silver eels (*Anguilla anguilla* L.) from the continental shelf to the deep sea. Aquat. Living Resour. 2: 9–20.

Tesch, F.W. 1991. Anguillidae. *In*: H. Hoestlandt [ed]. The Freshwater Fishes of Europe. AULA-Verlag, Wiesbaden, Germany.

Tesch, F.W. and N. Rohlf. 2003. Migration from continental waters to the spawning grounds. *In*: Eel Biology, Springer, Tokyo, Japan. pp. 223–234.

Todo, T., S. Adachi and K. Yamauchi. 1996. Molecular cloning and characterization of Japanese eel estrogen receptor cDNA. Mol. Cell. Endocrinol. 119: 37–45.

Todo, T., T. Ikeuchi, T. Kobayashi and Y. Nagahama. 1999. Fish androgen receptor: cDNA cloning, steroid activation of transcription in transfected mammalian cells, and tissue mRNA levels. Biochem. Biophys. Res. Commun. 254: 378–383.

Todd, P.R. 1979. Hormone-induced maturation of New Zealand freshwater eels. Rapport et Procès-verbaux Réunions. Conseil Permanent International pour l'Exploration de la Mer 174: 91–97.

Tomkiewicz, J. and R.S. Soerensen. 2008. Project Report: Artificial Reproduction of Eels III: Larval Culture (ROE III-LC). National Institute for Aquatic Resources, Technical University of Denmark (In Danish).

Tostivint, H. 2011. Evolution of the gonadotropin-releasing hormone (GnRH) gene family in relation to vertebrate tetraploidizations. Gen. Comp. Endocrinol. 170: 575–81.

Tsukamoto, K. 1992. Discovery of the spawning area for Japanese eel. Nature 356: 789–791.

Tsukamoto, K. 2006. Oceanic biology: spawning of eels near a seamount. Nature 439: 929.

Tsukamoto, K., T. Otake, N. Mochioka, T.-W. Lee, H. Fricke, T. Inagaki, J. Aoyama, S. Ishikawa, S. Kimura, M.J. Miller, H. Hasumoto, M. Oya and Y. Suzuki. 2003. Seamounts, new moon and eel spawnin: the search for the spawning site of the Japanese eel. Environ. Biol. Fish. 66: 221–229.

Tsukamoto, K., S. Chow, T. Otake, H. Kurogi, N. Mochioka, M.J. Miller, J. Aoyama, S. Kimura, S. Watanabe, T. Yoshinaga, A. Shinoda, M. Kuroki, M. Oya, T. Watanabe, K. Hata, S. Ijiri, Y. Kazeto, K. Nomura and H. Tanaka. 2011. Oceanic spawning ecology of freshwater eels in the western North Pacific. Nat. Commun. 2: 179.

Tsutsui, K., G.E. Bentley, L.J. Kriegsfeld, T. Osugi, J.Y. Seong and H. Vaudry. 2010. Discovery and evolutionary history of gonadotrophin-inhibitory hormone and kisspeptin: new key neuropeptides controlling reproduction. J. Neuroendocrinol. 22: 716–27.

Tzchori, I.G. Degani, A. Hurvitz and B. Moav. 2004. Cloning and developmental expression of the cytochrome P450 aromatase gene (CYP19) in the European eel (*Anguilla Anguilla*). Gen. Comp. Endocrinol. 138: 271–280.

Van den Thillart, G., V. Van Ginneken, F. Korner, R. Heijmans, R. Van der Linden and A. Gluvers. 2004. Endurance swimming of European eel. J. Fish Biol. 65: 312–318.

Van Ginneken, V. and G. Van den Thillart. 2000. Eel fat stores are enough to reach the Sargasso. Nature 403: 156–157.

Van Ginneken, V.J.T. and G.E. Maes. 2005. The European eel (*Anguilla Anguilla*, Linnaeus), its lifecyle, evolution and reproduction: a literature review. Rev. Fish Biol. Fisheries 15: 367–398.

Van Ginneken, V., C. Durif, S.P. Balm, R. Boot, M.W.A. Verstegen, E. Antonissenn and G. Van den Thillart. 2007a. Silvering of European eel (*Anguilla anguilla* L.): seasonal changes of morphological and metabolic parameters. Anim. Biol. 57: 63–77.

Van Ginneken, V., S. Dufour, M. Sbaihi, P. Balm, K. Noorlander, M. de Bakker, J. Doornbos, A. Palstra, E. Antonissen, I. Mayer and G. Van den Thillart. 2007b. Does a 5500-km swim trial stimulate early sexual maturation in the European eel (*Anguilla anguilla* L.)? Comp. Biochem. Physiol. 147A: 1095–1103.

Vettier, A., C. Labbé, A. Amérand, G. Da Costa, E. Le Rumeur, C. Moisan and P. Sébert. 2006. Hydrostatic pressure effects on eel mitochondrial functioning and membrane fluidity. Undersea Hyperbaric Medical Society 33: 149–156.

Vidal, B., C. Pasqualini, N. Le Belle, M.C. Holland, M. Sbaihi, P. Vernier, Y. Zohar and S. Dufour. 2004. Dopamine inhibits luteinizing hormone synthesis and release in the juvenile European eel: a neuroendocrine lock for the onset of puberty. Biol. Reprod. 71: 1491–500.

Vollestad, L.A. 1988. Tagging experiments with yellow eel, *Anguilla anguilla* (L.) in brackish water in Norway. Sarsia 73: 157–161.

Vollestad, L.A. 1992. Geographic variation in age and length at metamorphosis of maturing European eel: environmental effects and phenotypic plasticity. J. Anim. Ecol. 61: 41–48.

Vollestad, L.A. and B. Jonsson. 1986. Life-history characteristics of the European eel *Anguilla anguilla* in the Imsa River, Norway. Trans. Am. Fish. Soc. 115: 864–871.

Weil, C., P.Y. Le Bail, N. Sabin and F. Le Gac. 2003. *In vitro* action of leptin on FSH and LH production in rainbow trout (*Onchorynchus mykiss*) at different stages of the sexual cycle. Gen. Comp. Endocrinol. 130: 2–12.

Weltzien, F.A., C. Pasqualini, P. Vernier and S. Dufour. 2005. A quantitative real-time RT-PCR assay for European eel tyrosine hydroxylase. Gen. Comp. Endocrinol. 142: 134–42.

Weltzien, F.A., C. Pasqualini, M.-E. Sébert, B. Vidal, N. Le Belle, O. Kah, P. Vernier and S. Dufour. 2006. Androgen-dependent stimulation of brain dopaminergic systems in the female European eel (*Anguilla anguilla*). Endocrinology 147: 2964–2973.

Yada, T., A. Urano and T. Hirano. 1991. Growth hormone and prolactin gene expression and release in the pituitary of rainbow trout in serum-free culture. Endocrinology 129: 1183–1192.

Yam, K.M., K.L. Yu and W. Ge. 1999. Cloning and characterisation of goldfish activin betaA subunit. Gen. Comp. Endocrinol. 154: 45–54.

Yamada, Y., H. Zhang, A. Okamura, S. Tanaka, N. Horie, N. Mikawa, T. Utoh and H.P. Oka. 2000. Morphological and histological changes in the swim-bladder during maturation of the japanese eel. J. Fish Biol. 58: 804–814.

Yamamoto, K. and K. Yamauchi. 1974. Sexual maturation of Japanese eel and production of eel larvae in the aquarium. Nature 251: 220–222.

Yamauchi, K., M. Nakamura, H. Takahashi and K. Takano. 1976. Cultivation of larvae of Japanese eel. Nature 263: 412.

Yaron, Z., G. Gur, P. Melamed, H. Rosenfeld, B. Levavi-Sivan and A. Elizur. 2001. Regulation of gonadotropin subunit genes in tilapia. Comp. Biochem. Physiol. 129B: 489–502.

Ying, S.Y. 1988. Inhibins, activins, and follistatins: gonadal proteins modulating the secretion of follicle-stimulating hormone. Endocrine Rev. 9: 267–293.

Yoshiura, Y., H. Suetake and K. Aida. 1999. Duality of gonadotropin in a primitive teleost, Japanese eel (*Anguilla japonica*). Gen. Comp. Endocrinol. 114: 121–131.

Zachmann, A.J. Falcon, S.C. Knijff, V. Bolliet and M.A. Ali. 1992. Effects of photoperiod and temperature on rhythmic melatonin secretion from the pineal organ of the white sucker (*Catostomus commersoni*) *in vitro*. Gen. Comp. Endocrinol. 86: 26–3.

Zhang, H., F. Futami, N. Horie, A. Okamura, T. Utoh, N. Mikawa, Y. Yamada, S. Tanaka and N. Okamoto. 2000. Molecular cloning of fresh water and deep-sea rod opsin genes from Japanese eel *Anguilla japonica* and expressional analyses during sexual maturation. FEBS Lett. 469: 39–43.

The Swimbladder

Bernd Pelster

Introduction

The eel is known as a bottom dwelling fish, but it has a well-developed swimbladder, which is an established model for the analysis of swimbladder function. The swimbladder originates as an unpaired dorsal outgrowth of the esophagus and in the adult stage is a gas-filled sac located below the vertebral column. Although the swimbladder of some teleost fishes may serve respiratory functions or can be important for sound production, it primarily is seen as a buoyancy organ. As a buoyancy organ this structure has attracted attention since more than two hundred years (see Alexander 1966, Dorn 1961). Although the eel swimbladder has a somewhat peculiar structure, the presence of a particularly impressive 'Wundernetz', the *rete mirabile*, which is a fantastic countercurrent system allowing access to both sides of the *rete*, may explain why it has been extensively studied in the European eel *Anguilla anguilla*. Its structure had already been delineated in 1882 (Pauly 1882), but it was described in detail almost 100 years later (Dorn 1961).

While most bottom dwelling fish are characterized by a reduced swimbladder or even lack a swimbladder at all, the well-developed eel swimbladder apparently is connected to the life cycle of this teleost. The eel

Institute for Zoology, University of Innsbruck, Technikerstr. 25, Austria and Center for Molecular Biosciences, University Innsbruck.
E-mail: bernd.pelster@uibk.ac.at

is a catadromous fish. The leptocephalus larva, the larval stage of the eel, after a one to three years journey from the Sargasso Sea, enters European freshwater and develops to a bottom dwelling adult fish, but the gonads are not yet developed. In preparation of sexual maturity the yellow eel metamorphoses into a so-called silver eel, returning to the sea and migrating back to the Sargasso Sea. This maturation is connected to a significant improvement in swimbladder function, and during this migratory phase the swimbladder most likely becomes an important structure. Consequently, various aspects of eel swimbladder structure and function have been addressed, from its initial inflation to its role during the migratory phase. Within the last decades parasitological aspects have been added, because the nematode *Anguillicoloides crassus* was found in European and American eels and seriously affected swimbladder function.

Initial Swimbladder Inflation

The time of swimbladder development varies from species to species. To be able to drift in the water pelagic fish larvae often use oil and lipid droplets to achieve a status of neutral or near neutral buoyancy. Therefore in many species the swimbladder starts functioning at the time of yolk and oil depletion (Govoni and Hoss 2001, Pelster 1997a). In the leptocephalus larvae of the European eel, however, the swimbladder becomes visible as 'Schwimmblasenanlage' shortly after the larvae reach the European coast and enter the freshwater system after a journey of one to three years from the Sargasso Sea (Als et al. 2011). This is the time of the metamorphosis of the leptocephalus to the glass eel. During the following weeks and months the swimbladder of the European eel develops and becomes functional (Zwerger et al. 2002). This development is characterized by distinct changes in the size of the swimbladder as well as in the histology of the swimbladder tissue. In the initial stage the swimbladder is filled with surfactant, which in later stages covers the surface of the swimbladder epithelium (Prem et al. 2000), and no gas space is detectable (Fig. 1).

In the young glass eel, the swimbladder epithelial cells are columnar. These cells become the so-called gas gland cells, but they do not yet have the typical enlargement of basolateral membranes, the basolateral labyrinth, which is present in well-developed gas gland cells in adult eels. At this stage the vascularization of the swimbladder tissue is poorly developed, and the submucosa is present as a thick layer of connective tissue, giving a large diffusion distance between blood vessels and the swimbladder lumen. Within the next two or three months of development, gas gland cells develop their typical basolateral labyrinth, and the thickness of the submucosa is significantly reduced, resulting in a short diffusion distance between blood vessels and the swimbladder lumen. During this time

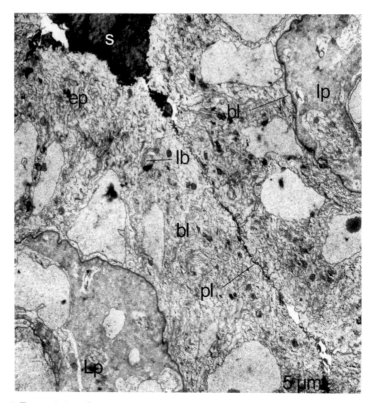

Figure 1. Transmission electron microscopical image of the epithelium of a very young glass eel (early April). The prospective lumen of the swimbladder is filled with surfactant (s). Within the epithelial cells (ep), which differentiate to gas gland cells, lamellar bodies (lb) filled with surfactant are present. The baso-lateral membrane (bl) does not yet show the typical fouldings, established in the gas gland cells of adult eels. Lp, lamina propria mucosae; pl, prospective lumen. Taken from Zwerger et al. (2002) with permission.

the size of the swimbladder increases remarkably. The first filling of the swimbladder with gas is observed while the gas gland cells are still in a poorly differentiated status. The eel is a physostome fish, i.e., the *ductus pneumaticus*, the connection between the esophagus and the swimbladder, persists until adulthood. It is generally assumed that physostome fish initially fill their swimbladder by gulping air at the water surface. Quite surprisingly, however, a large number of glass eels without access to the water surface were able to fill the swimbladder (Zwerger et al. 2002). The authors assumed that glass eels are able to pick up tiny gas bubbles from the water and transfer them to the swimbladder.

At the molecular level the initial epithelial budding of the zebrafish swimbladder from the foregut endoderm requires hedgehog (Hh) signaling and involves Wnt signaling with crosstalk between Wnt/ß-catenin and

Hh signaling, supporting the notion that the swimbladder and lungs are homologous structures (Winata et al. 2009, Yin et al. 2011). Unfortunately the molecular background for the development of the swimbladder in the leptocephalus larva remains unknown so far.

Recent experiments on the zebrafish swimbladder also revealed that the swimbladder epithelium and also blood flow through the swimbladder are necessary for a proper development of the swimbladder (Winata et al. 2010).

Structure of the Adult Swimbladder

The eel is a physostome fish and the *ductus pneumaticus* persists, but it is functionally closed. Thus, the adult eel cannot gulp air at the water surface. In the adult eel the swimbladder can only be filled by gas secretion, induced by the metabolic activity of the gas gland cells. The *ductus pneumaticus* develops into the resorbing section of the swimbladder (Fig. 2). A muscular sphincter connects this resorbing part of the swimbladder with the secretory section. Looking at the swimbladder the resorbing section is a thin walled,

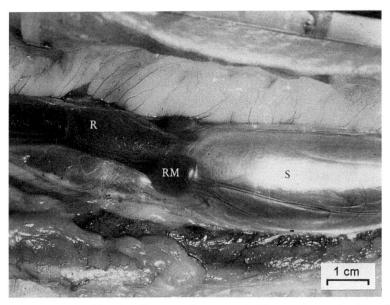

Figure 2. The swimbladder of a yellow stage European eel with a secretory section (S) and the pneumatic duct, which serves as a resorbing section (R) of the swimbladder. Two *retia mirabilia* (RM) are located next to the sphincter muscle separating the secretory and the resorbing section. The ductus pneumaticus is functionally closed, so that the eel cannot gulp air at the water surface to fill the swimbladder. Taken from Pelster (Pelster 2011) with permission.

Color image of this figure appears in the color plate section at the end of the book.

translucent air sack with an extensive vascularization. In contrast, the secretory section has a silvery appearance and looks more robust. The silvery appearance is due to the presence of guanine crystals encrusted in an outer layer of the swimbladder wall, the submucosa (Kleckner 1980a, Lapennas and Schmidt-Nielsen 1977). This layer significantly reduces the gas permeability of the secretory section and prevents or at least reduces the loss of gas (Denton et al. 1972, Kleckner 1980a, Kutchai and Steen 1971, Lapennas and Schmidt-Nielsen 1977).

A cross section through the wall of the secretory part of the swimbladder shows the presence of several tissue layers. The outer layer is the tunica externa (the serosa according to (Dorn 1961)), a thin layer of dense connective tissue, forming an outer capsule for the swimbladder. Below the tunica externa is the submucosa with the guanine crystals. A thin muscularis mucosa with smooth muscle cells and the single layered swimbladder epithelium complete the structure (Dorn 1961, Fänge 1953). The contractile muscularis mucosa contains mostly circular oriented smooth muscle fibers near the swimbladder lumen, while the cells in the outer layers are longitudinally oriented (Dorn 1961). Contraction of the muscularis mucosa reduces the volume of the secretory part of the swimbladder and allows for the movement of gas from this secretory part of the swimbladder to the resorbing section.

Gas Gland Cells

In the eel the single layered swimbladder epithelium forms the so-called gas gland cells, essential for the process of gas secretion. Gas gland cells are cubical or cylindrical with a size ranging from 10 to 25 µm. Gas gland cells are highly polarized with some small microvilli on the luminal side. The basal side often is more densely vacuolated and the membrane shows a large number of infoldings, the basolateral labyrinth (Fig. 3). Extensive membrane foldings typically are found in highly transport active sites. In gas gland cells, however, the distribution of membrane transport proteins has not yet been analyzed in detail and the meaning of these membrane enlargements therefore is not yet clear (Dorn 1961, Morris and Albright 1975).

The variable density of the granulated cytoplasma of gas gland cells with 'light' and 'dark' cells may represent variable functional states, and not necessarily indicates presence of different cell types (Dorn 1961, Morris and Albright 1975). The basolateral labyrinth in the dark cells is extensively expressed in contrast to light cells. This may suggest that these cells represent the more active state. Due to the high metabolic and ion transport activities in secretory cells typically a large number of mitochondria are found. Gas gland

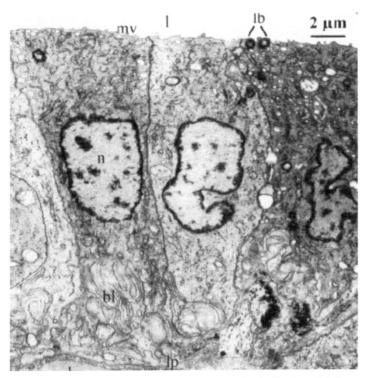

Figure 3. Transmission electron microscopical picture of gas gland cells of an adult European eel (yellow eel). The cells are cuboidal, with a typical basolateral labyrinth (bl) near blood vessels (bv) and some small microvilli (mv) at the apical membrane in contact with the swimbladder lumen (l). Electron dense lamellar bodies (lb) are present near the apical membrane. lp, lamina propria; n, nucleus. Taken from Prem et al. (2000) with permission.

cells, however, are characterized by the presence of only few filamentous or elongated mitochondria with few tubular cristae (Dorn 1961, Prem et al. 2000, Würtz et al. 1999).

Metabolic studies on gas gland cells revealed enzyme activities of the anaerobic glycolytic pathway that were comparable to the activities measured in white muscle tissue, while enzymes of the aerobic metabolism were very low (Pelster and Scheid 1991). Surprisingly, activities of the pentose phosphate shunt were detected in gas gland cells, suggesting that the activity of the pentose phosphate shunt in this tissue exceeded the activity established in other tissues. Using ^{14}C labeled glucose it could be shown that about 20% of the glucose taken up from the blood is metabolized via the pentose phosphate shunt in gas gland cells of the American eel, which results in the production of CO_2 without concomitant consumption of oxygen (Pelster et al. 1994). Calculation of the contribution of the various pathways to glucose metabolism and CO_2 production in gas gland cells of the European eel indicated that only 1% of the glucose taken up from the blood was metabolized aerobically. Aerobic metabolism thus contributed 20% to total CO_2 production. The bulk part of the CO_2 produced in gas gland cells was formed by decarboxylation in the pentose phosphate shunt (Pelster 1995b). About 80% of the glucose was converted to lactate, resulting in an equal amount of protons produced in the anaerobic metabolism (Pelster and Scheid 1993). This lactate production is observed even though the tissue is very well oxygenated, and hyperbaric oxygen partial pressures can be expected. In swimbladder tissue of *Sebastodes miniatus* lactate production has been recorded under an oxygen pressure of about 50 atmospheres (D'Aoust 1970).

Metabolic activity of gas gland cells thus produces CO_2, which can diffuse out of the cells along the partial pressure gradient, and it produces lactate with a concomitant proton, and both are released into the blood stream. Lactate transport has not yet been studied in detail, but it can be assumed that monocarboxylate transporters are involved. Several pathways may contribute to the transport of protons from the gas gland cells to the blood stream (Pelster 1995a, Pelster 2004). Besides sodium proton exchange (Na^+/H^+ exchange) a V-ATPase has been identified in gas gland cells assuring the efficient secretion of protons even against a proton gradient, i.e., in a situation when blood pH is significantly lower than intracellular pH (Boesch et al. 2003, Niederstätter and Pelster 2000, Pelster 2004, Pelster and Niederstätter 1997).

The diffusion of CO_2 into the blood also will acidify the blood. In gas gland cells this pathway is supported by the presence of a membrane bound carbonic anhydrase, facing the blood (Pelster 1995a, Würtz et al. 1999). Membrane bound carbonic anhydrase rapidly converts CO_2 entering the extracellular space and the blood into HCO_3^-, so that the partial pressure gradient driving CO_2 out of the cell is sustained.

Blood Supply

Crucial for the functioning of the swimbladder is the blood supply. The swimbladder artery supplying the secretory part of the swimbladder originates from the dorsal artery and in the eel is characterized by the presence of a paired countercurrent system, the *rete mirabile*, also called the red body or the 'Wundernetz' (Fig. 2 (Dorn 1961, Müller 1840, Woodland 1911)). The two *retia mirabilia* are located next to the sphincter separating the resorbing and the secretory section of the swimbladder. A detailed description of the eel *rete mirabile* was provided by Wagner et al. (Wagner et al. 1987). At the entrance of the *rete mirabile* the swimbladder artery splits into several ten thousand arterial capillaries running parallel to each other for the length of several millimeters. After leaving the *rete mirabile* the arterial capillaries at the swimbladder pole reconvene to only two or three larger arterial vessels, which then give raise to an additional capillary system supplying the secretory swimbladder epithelium, the gas gland cells. From the gas gland cells the venous capillaries rejoin forming two or three larger veins returning to the *rete mirabile*. At the entrance of the rete these veins again split into several ten-thousand venous capillaries running exactly parallel to the arterial capillaries of the rete. Thus, in the countercurrent system each arterial capillary is surrounded by several venous capillaries, and vice versa. The diffusion distance between arterial and venous vessels is about 1–2 μm (Stray-Pedersen and Nicolaysen 1975). With a capillary length of several millimeters measured in the rete of a yellow eel the *rete mirabile* represents a fantastic countercurrent structure with a large surface area and very small diffusion distances for the exchange of permeable molecules between venous and arterial capillaries. During metamorphosis to the silver eel, the *rete mirabile* even increases in size and the length of the capillaries increases. The length of the capillaries correlates with the countercurrent concentrating ability. This suggests an increased importance of this structure during the subsequent spawning migration to the Sargasso Sea, because it is known that the length of the rete capillaries correlates with depth of occurrence of a species (Marshall 1972).

This bipolar structure of the eel rete allows access to larger blood vessels entering and leaving the *rete mirabile* on the arterial side, and also entering and leaving the rete on the venous side, and therefore is ideal for the analysis of the exchange phenomena occurring within the *rete mirabile* (Kobayashi et al. 1990a, Kobayashi et al. 1989a, Kobayashi et al. 1989b). Accordingly, most of our knowledge on the physiology and functioning of the countercurrent system of the teleost swimbladder is based on data collected from the eel swimbladder. In an unipolar *rete mirabile*, blood vessels after passing the arterial part of the *rete mirabile* immediately return to the venous side, leaving no space for micropuncture experiments, which are essential for the functional analysis of this vascular arrangement.

An important feature of the swimbladder vascular system is that the resorbing part of the bladder, the modified ductus pneumaticus in the eel, has a separate blood supply, independent of the vascular system of the secretory bladder. This allows for a functional separation of these two swimbladder sections, and assures that the resorbing and the secretory parts operate independent of each other.

Swimbladder Innervation

Nervous control of swimbladder function has been studied in several species, mainly cod and perch, but also the eel. Swimbladder function is under the control of the autonomic nervous system. The swimbladder is innervated by the splanchnic nerve, originating from the celiac ganglion, and the vagosympathetic trunk. Several studies demonstrated that vagotomy or cholinergic blocking agents significantly impair gas deposition (Fänge et al. 1976, Lundin and Holmgren 1991). Adrenergic nerve cells could not be detected in proximity of gas gland cells, so that metabolic and ion regulatory activity of gas gland cells appear to be under cholinergic, parasympathetic control (McLean and Nilsson 1981).

Control of blood flow, however, does include adrenergic effects. Electrical stimulation of the splanchnic nerve of the vagosympathetic trunk elicits vasoconstriction, and both nerves stain for catecholamines (McLean and Nilsson 1981, Nilsson 1972, Wahlqvist 1985, Schwerte et al. 1997). Perfused eel or cod swimbladder preparations show a clear responsiveness to adrenergic agonists and antagonists. In a blood-perfused swimbladder preparation of the European eel α-adrenergic stimulation resulted in a vasoconstriction close to the *rete mirabile* (Pelster 1994). In addition, VIP and substance P responsiveness has been demonstrated (Lundin and Holmgren 1991, Schwerte et al. 1999).

Muscle cells within the swimbladder wall have been shown to be under autonomic control, with contribution of α- and β-adrenergic reactivity (Fänge et al. 1976, Stray-Pedersen 1970). These muscle cells are responsible for shifting gas between secretory and resorbing sections of the bladder. Similarly, the oval appears to be under adrenergic control (Nilsson 1971, Ross 1978, Wahlqvist 1985).

Mechanisms of Gas Deposition

The Single Concentrating Effect

The term 'secretion' typically implies an active process, primary or secondary active, but gas secretion into the fish swimbladder is passive and is achieved by diffusion of gas molecules from the blood or from gas gland cells into the swimbladder along a partial pressure gradient. Gas transfer into the swimbladder does not involve primary or secondary active transport phenomena. Therefore the term 'gas deposition' may in fact be more adequate. The high gas partial pressures necessary to establish a diffusion gradient between the blood and the swimbladder lumen are established by two mechanisms, the reduction of the effective gas-carrying capacity of blood in swimbladder vessels (the single concentrating effect) and subsequent countercurrent concentration of gases in the countercurrent system, the *rete mirabile* (Kuhn et al. 1962, Kuhn et al. 1963).

Crucial for the reduction of the effective gas-carrying capacity of swimbladder blood is the metabolic activity of the epithelial gas gland cells (Fig. 4).

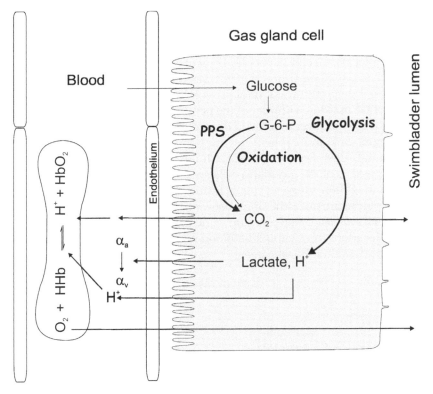

Figure 4. Present concept of the metabolism of swimbladder gas gland cells and its relation to the gas-carrying capacity in the blood. Glucose is taken up from the blood and mainly converted into lactate, which results in the production of an equal amount of protons. This lactate production is not inhibited by hyperbaric oxygen tensions, i.e., gas gland cells do not show a Pasteur effect. A significant fraction of glucose is converted to CO_2 in the pentose phosphate shunt (PPS), while only a very small fraction of glucose is oxidized. Due to this CO_2 production in the cell CO_2 diffuses down the partial pressure gradient into the swimbladder lumen as well as into the blood, where it contributes to blood acidification. Lactate and protons are also secreted into the blood. Protons and the increase in blood PCO_2 acidify the erythrocytes, and thus reduce the oxygen-carrying capacity of the hemoglobin. Oxygen is released from the hemoglobin (Hb) *via* the Root effect and diffuses down the partial pressure gradient through the gas gland cells into the swimbladder lumen. The release of lactate increases the overall salt concentration in the blood and therefore reduces the solubility (αa, αv) of gases in the blood (salting-out effect).

Although the oxygen partial pressure usually is high in the swimbladder epithelium, gas gland cells are specialized for the anaerobic production of acidic metabolites (see above; Pelster 1995b). In the eel about 80% of the glucose removed from the blood is converted to lactic acid, and the release of lactate and of protons significantly acidifies the blood. Due to the production of CO_2 in the pentose phosphate shunt a high PCO_2 has to be expected in gas gland cells, driving the outward diffusion of CO_2 into the blood. This also contributes to the acidification of blood during passage of the gas gland cells. Accordingly, in the European eel *Anguilla anguilla* blood returning to the *rete mirabile* after passage of metabolically active gas gland cells is significantly acidified (Fig. 5) (Kobayashi et al. 1990b, Steen 1963b). In fish blood an increase in proton concentration not only provokes a Bohr effect, which describes the decrease in oxygen affinity of the hemoglobin with decreasing pH, but also it typically induces a so-called Root effect, i.e., an acidification decreases the oxygen carrying capacity of the hemoglobin (Pelster 2001, Pelster and Randall 1998, Root 1931). Accordingly, the acidification of the blood during passage of the gas gland cells switches on the Root effect and part of the hemoglobin releases oxygen, resulting in a significant increase in the oxygen partial pressure (the single concentrating effect for oxygen). Figure 5 shows pH and gas partial pressure values which have been recorded at the various locations in swimbladder blood vessels in an active, gas depositing swimbladder of the European eel.

Rete mirabile

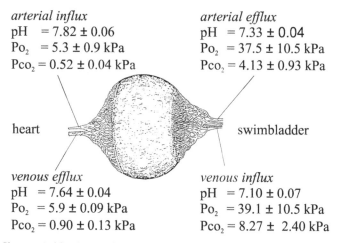

arterial influx
pH   = 7.82 ± 0.06
Po_2 = 5.3 ± 0.9 kPa
Pco_2 = 0.52 ± 0.04 kPa

arterial efflux
pH = 7.33 ± 0.04
Po_2 = 37.5 ± 10.5 kPa
Pco_2 = 4.13 ± 0.93 kPa

heart

swimbladder

venous efflux
pH   = 7.64 ± 0.04
Po_2 = 5.9 ± 0.09 kPa
Pco_2 = 0.90 ± 0.13 kPa

venous influx
pH = 7.10 ± 0.07
Po_2 = 39.1 ± 10.5 kPa
Pco_2 = 8.27 ± 2.40 kPa

Figure 5. Changes in blood pH, PO_2 and PCO_2 recorded in swimbladder vessels entering or leaving the *rete mirabile* of the European eel during gas deposition (data from Kobayashi et al. 1990b). Note the significant decrease in blood pH and the significant increase in gas partial pressures during passage of the *rete mirabile*. Only small changes are detected following passage of the gas gland cells, which can be explained by the deposition of gas into the swimbladder. During subsequent passage of the venous vessels of the *rete mirabile* the changes observed during arterial passage are largely reversed. This confirms that these changes are mostly due to back-diffusion from venous to arterial vessels in the rete. In consequence, low pH values and high gas partial pressures are recorded at the swimbladder pole of the *rete mirabile*, while at the heart pole values for arterial and venous blood are comparable to influx and efflux to most other organs. The drawing of the *rete mirabile* has been adopted from Wagner et al. (Wagner et al. 1987).

Although part of the CO_2 diffusing into the blood from gas gland cells is converted into HCO_3^-, the PCO_2 of the blood increases during passage of the gas gland cells (the single concentrating effect for CO_2). Finally, the secretion of lactate causes an increase in the total solute concentration of the blood, and the physical solubility of gas in blood decreases with increasing solute concentration. Accordingly, the increase in lactate concentration in blood will induce the so-called salting out effect and increase the gas partial pressure of all gases in blood (the single concentrating effect for all gases including nitrogen). Given the measured changes in lactate concentration in swimbladder blood this effect is very small compared to the single concentrating effect for oxygen and CO_2 and probably of minor importance (Pelster et al. 1988, Steen 1963c).

These considerations show that during passage of the gas gland cells the metabolic activity of these cells induces an initial increase in gas partial pressure of all gases in the blood. Depending on the rate of acidification and on the hemoglobin concentration this effect may be very large for oxygen (Pelster 2001). For PCO_2 also a significant increase can be expected, depending on the activity of the pentose phosphate shunt, and this is confirmed by actual measurements performed in *in situ* preparations of the European eel (Fig. 5) (Kobayashi et al. 1990b, Steen 1963b). For inert gases including nitrogen the increase in gas partial pressure probably is only in the range of 1% or 2% (Pelster et al. 1988).

The increase in gas partial pressure in blood and the increase in PCO_2 in gas gland cells due to metabolic production will generate a pressure head for the diffusion of gas molecules into the swimbladder. Nevertheless, the gas partial pressure for all gases in the venous blood returning to the venous vessels of the *rete mirabile* will be elevated compared to the partial pressures measured in blood that leaves the arterial vessels of the rete on the swimbladder pole. This provides the basis for the next step in gas deposition, the multiplication of this single concentrating effect in a countercurrent system.

Countercurrent Concentration

Countercurrent arrangement of arterial and venous blood vessels or of blood flow and respiratory water or air flow allows for an efficient exchange between the two different compartments. Countercurrent exchangers in nature typically are passive structures that do not include active or secondary active transport. The efficiency of this exchange mechanism therefore is dependent on the diffusion distance between the two compartments and the contact area (Kobayashi et al. 1990a, Kuhn et al. 1963, Pelster 2001). The eel *rete mirabile* probably is one of the most efficient exchange systems that have been studied in great detail. These studies revealed that in addition to the geometric parameters describing the countercurrent exchanger the permeability of the capillary membranes is of importance. While it was expected that rete membranes are permeable for gas molecules, a detailed analysis of the eel countercurrent system demonstrated that water and small metabolites like lactate can also be shifted between venous and arterial vessels in the rete along a concentration gradient (Kobayashi et al. 1989a, Kobayashi et al. 1989b). This results in a countercurrent concentration of small solutes, which even enhances the countercurrent concentrating ability of the rete.

Based on geometric estimates, gas solubility's or transport capacities and diffusion coefficients the concentrating ability of a countercurrent exchanger can be modeled (Sund 1977). Taking into account the hemoglobin concentration and the Root effect, maximum gas partial pressures that can be achieved can be estimated. These calculations revealed that the back-diffusion of CO_2 in the *rete mirabile*, by causing an acidification of blood during arterial passage of the rete, enhances the countercurrent concentration of oxygen (Kobayashi et al. 1990b). Due to the passive nature of the countercurrent system, the concentrating effect achieved during passage of the *rete mirabile* is related to the magnitude of the single concentrating effect. Therefore the highest partial pressures can be achieved for oxygen, followed by CO_2. Because the salting out effect that can be expected on the basis of measured lactate and salt concentrations in eel swimbladder blood is so small, the maximum partial pressures for inert gases achieved in the rete are in the lower range. Depending on the properties of the *rete mirabile*, however, these calculations demonstrated that partial pressures of several hundred atmospheres can be achieved by countercurrent concentration, explaining the presence of fish with a gas filled swimbladder at depth of several 1000 m (Fig. 6).

Gas Resorption

The partial pressure of gases in the blood within the gills ideally is equilibrated with water gas partial pressures, and gas partial pressures in the water hardly change with depth (Enns et al. 1967, Pelster 1997b). Therefore the partial pressure of swimbladder gas typically exceeds the partial pressure of dissolved gases in the systemic circulation, and with increasing water depth the pressure head for the gases to leave the swimbladder increases. To avoid or at least reduce diffusional loss of gas through the swimbladder wall guanine and hypoxanthine crystals are incorporated in the submucosa, which significantly reduces its gas permeability (Denton et al. 1972, Kleckner 1980a, Lapennas and Schmidt-Nielsen 1977).

For the secretory part of the swimbladder the *rete mirabile* acts as a barrier which prevents gas from leaving the swimbladder. Venous blood

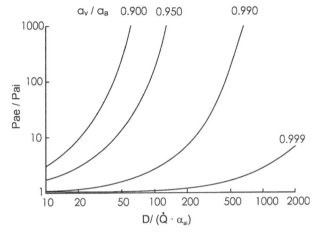

Figure 6. Efficiency of the rete in enhancing inert gas partial pressure, calculated as the ratio of partial pressure in the arterial efflux and influx (Pae/Pai), in relation to the conductance ratio (D/Q • α_{ai}), where D is the diffusion coefficient, Q is blood flow and α_{ai} is the physical solubility of gas in blood. The magnitude of the salting-out effect is calculated as the ratio of venous and arterial gas solubility in (α_v/α_a). Based on measured changes in lactate and salt concentrations in blood of the European eel a salting out effect in the range of 1 or 2% can be expected. This calculation assumes that no gas is deposited into the swimbladder. The deposition of gas into the swimbladder would reduce the gas partial pressures in blood returning to the rete and therefore reduce the countercurrent concentrating effect. Adapted from Kobayashi et al. (Kobayashi et al. 1989a).

leaving the secretory section of the swimbladder must pass through the *rete mirabile*. Analogous to a diffusion shunt, gases with an elevated gas partial pressure are shunted from the venous to the arterial rete capillaries and returned to the swimbladder. The rete thus not only plays a role in gas deposition, but also in its maintenance within the bladder.

Therefore gas can be effectively resorbed only in the resorbing part of the bladder, which has its own blood supply. By opening the sphincter between the two swimbladder sections a bolus of gas can be released into the resorbing part, which is a well vascularized air sack. Because the gas partial pressures in this bolus, depending on the water depth, exceed the partial pressure in the systemic circulation, gases are resorbed and distributed in the general circulation. The reabsorption of gases is correlated to the partial pressure gradient and their transport capacity in blood, and the more soluble gases are more readily absorbed. This explains why CO_2, which typically is present in newly deposited gas, is hardly found under steady-state conditions. This also contributes to the accumulation of the less-soluble inert gases in the swimbladder (Pelster 1997b, Piiper 1965, Steen 1963a). Given that oxygen most likely represents the main gas in the bladder, this oxygen could be resorbed for the benefit of other organs. There is, however, one caveat that must be taken into account. Blood returning from the swimbladder is directed towards the heart and subsequently passes the gills. If the partial pressure of oxygen in the blood is significantly elevated at this point, oxygen would be lost to the surrounding water. Opening the sphincter and resorbing gas from the swimbladder therefore can be expected to be mainly for buoyancy reasons, i.e., to reduce overall buoyancy when ascending. In this situation the swimbladder probably is only of minor importance as a respiratory organ.

Swimbladder Function During Spawning Migration

The ~5000 km spawning migration of the European eel from European freshwater streams to the Sargasso Sea fascinates biologists since many decades and remains a mystery until today (Van den Thillart et al. 2009, Sebert et al. 2009). Several attempts have been undertaken to tag eels and follow them by boat, with limited success rates. Lost tags or signals, eels apparently eaten by larger predators, many aggravating circumstances prevented fully successful studies (Fricke and Kaese 1995, McCleave and Arnold 1999). A recent study used miniaturized pop-up satellite archival transmitters (PSAT) to follow the spawning migration. The experiment again failed to follow eels all the way to the Sargasso Sea, but some eels could be traced for up to 1300 km from release and provided unique behavioral insights (Aarestrup et al. 2009). Eels released on the West coast of Ireland in October and November did not travel straight to the Sargasso, but headed more South towards the Canaries and Azores. Mean travelling speed was 13.8 km per day, which was less than half the expected value. This could confirm the hypothesis proposed by Fricke and Kaese (Fricke and Kaese 1995) that eels use the south- and west-flowing currents that begin in west of Africa and continue as part of the subtropical gyre system to the Caribbean. These currents may allow for much higher travelling speeds so that the spawning grounds may be reached in time around April. Swim tunnel experiments revealed that the pop-up tags, although miniaturized, impair the swimming performance of the eels, so that the travelling speed measured with the tagged eels may be significantly lower than the speed an untagged eel could achieve during the spawning migration (Methling et al. 2011, Burgerhout et al. 2011).

Another observation of these studies was extremely exciting with respect to swimbladder function. The eels performed daily migrations predominantly between the depth of about 200 m and 1000 m. At night eels were detected in warmer water layers around 150–300 m at a temperature of $11.68° \pm 0.48°C$, during day times they steeply descended down to 800–1000 m with a water temperature of $10.12°C \pm 0.89°C$ (Fig. 7) (Aarestrup et al. 2009).

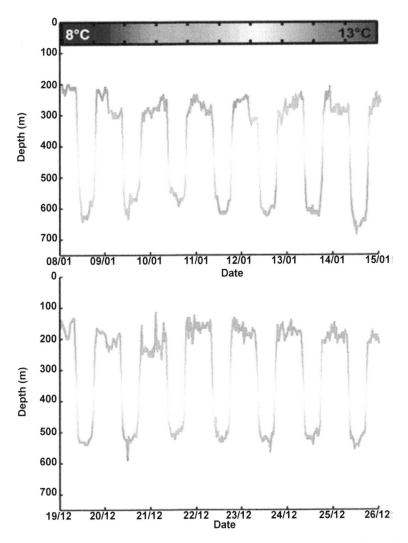

Figure 7. Oceanic daily vertical migration of two eels recorded by pop-up satellite archival transmitters attached to silver eels. Eels were released near the Irish west coast and followed for several days during their spawning migration to the Sargasso Sea. Depth values are colored to indicate the temperature encountered at the respective depth. Taken from Aarestrup et al. (Aarestrup et al. 2009), with permission.

Color image of this figure appears in the color plate section at the end of the book.

At the water depth of 200 m hydrostatic pressure is 21 atm, at a depth of 1000 m it is increased to 101 atm. If a gas deposition rate of about 1 or 2 swimbladder volumes per day is assumed, this daily migration cannot be performed in a status of neutral buoyancy, because the rate of gas deposition is far too low. Even if the secretion rate of silver eels would be 5-times higher than in yellow eels (Kleckner 1980b), it would be too low to achieve full compensation. It appears most likely that eels have a status of near neutral buoyancy in the upper or lower level of the travelling range, and compensate for the differences by swimming activity when they leave this depth. This is known from several deep sea species which perform daily migrations between shallow and deeper water layers (Gee 1983, Kalish et al. 1986, Vent and Pickwell 1977). With respect to swimbladder function these data demonstrate that we are far from understanding the functioning of the swimbladder. It would be very interesting to measure the actual buoyancy status of the eels during these daily migrations.

The reason for these daily vertical migrations is not obvious. During metamorphosis to the silver eel the alimentary tract degenerates (Pankhurst and Sorensen 1984, Van Ginneken et al. 2007) and eels don't feed during spawning migration. Accordingly, feeding activities cannot explain these daily migrations. A possible explanation might be the avoidance of predators in the dim light in the upper, warmer water layers during day time and the temperature changes (Aarestrup et al. 2009). Cold temperatures may be advantageous to avoid an early development of the gonads, while warmer temperatures may be advantageous for swimming activity. The mean temperature changes encountered at the two different depth ranges, however, is only 1.5°C. Therefore alternative explanations may come up in the near future.

Infection of the Swimbladder with the Nematode *Anguillicoloides crassus*

The nematode *Anguillicoloides crassus* (Kuwahara et al. 1974) was originally described as a parasite of the Japanese eel *Anguilla japonica* (Moravec 1992, Moravec and Taraschewski 1988). In the Japanese eel the infection rate apparently is low, and the viability of the eel is not seriously affected. In the early 80-ties the first infections of the European eel were described (Bernies et al. 2011, Combes 2001, Moravec 1992, Schabuss et al. 2005). Within only one decade the parasite spread over Europe, and in some regions infection rates of almost 100% were reported. Within recent years the infection rates appear to level off, or even decrease slightly (Lefebvre and Crivelli 2004, Schabuss et al. 2005).

Eels are infected with *Anguillicoloides crassus* L3 larval stage by eating infected copepods (Haenen et al. 1989) Dracunculoidea. This infection can even occur at the glass eel stage, almost immediately after the initial inflation of the swimbladder (Nimeth et al. 2000). From the gut the L3 larvae migrate into the swimbladder by penetrating the swimbladder wall. The sanguivorous, histotrophic L3 larvae develop to the L4 stage within the swimbladder wall, and the pre-adult form enters the swimbladder lumen. Due to the sanguivorous, histotrophic lifestyle of the nematode larvae the swimbladder wall is excessively damaged, the connective tissue of the swimbladder wall is significantly thickened and the histology is completely altered (Molnár et al. 1995, Nimeth et al. 2000, Würtz and Taraschewski 2000). Histological examination of the swimbladder epithelium revealed that the unicellular epithelium in heavily infected eels became hyperplastic and developed into a multi-layered epithelium (Fig. 8).

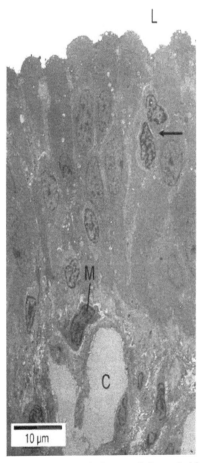

Figure 8. Transmission electron microscopic image of the swimbladder epithelium of the European eel infected with the nematode *Anguillicoloides crassus*. The unicellular epithelium becomes hyperplastic and multi-layered. The contact to blood vessels (C) is significantly reduced, and macrophages (M) invade the tissue. L, swimbladder lumen; the arrow indicates modifications in the shape of the nuclei and in heterochromatin distribution. Taken from Würtz and Taraschewski (Würtz and Taraschewski 2000), with permission.

Presence of macrophages indicated activation of immune responses (Würtz and Taraschewski 2000). The basolateral labyrinth of gas gland cells was reduced, suggesting an impairment of gas gland cell function. In heavily infected eels the swimbladder may almost completely be filled with nematodes, leaving almost no space for any swimbladder gas. Accordingly, the gas composition of infected swimbladders was significantly different from the composition of an uninfected eel, and especially the oxygen content of infected swimbladders was significantly reduced (Würtz et al. 1996). In spite of this severe damage to the swimbladder the eel may recover from an infection, since eels without parasites have been found in which the modified histology of the swimbladder confirms a previous infection (Würtz and Taraschewski 2000).

While the behavior of the bottom dwelling freshwater eels is not obviously affected by an infection of the swimbladder with *Anguillicoloides* , it was speculated that the spawning migration may be impaired (Würtz et al. 1996). Experimental studies revealed that in glass eels an infection of the swimbladder does not immediately impair swimming performance (Nimeth et al. 2000), and in yellow eels short term swimming against a current was not severely affected (Münderle et al. 2004). Looking at the swimming speed of infected and uninfected yellow eels an 18% reduction was recorded in eels infected with more than 10 nematodes (Sprengel and Lüchtenberg 1991). In silver eels an infection caused lower cruising speeds and a higher cost of transport. Eels with an infected or damaged swimbladder at all tested swimming speeds had a higher oxygen consumption than uninfected eels (Palstra et al. 2007a). Eels without parasites in the swimbladder but with a damaged swimbladder as a result of a previous infection showed similar effects. Almost half of the eels that contained damaged swimbladders stopped swimming at low aerobic swimming speeds. Simulated migration trials confirmed that eels with a high parasite level or with damaged swimbladder show early migration failure. These results suggest that migrating silver eels with severely infected or damaged swimbladders will not be able to reach the spawning grounds (Palstra et al. 2007a, Palstra et al. 2007b, Sebert et al. 2009). Given the high infection rates recorded in various rivers and streams of the European freshwater system this parasite appears to be a major threat to the survival of European eels.

Perspectives

The eel swimbladder is a buoyancy organ, which has fascinated biologists for more than a century. Nevertheless, its manifold physiological aspects are far from being understood. The countercurrent exchanger, the *rete mirabile*, has been identified as a passive exchange system. The metabolism and the secretory activity of the gas gland cells are crucial for the initiation of the single concentrating effect. Membrane transport characteristics, which are essential for the decrease in gas solubility in the blood, have not been analyzed in detail. Another important aspect is the low rate of aerobic metabolism, in spite of the fact that the cells are under hyperoxic conditions. The production of reactive oxygen species, which usually are formed under hyperoxic conditions, has not been addressed. Finally, the role of an infection with *Anguillicoloides* is far from clear. How can the yellow eel cope with an infection, how does a persistent infection or a previous infection influence metamorphosis to a silver eel and the spawning migration? Even after many decades of research, the eel swimbladder remains a mystery and promises to remain an interesting object of research.

Acknowledgements

The author would like to thank Drs. Francesca Trischitta and Thorsten Schwerte for critical comments on the manuscript.

References

Aarestrup, K., F. Okland, M.M. Hansen, D. Righton, P. Gargan, M. Castonguay, L. Bernatchez, P. Howey, H. Sparholt, M. Pedersen and R.S. McKinley. 2009. Oceanic Spawning Migration of the European Eel (*Anguilla anguilla*). Science 325: 1660.

Alexander, R.M. 1966. Physical aspects of swimbladder function. Biol. Rev. 41: 141–176.

Als, T.D., M. Hansen, G.E. Gregory, M. Castonguay, L. Riemann, K. Aarestrup, P. Munk, H. Sparholt, R. Hanel and L. Bernatchez. 2011. All roads lead to home: panmixia of European eel in the Sargasso Sea. Molec. Ecol. 20(7): 1333–1346.

Bernies, D., A. Brinker and A. Daugschies. 2011. An invasion record for the swimbladder nematode *Anguillicoloides crassus* in European eel *Anguilla anguilla* in a deep cold-monomictic lake, from invasion to steady state. J. Fish Biol. 79(3): 726–746.

Boesch, S.T., H. Niederstätter and B. Pelster. 2003. Localization of the vacuolar-type ATPase in swimbladder gas gland cells of the European eel *(Anguilla anguilla)*. J. Exp. Biol. 206: 469–475.

Burgerhout, E., R. Manabe, S.A. Brittijn, J. Aoyama, K. Tsukamoto and G. Van den Thillart. 2011. Dramatic effect of pop-up satellite tags on eel swimming. *Naturwissenschaften* 98: 631–634.

Combes, C. 2001. Parasitism: The Ecology and Evolution of Intimate Interactions. University of Chicago Press, Chicago, USA.

D'Aoust, B.G. 1970. The role of lactic acid in gas secretion in the teleost swimbladder. Comp. Biochem. Physiol. 32: 637–668.

Denton, E.J., J.D. Liddicoat and D.W. Taylor. 1972. The permeability to gases of the swimbladder of the conger eel (*Conger conger*). J. Mar. Biol. Ass. UK. 52: 727–746.

Dorn, E. 1961. Über den Feinbau der Schwimmblase von *Anguilla vulgaris* L. Licht- und Elektronenmikroskopische Untersuchungen. Zeitschrift für Zellforschung. 55: 849–912.

Enns, T., E. Douglas and P.F. Scholander. 1967. Role of the swimbladder rete of fish in secretion of inert gas and oxygen. Adv. Biol. Med. Phys. 11: 231–244.

Fänge, R. 1953. The mechanisms of gas transport in the euphysoclist swimbladder. Acta Physiol. Scand. 30: 1–133.

Fänge, R., S. Holmgren and S. Nilsson. 1976. Autonomic nerve control of the swimbladder of the goldsinny wrasse, *Ctenolabrus rupestris*. Acta Physiol. Scand. 97: 292–303.

Fricke, H. and R. Kaese. 1995. Tracking of artificially matured eels (*Anguilla anguilla*) in the Sargasso Sea and the problem of the eel's spawning site. Die Naturwissenschaften. 82: 32–36.

Gee, J.H. 1983. Ecologic implications of buoyancy control in fish. *In*: P.W. Webb [ed]. Fish Biomechanics. Praeger Scientific. New York, USA. pp. 140–176.

Govoni, J.J. and D.E. Hoss. 2001. Comparison of the Development and Function of the Swimbladder of *Brevoortia tyrannus* (Clupeidae) and *Leiostomus xanthurus* (Sciaenidae). Copeia. 2: 430–442.

Haenen, O.L.M., L. Grisez, D. DeCharleroy, C. Belpaire and F. Ollevier. 1989. Experimentally induced infections of European eel *Anguilla anguilla* with *Anguillicola crassus* (Nematoda, Dracunculoidea) and subsequent migration of larvae. Dis. Aquatic Org. 7: 97–101.

Kalish, J.M., C.F. Greenlaw, W.G. Pearcy and D. Van Holliday. 1986. The biological and acoustical structure of sound scattering layers off oregon. Deep Sea Res. Part A. Oceanographic Research Papers 33(5): 631–653.

Kleckner, R.C. 1980a. Swimbladder wall guanine enhancement related to migratory depth in silver phase *Anguilla rostrata*. Comp. Biochem. Physiol. 65A: 351–354.

Kleckner, R.C. 1980b. Swim bladder volume maintenance related to initial oceanic migratory depth in silver phase *Anguilla rostrata*. Science 208: 1481–1482.

Kobayashi, H., B. Pelster and P. Scheid. 1989a. Solute back-diffusion raises the gas concentrating efficiency in counter-current flow. Respir. Physiol. 78: 59–71.

Kobayashi, H., B. Pelster and P. Scheid. 1989b. Water and lactate movement in the swimbladder of the eel, *Anguilla anguilla*. Respir. Physiol. 78: 45–57.

Kobayashi, H., B. Pelster, J. Piiper and P. Scheid. 1990a. Counter-current blood flow in tissues: protection against adverse effects. *In*: J. Piiper, T.K. Goldstick and M. Meyer [eds]. Oxygen Transport to Tissue. Plenum Press, New York and London. pp. 3–11.

Kobayashi, H., B. Pelster and P. Scheid. 1990b. CO_2 back-diffusion in the rete aids O_2 secretion in the swimbladder of the eel. Respir. Physiol. 79: 231–242.

Kuhn, H.J., P. Moser and W. Kuhn. 1962. Haarnadelgegenstrom als Grundlage zur Erzeugung hoher Gasdrücke in der Schwimmblase von Tiefseefischen. Nachweis der Sekretion kleiner Mengen von Milchsäure am Scheitel der Haarnadel als Ursache des Einzeleffektes. Pflügers Archiv. 275: 231–237.

Kuhn, W., A. Ramel, H.J. Kuhn and E. Marti. 1963. The filling mechanism of the swimbladder. Generation of high gas pressures through hairpin countercurrent multiplication. Experientia. 19: 497–511.

Kutchai, H. and J.B. Steen. 1971. The permeability of the swimbladder. Comp. Biochem. Physiol. 39A: 119–123.

Kuwahara, A., A. Niimi and H. Itagaki. 1974. Studies on a nematode parasitic in the air bladder of the eel. I. description of *A. crassus* n.sp. (Philometridea, Anguillicolidae). Japanese J. Parasitol. 23: 275–279.

Lapennas, G.N. and K. Schmidt-Nielsen. 1977. Swimbladder permeability to oxygen. J. Exp. Biol. 67: 175–196.

Lefebvre, F.S. and A.J. Crivelli. 2004. Anguillicolosis: dynamics of the infection over two decades. Dis. Aquatic Org. 62: 227–232.

Lundin, K. and S. Holmgren. 1991. An x-ray study of the influence of vasoactive intestinal polypeptide and substance P on the secretion of gas into the swimbladder of a teleost *Gadus morhua*. J. Exp. Biol. 157: 287–298.

Marshall, N.B. 1972. Swimbladder organization and depth ranges of deep-sea teleosts. Symposium of Society of the experimental Biology. 26: 261–272.

McCleave, J.D. and G.P. Arnold. 1999. Movements of yellow- and silver-phase European eels (*Anguilla anguilla* L.) tracked in western North Sea. ICES J. Mar. Sci. 56: 510–536.

McLean, J.R. and S. Nilsson. 1981. A histochemical study of the gas gland innervation in the Atlantic cod, *Gadus morhua*. Acta Zool. 62: 187–194.

Methling, C., C. Tudorache, P.V. Skov and J.F. Steffensen. 2011. Pop-Up Satellite Tags Impair Swimming Performance and Energetics of the European Eel (*Anguilla anguilla*). PLoS One 6: e20797.

Molnár, K., J. Szakolczai and F. Vetési. 1995. Histological changes in the swimbladder wall of eels due to abnormal location of adults and second stage larvae of *Anguillicola crassus*. Acta Vet. Hung. 43: 125–137.

Moravec, F. 1992. Spreading of the nematode *Anguillicola crassus* (Dracunculoidea) among eel populations in Europe. Folia Parasitol. 39: 247–248.

Moravec, F. and H. Taraschewski. 1988. Revision of the genus *Anguillicola* Yamaguti 1935 (Nemataoda: Anguillicolidae) of the swimbladder of eels, including descriptions of two new species, *A. novaezelandiae* sp. n. and *A. papernai* sp. n. Folia Parasitol. 35: 125–156.

Morris, S.M. and J.T. Albright. 1975. The ultrastructure of the swimbladder of the toadfish, *Opsanus tau* L. Cell and Tissue Res. 164: 85–104.

Müller, J. 1840. Über Nebenkiemen und Wundernetze. Archiv für Anatomie, Physiologie und wissenschaftliche Medizin. 1840: 101–142.

Münderle, M., B. Sures and H. Taraschewski. 2004. Influence of *Anguillicola crassus* (Nematoda) and *Ichthyophthirius multifiliis* (Ciliophora) on swimming activity of European eel *Anguilla anguilla*. Dis. Aquatic Org. 60: 133–139.

Niederstätter, H. and B. Pelster. 2000. Expression of two vacuolar-type ATPase B subunit isoforms in swimbladder gas gland cells of the European eel: nucleotide sequences and deduced amino acid sequences. Biochim. Biophys. Acta Struct. Express. 1491(1-3): 133–142.

Nilsson, S. 1971. Adrenergic innervation and drug responses of the oval sphincter in the swimbladder of the cod (*Gadus morhua*). Acta Physiol. Scand. 83: 446–453.

Nilsson, S. 1972. Autonomic vasomotor innervation in the gas gland of the swimbladder of a Teleost (*Gadus morhua*). Comp. Gen. Pharmacol. 3: 371–375.

Nimeth, K., P. Zwerger, J. Würtz, M. Salvenmoser and B. Pelster. 2000. Infection of the glass-eel swimbladder with the nematode *Anguillicola crassus*. Parasitol. 121: 75–83.

Palstra, A.P., D.F.M. Heppener, V.J.T. Van Ginneken, C. Szekely and G.E.E.J. Van den Thillart. 2007a. Swimming performance of silver eels is severely impaired by the swim-bladder parasite *Anguillicola crassus*. J. Exp. Mar. Biol. Ecol. 352(1): 244–256.

Palstra, A., D. Curiel, M. Fekkes, M. de Bakker, C. Szekely, V. Van Ginneken and G. Van den Thillart. 2007b. Swimming stimulates oocyte development in European eel. Aquacult. 270: 321–332.

Pankhurst, N.W. and P.W. Sorensen. 1984. Degeneration of the alimentary tract in sexually maturing European *Anguilla anguilla* (L.) and American eels *Anguilla rostrata* (LeSueur). Can. J. Zool. 62(6): 1143–1149.

Pauly, A. 1882. Beitrag zur Anatomie der Schwimmblase des Aals (*Anguilla fluviatilis*). Habilitationsschrift, Universität München, Germany.

Pelster, B. 1994. Adrenergic control of swimbladder perfusion in the European eel *Anguilla anguilla*. J. Exp. Biol. 189: 237–250.

Pelster, B. 1995a. Mechanisms of acid release in isolated gas gland cells of the European eel *Anguilla anguilla*. Amer. J. Physiol. 269: R793–R799.

Pelster, B. 1995b. Metabolism of the swimbladder tissue. Biochem. Mol. Biol. Fishes. 4: 101–118.

Pelster, B. 1997a. Buoyancy. *In*: D.H. Evans [ed]. The Physiology of Fishes. CRC Press, Boca Raton and New York, USA. pp. 25–42.

Pelster, B. 1997b. Buoyancy at depth. *In*: D. Randall and A.P. Farrell [eds]. Fish Physiology, Vol. 16: Deep-Sea Fish. Academic Press, San Diego, USA. pp. 195–237.

Pelster, B. 2001. The generation of hyperbaric oxygen tensions in fish: News Physiol. Sci. 16: 287–291.

Pelster, B. 2004. pH regulation and swimbladder function in fish. Respir. Physiol. & Neurobiol. 144(2-3): 179–190.

Pelster, B. 2011. Swimbladder Function and Buoyancy Control in Fishes. *In*: A.P. Farrell [ed]. Encyclopedia of Fish Physiology: From Genome to Environment. Academic Press. San Diego, USA. pp. 526–534.

Pelster, B., J. Hicks and W.R. Driedzic. 1994. Contribution of the pentose phosphate shunt to the formation of CO_2 in swimbladder tissue of the eel. J. Exp. Biol. 197: 119–128.

Pelster, B., H. Kobayashi and P. Scheid. 1988. Solubility of nitrogen and argon in eel whole blood and its relationship to pH. J. Exp. Biol. 135: 243–252.

Pelster, B. and H. Niederstätter. 1997. pH-dependent proton secretion in cultured swim bladder gas gland cells. Amer. J. Physiol. 273: R1719–R1725.

Pelster, B. and D.J. Randall. 1998. The physiology of the Root effect. *In*: S.F. Perry and B.L. Tufts [eds]. Fish Physiology Vol. 17: Fish Respiration. Academic Press, San Diego, USA. pp. 113–139.

Pelster, B. and P. Scheid. 1991. Activities of enzymes for glucose catabolism in the swimbladder of the European eel *Anguilla anguilla*. J. Exp. Biol. 156: 207–213.

Pelster, B. and P. Scheid. 1993. Glucose metabolism of the swimbladder tissue of the European eel *Anguilla anguilla*. J. Exp. Biol. 185: 169–178.

Piiper, J. 1965. Physiological equilibria of gas cavities in the body. *In*: W.O. Fenn and H. Rahn [eds]. Handbook of Physiology, Respiration. American Physiological Society, Bethesda, Maryland, USA. pp. 1205–1218.

Prem, C., W. Salvenmoser, J. Würtz and B. Pelster. 2000. Swimbladder gas gland cells produce surfactant: *in vivo* and in culture. Am. J. Physiol. Regul. Integ. Comp. Physiol. 279: 2336–2343.

Root, R.W. 1931. The respiratory function of the blood of marine fishes. Biol. Bull. 61: 427–456.

Ross, L.G. 1978. The innervation of the resorptive structures in the swimbladder of a physoclist fish, *Pollachius virens* (L.). Comp. Biochem. Physiol. 61C: 385–388.

Schabuss, M., C.R. Kennedy, R. Konecny, B. Grillitsch, W. Reckendorfer, F. Schiemer and A. Herzig. 2005. Dynamics and predicted decline of *Anguillicola crassus* infection in European eels, *Anguilla anguilla*, in Neusiedler See, Austria. J. Helminthol. 79: 159–167.

Schwerte, T., M. Axelsson, S. Nilsson and B. Pelster. 1997. Effects of vagal stimulation on swimbladder blood flow in the European eel *Anguilla anguilla*. J. Exp. Biol. 200: 3133–3139.

Schwerte, T., S. Holmgren and B. Pelster. 1999. Vasodilation of swimbladder vessels in the European eel (*Anguilla anguilla*) induced by vasoactive intestinal polypeptide, nitric oxide, adenosine and protons. J. Exp. Biol. 202: 1005–1013.

Sebert, P., A. Vettier, A. Amerand and C. Moisan. 2009. High pressure resistance and adaptation of European eels. *In*: G. Van den Thillart, S. Dufour and J.C. Rankin, [eds]. Spawning Migration of the European Eel. Springer-Verlag, New York, USA. pp. 99–127.

Sprengel, G. and H. Lüchtenberg. 1991. Infection by endoparasites reduces maximum swimming speed of European smelt *Osmerus eperlanus* and European eel *Anguilla anguilla*. Dis. Aquatic Organ. 11: 31–35.

Steen, J.B. 1963a. The physiology of the swimbladder in the eel *Anguilla vulgaris*. II. The reabsorption of gases. Acta Physiol. Scand. 58: 138–149.

Steen, J.B. 1963b. The physiology of the swimbladder in the eel *Anguilla vulgaris*. III. The mechanism of gas secretion. Acta Physiol. Scand. 59: 221–241.

Steen, J.B. 1963c. The physiology of the swimbladder of the eel *Anguilla vulgaris*. I. The solubility of gases and the buffer capacity of the blood. Acta Physiol. Scand. 58: 124–137.

Stray-Pedersen, S. 1970. Vascular responses induced by drugs and by vagal stimulation in the swimbladder of the eel, *Anguilla vulgaris*. Comp. Gen. Pharmacol. 1: 358–364.

Stray-Pedersen, S. and A. Nicolaysen. 1975. Qualitative and quantitative studies of the capillary structure in the *rete mirabile* of the eel, *Anguilla vulgaris* L. Acta Physiol. Scand. 94: 339–357.

Sund, T. 1977. A mathematical model for counter-current multiplication in the swimbladder. J. Physiol. 267: 679–696.

Van den Thillart, G., S. Dufour, J.C. Rankin. 2009. Spawning Migration of the European Eel: Reproduction index, a useful tool for conservation management. Springer Verlag. Fish & Fisheries Series.

Van Ginneken,V., C. Durif, S.P. Balm, R. Boot, M.W.A. Verstegen, E. Antonissen and G. Van den Thillart. 2007. Silvering of European eel (*Anguilla anguilla* L.): seasonal changes of morphological and metabolic parameters. Animal Biol. 57(1): 63–77.

Vent, R.J. and G.V. Pickwell. 1977. Acoustic volume scattering measurements with related biological and chemical observations in the northeastern tropical Pacific. *In*: N.R. Andersen and B.J. Zahuranec [eds]. Oceanic Sound Scattering Prediction. Plenum Press, New York, USA. pp. 697–716.

Wagner, R.C., R. Froehlich, F.E. Hossler and S.B. Andrews. 1987. Ultrastructure of capilaries in the red body (*rete mirabile*) of the eel swim bladder. Microvasc. Res. 34: 349–362.

Wahlqvist, I. 1985. Physiological evidence for peripheral ganglionic synapses in adrenergic pathways to the swimbladder of the Atlantic cod, *Gadus morhua*. Comp. Biochem. Physiol. 80C: 269–272.

Winata, C., S. Korzh, I. Kondrychyn, V. Korzh and Z. Gong. 2010. The role of vasculature and blood circulation in zebrafish swimbladder development. BMC Develop. Biol. 10(1): 3.

Winata, C.L., S. Korzh, I. Kondrychyn, W. Zheng, V. Korzh and Z. Gong. 2009. Development of zebrafish swimbladder: The requirement of Hedgehog signaling in specification and organization of the three tissue layers. Develop. Biol. 331(2): 222–236.

Woodland, W.N.F. 1911. On the structure and function of the gas glands and *retia mirabilia* associated with the gas bladder of some teleostean fishes. Proc. Zool. Soc. London. 1: 183–248.

Würtz, J., W. Salvenmoser and B. Pelster. 1999. Localization of carbonic anhdrase in swimbladder tissue of European eel (*Anguilla anguilla*) and perch (*Perca fluviatilis*). Acta Physiol. Scand. 165: 219–224.

Würtz, J. and H. Taraschewski. 2000. Histopathological changes in the swimbladder wall of the European eel *Anguilla anguilla* due to infections with *Anguillicola crassus*. Dis. Aquat. Org. 39: 121–134.

Würtz, J., H. Taraschewski and B. Pelster. 1996. Changes in gas composition in the swimbladder of the European eel (*Anguilla anguilla*) infected with *Anguillicola crassus* (Nematoda). Parasitol. 112: 233–238.

Yin, A., S. Korzh, C.L. Winata, V. Korzh and Z. Gong. 2011. Wnt Signaling Is Required for Early Development of Zebrafish Swimbladder. PLoS One. 6(3): e18431.

Zwerger, P., K. Nimeth, J. Würtz, W. Salvenmoser and B. Pelster. 2002. Development of the swimbladder in the European eel (*Anguilla anguilla*). Cell Tissue Res. 307: 155–164.

Morpho-Functional Design of the Eel Heart

Daniela Amelio,[a,]* Filippo Garofalo,[b] Sandra Imbrogno[c]
and Bruno Tota[d]

Introduction

Eels (genus *Anguilla*) are fascinating animals, which for their unique features have become an interesting laboratory model. This is due not only to their exceptional resistance to experimental conditions, but also to the incomparable capacities to face the severe environmental challenges imposed by their complex life history (e.g., Hyde et al. 1987). For example, following metamorphosis, the eels travel a long distance to spawn, but because of their swimming behaviour ("anguilliform locomotion") they are relatively inefficient swimmers (van Ginneken et al. 2005).

To match such high levels of locomotor endurance, eel have evolved relevant metabolic and cardio-respiratory capacities (van Ginneken and van den Thillart 2000, Ellerby et al. 2001). As first documented by Steffensen and Lomholt (1990) in European eel farms, this fish exhibits exceptional tolerance to very severe chronic hypercapnia, greater than has been documented

University of Calabria, Dept. of Cell Biology, Arcavacata di Rende, 87036 (Cosenza), Italy.
[a]E-mail: daniela.amelio@unical.it
[b]E-mail: filigaro@libero.it
[c]E-mail: sandra.imbrogno@unical.it
[d]E-mail: tota@unical.it
*Corresponding author

in any other teleost (Larsen and Jensen 1997). An elevated capacity to accumulate plasma HCO_3^- may be implicated in the eel compensation for acidosis (Farrell and Lutz 1975) and hypercapnic acidosis *per se* (McKenzie et al. 2002). However, despite a consequent reduction in arterial pH and 80% decline in arterial blood total O_2 content, no effects on cardiac output or whole animal O_2 uptake were observed (McKenzie et al. 2002, 2003). Moreover, the eel ability to regulate the aerobic metabolism during hypoxia, despite quite profound hypoxaemia, is consistent with an extraordinary capacity of their respiratory and cardio-circulatory systems to satisfy tissue demands (McKenzie et al. 2003). Of note, during migration both *A. japonica* and *A. anguilla* face hydrostatic pressure challenges of at least 20 ± 30 atm (200 ± 300 m depth) (Aoyama et al. 1999, McCleave and Arnold 1999). In the eel, unlike trout (*Oncorhynchus mykiss*), a short exposure to high hydrostatic pressure (over 100 atm) did not induce significant alterations in aerobic metabolism (for review see: Sébert 2001). This is related to a low sensitivity of mitochondrial red myotomal muscle to high hydrostatic pressure, evaluated in terms of oxidative phosphorylation (Sébert and Theron 2001). In addition, the eel has an efficient anaerobic metabolism, which, conceivably, might balance the deleterious effects observed during the first hours of pressure exposure (Sébert 1997).

For these considerations, the eel cardiovascular system may be expected to respond with a remarkable multilevel morpho-functional plasticity and adaptability to a variety of intrinsic (body size and shape, fibre type distribution in the myotomal muscle, mode of swimming, etc.) and extrinsic (temperature, partial pressure of oxygen, pH, environmental pollutants, etc.) factors associated with the mode of the animal life (Peyraud-Waitzenegger and Soulier 1989, Sancho et al. 2000, Bailey et al. 2000, McKenzie et al. 2003). In the present chapter we will consider the following aspects of the eel heart: morphology, cardiac allometry and morphodynamics, intrinsic and extrinsic regulatory mechanisms of pumping performance, including autocrine-paracrine modulation and the nitric oxide synthases (NOS)/nitric oxide (NO) signalling, and the influences of exercise and temperature.

Cardiac Morphology

The eel heart, like that of other teleosts, comprises three contractile chambers, i.e., the sinus venosus, the atrium and the ventricle, and an elastic non-contractile structure, the bulbus arteriosus (Fig. 1a). The *sinus venosus* is a large thin-walled chamber located dorsally and separated from the atrium by the sinus valves (Yamauchi 1980). Its structure varies between species, being formed by variable amounts of muscle and connective tissue. In *Anguilla anguilla* it is mostly made up of myocardium (Farrell and Jones 1992). The single chambered *atrium*, which shows considerable size and

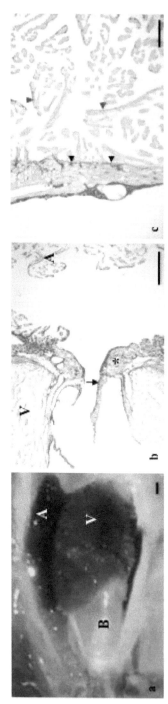

Figure 1. (a) Ventral view of *Anguilla anguilla* heart (A=atrium, B=bulbus, V=ventricle). Scale bars: 0,1 cm. (b) and (c): Sirius red. (b) shows the AV region; note the continuity between the AV and the atrial muscle; AV valves (arrows) anchor to a ring of compact, vascularized myocardium (asterisks). In (c) collagen at the *compacta/spongiosa* boundary (black arrowheads) and ventricular trabeculae are evident; note the collagen bundles localized at the subendocardial level (blue arrowheads). Scale bars: 100 μm. (Unpublished data).

Color image of this figure appears in the color plate section at the end of the book.

shape variability between teleost species (Santer 1985, Farrell and Jones 1992), is formed externally by a rim of myocardium enveloping a complex system of thin trabeculae in the eel. The myocardial tissue is surrounded by a subepicardial layer containing a large amount of collagen fibers (Fig. 1c), also evident at the subendocardial level of the trabeculae, where it contributes to support the atrial structure (Icardo 2012).

There are two main types of fish ventricle organization: the trabeculate type and the mixed type. The former is made up of a crisscrossed array of myocardial bundles (*trabeculae*) that give the ventricular wall a spongy appearance, hence named *spongiosa*. It is supplied by the superfusing venous blood circulating through the intertrabecular spaces (*lacunae*). In the mixed type of ventricle the *spongiosa* is invested in an outer layer of orderly and densely arranged myocardial bundles, thence named *compacta*, which is perfused by an arterial supply (Tota 1989). The *mixed type ventricle* of *A. anguilla* (Figs. 1c, 2a) shows an external tubular shape (Icardo 2012). The *compacta* is formed by bundles of muscle tissue variously oriented (outer longitudinal and inner mainly circular) and vascularized by coronary vessels (Fig. 2a,b,d,e). The *spongiosa* is avascular and the numerous *trabeculae* are lined by a thin layer of endocardial endothelium (EE), separating the ventricular lumen into an extensive network of lacunae of different size (Fig. 2a,d).

Of note, the EE possesses a variety of receptors for endocrines/growth factor substances, which is consistent with the view that it may act as an autocrine–paracrine and sensor-integrator device (e.g., *via* EE-NOS/NO-mediated signaling, as detailed below). Therefore, the extensive surface area, typical of fish heart (Tota et al. 2010, Imbrogno et al. 2011), may be important not only as an exchange boundary between blood-born stimuli and subjacent myocardium, but also for an intracavitary regulation of cardiac performance (Tota et al. 2010). Recently, Icardo (2012) reported the presence of a large collagen network, extending between the subepicardium and the subendocardium, which mimics the distribution of both the myocardial bundles in the *compacta* and the trabeculae in the *spongiosa*. In particular, the subendocardial localization of the collagen in the trabeculae (Fig. 1c) points to an important mechanical role in the maintenance of the ventricular structure and dynamics. Like in mammalian hearts, this functional role of the collagen network is strongly supported by the morpho-funtional evidence showing that in the growing eels the increased amount of collagen in the ventricle (Fig. 2e) parallels its improved mechanical performance (Cerra et al. 2004). Likely, the collagen fibers located at the boundary between *compacta* and *spongiosa* can also act as a mechanical link of two differently oriented myocardial layers (Poupa et al. 1974, Tota 1978, Icardo et al. 2005).

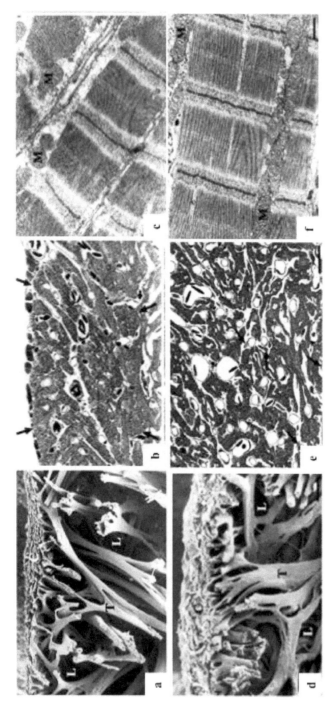

Figure 2. (a,d) Scanning, (b,e) light and (c,f) transmission-electron microscope images of small (a,b,c) and large (d,e,f) eels. (a,d) the thickness of the *compacta* and that of the trabeculae increases with growth. (b,e) the entire *compacta* thickness (arrows) of the small eel is included in (b). At the same magnification, only part of the *compacta* appears in (e). Note the increased vascularization of the latter. Arrows in (e) indicate accumulations of collagen; no such accumulations appear in (b). (c,f) myofibrils are developed at both ages. Note in (f) the alignment of the mitochondrial and myofibril axis. L, lacunary spaces; T, trabeculae; M, mitochondria. Bars, 25 μm (a,d); 25 μm (b,e); 600 nm (c,f) from Cerra et al. 2004.

As in other teleosts, two distinct structures have been described in the eel heart, namely the atrio-ventricular (AV) region and the conus arteriosus (Icardo 2012). The AV region (Fig. 1b), which supports the AV valves, is made up by a ring of compact and vascularized myocardium, containing a variable amount of collagen and elastin (Santer and Cobb 1972, Farrell and Jones 1992, Icardo and Colvee 2011, Icardo 2012). A ring of connective tissue contributes to discriminate the AV muscle from that of the atrium and the ventricle. Contradicting the commonly accepted view that the conus arteriosus of teleosts has been lost during evolution, Icardo (2006) demonstrated in the modern teleosts, including the eel, the conservation of this region as a distinct heart segment interposed between the ventricle and the bulbus arteriosus. Formed by compact myocardium, which is generally vascularized and supports the conus valves (Schib et al. 2002), this region is easily identifiable in completely trabeculated ventricles, being more difficult to be discerned in those possessing a compact layer.

As a consequence of the more caudal position, the eel *bulbus arteriosus* appears comparatively smaller (Fiedler 1991). Its inner wall shows an irregular surface due to the presence of branching ridges, covered by flattened EE cells that contain dense bodies with secretory function (Icardo et al. 2000). The middle layer is formed by smooth muscle cells, embedded in a meshwork matrix containing thin and thick filaments (Icardo et al. 2000). The stretching of this meshwork suggests an active role as smooth muscle in wall dynamics. Furthermore, large areas of the extracellular space are occupied by elastin-like materials, whose amount decreases toward the external layer (Icardo et al. 2000). On the contrary, the collagen, although present across the entire wall thickness, increases from the inner toward the outer bulbus surface. Conceivably, such gradient of collagen matrix may increase wall strength, maintaining bulbus dilation within safe physiological changes.

The myocardium is covered by an external mesothelial layer, the *epicardium*. It is formed by flattened cells containing abundant pinocytotic vesicles, which is consistent with an active solute interchange with the pericardial cavity (Icardo et al. 2000). As in other teleosts, the detection in the epicardial cells of natriuretic peptide receptors (Cerra et al. 1997) suggests a NP-dependent control of pericardial activity. Moreover, the presence of eNOS at the epicardial level (Tota et al. 2005, Amelio et al. 2006, 2008, Imbrogno et al. 2011) indicates an autocrine-paracrine function in relation to both interstitial fluid balance and myocardial remodeling/regeneration (for more details see Poss et al. 2002, Lepilina et al. 2006).

Body Growth and Cardiac Morphodynamics

As a rule in all animals, the eel heart responds to developmental (genetic programs) and often striking environmental (e.g., hemodynamic forces) stimuli, and grows by either cell division (hyperplasia), or cell enlargement (hypertrophy), or both. This growth obeys the scaling relationship (allometry) which describes how the organism changes as body size increases, that is, the variability in shape as a function of size (Hochachka and Somero 2002). In most fishes, the relative heart weight (RHW=heart mass/body mass x100) is bigger in active swimmers than in sedentary species (Poupa and Lindstrom 1983, Gamperl and Farrell 2004). A hallmark of cardiac performance is represented by cardiac output (CO), i.e., the product of heart rate (HR) and stroke volume (SV). Since the ventricle is the major pump of the heart, its growth and remodelling highlight the remarkable morpho-functional plasticity which can be attained at the organ level. In particular, a close link exists between the myoarchitetture of the ventricle and its mechanical performance, evaluated in terms of relative contribution of pressure and volume work to the stroke work (SW, i.e., the product of SV and the mean ventral aortic pressure) (Tota and Gattuso 1996). In many fish, this relationship permits to discern between ventricles producing mainly volume work and those producing mainly pressure work. To evaluate in the eel whether and how body growth can affect significant morpho-dynamic rearrangements of the ventricle, Cerra et al. (2004) compared these parameters between juvenile (small-size) and adult (large-size) eels. During growth, heart and ventricle mass showed a comparable increase: for a fourfold increase of body weigth from 138.53 ± 7.45 g (means ± S.E.M, n=22) in juveniles to 565.43 ± 51.57 g (n=22) in adults, an equivalent increase of absolute ventricular mass was documented from 0.084 ± 0.0077 g to 0.358 ± 0.043 g. This was confirmed by the similar values of the relative ventricular mass (RVM=ventricle mass x 100/body mass: 0.061 ± 0.0036 and 0.062 ± 0.0036 for small and large eels, respectively). The cardiac basal physiological parameters resulted lower in small eel hearts compared to the large counterparts. In fact, HR was 38.93 ± 2.82 and 52.7 ± 1.8 beats min^{-1} and SV was 0.27 ± 0.017 and 0.37 ± 0.016 ml kg^{-1}, in small and large eels respectively. Interestingly from a structural point of view, Cerra et al. (2004) found an augmented thickness of the *compacta* in the large eel ventricle, while in the *spongiosa* an increased trabecular diameter, paralleled by a decrease of the lacunary spaces, resulted in an increased layer density (Fig. 2a,d). The increased thickness in the *compacta* is achieved through the enlargements of both muscular and vascular compartments with a reduction of the interstitium space (Fig. 2b,e). In many growing fish species, an increased thickness of the *compacta* has been documented (*Ciprinus carpio*: Bass et al. 1973, *Salmo salar*: Poupa et al. 1974, *Thunnus thynnus*: Poupa et al. 1981, *Salmo gairdneri*: Farrell et al. 1988, Johansen et al. 2011). Of note, the *compacta* thickening with consequent enlargement of the epicardium-endocardium distance, may have also stimulated the increased vascularization to adequately supply blood to the innermost myocardial cells. In both *compacta* and *spongiosa* of large eels, the larger number of cardiomyocytes together with the decrement of both the cell cross-sectional area and the myofibrillar compartment indicates that the myocardial growth is achieved through hyperplasia. Hyperplastic ventricular growth is common in many fish species, although hypertrophy has also been shown to play an important role (Farrell et al. 1988, Bailey et al. 1997, Clark and Rodnick 1998). Moreover, during growth, the decrement of cellular cross-sectional area in both *compacta* and *spongiosa*, could be an advantage to face the limitations induced by the low area-to-volume ratio, due to the absence of a transverse tubule network for excitation/contraction coupling (EC) (Santer 1985, Rodnick and Sidell 1997, Harwood et al. 2002, and references therein). In the large eels, unlike the *spongiosa*, the *compacta* shows an increased mitochondrial compartment (Fig. 2d,f) and its association with the myofibrils represents a basic index of the potential cardiac work (Kayar et al. 1986, Barth et al. 1992). It is thus conceivable that the accelerated contractile rhythm and the higher pumping capacity, reported for large eels, can be achieved by the enlargement of the mitochondrial compartment to match adequately the energetic demands of the myofibril apparatus. Importantly from a hemodynamic viewpoint, the large eels possess an enhanced ability to maintain work against higher output pressure (PO). In fact, when filling pressure (preload) was increased, both small and large eels showed a similar response, whereas large eels were more able to tolerate increased output pressure (afterload) corresponding to increased aortic blood pressure. In particular, small eel hearts decreased SV at afterload greater than 3kPa, in contrast to larger hearts, which maintained constant SV up to 6kPa (Fig. 3).

Consequently, as a result of these growth-related morpho-functional changes, the small eel ventricle, with its large lacunary spaces and limited tolerance to increased afterloads, seems better adapted to generate volume work; in contrast, large eels are better suited to produce pressure work (for reference: Icardo et al. 2005).

Regulation of Cardiac Performance

In vertebrates, the control of cardiac performance, achieved by intrinsic and extrinsic (neuro-humoral) mechanisms, can be appreciated in terms of three temporal paradigms: beat-to-beat, length-dependent regulation (Starling's law of the heart); short-term regulation by biochemical changes within the cardiomyocyte (E–C coupling, myocardial contractility); long-

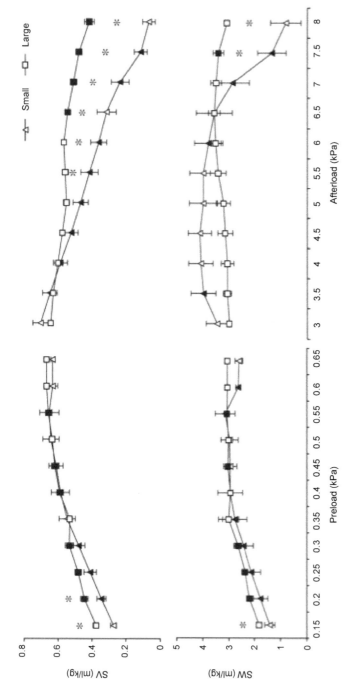

Figure 3. Effects of preload and afterload elevation on stroke volume (SV) and stroke work (SW) in isolated and perfused heart of small (triangles; n=7) and large (squares; n=7) eels. Values are means ± S.E.M. Comparison within group: open symbols, not significant; closed symbols, P<0.05. Comparison between groups: *P<0.05 (from Cerra et al. 2004).

term regulation by modified gene expression. Unlike mammals, fishes respond to different hemodynamic loads (filling pressure or preload and systemic aortic pressure or afterload) by increasing CO mainly through an increased SV rather than HR. The neuro-humoral control is mainly achieved by the cholinergic and adrenergic fibres, as well as a variety of hormones (e.g., the cardiac natriuretic peptides, angiotensin, etc.) and humoral agents, including the catecholamines (CA) released by both extracardiac and cardiac chromaffin cells (Randall and Perry 1992, Farrell and Jones 1992, Nilsson and Holmgren 1992).

Intrinsic Mechanism: Frank-Starling Response

The intrinsic control of cardiac function is exemplified by the length-dependent response to changing preload. In mammals, the stretched cardiac muscle modifies its contractility following a biphasic module characterized by an immediate increase in force (heterometric regulation or Frank-Starling response), followed, after 10–15 min, by a slower and persistent force increment (Anrep effect). The rapidly occurring stretch-related increase in developed force has been attributed to a length-dependent increase in cross-bridge formation and myofilament calcium responsiveness (Katz 2002), while the slow increase has been correlated with a corresponding increase in the amplitude of the intracellular Ca^{2+} transient (Calaghan et al. 1999, Casadei and Sears 2003). Although it is not yet clear whether in non-mammalian vertebrates the cardiac response to stretch assumes also the same biphasic module, it is commonly recognized that in fish, the end-diastolic volume and the consequent stretch-related increase in developed force is a major regulator of cardiac performance (Olson 1998). In fact, unlike mammals, fish respond to different hemodynamic loads by enhancing CO mainly through an increased SV rather than HR (Tota and Gattuso 1996, Olson 1998). Teleosts, including the eel, show an elevated sensitivity to the Frank-Starling response (icefish: Tota et al. 1991, gilthead seabream: Icardo et al. 2005, eel: Imbrogno et al. 2011), attributed to a greater myocardial extensibility of the highly trabeculated heart, coupled to a maintained increase in myofilament Ca^{2+} sensitivity over a large range of sarcomere lengths (Di Maio and Block 2008, Shiels and White 2008). In particular, the sensitivity of *A. anguilla* heart to the Frank-Starling response was significantly enhanced by a basal release of endogenous NO (Imbrogno et al. 2001). This occurs through a nonclassical cGMP-independent pathway involving an Akt-mediated activation of eNOS-dependent NO production, which in turn modulates the rate of Ca^{2+} reuptake by the sarcoplasmic reticulum through phospholamban (PLN) S-nitrosylation (Garofalo et al. 2009).

Extrinsic Regulation: Neurohumoral Control

The teleost (and eel) heart receives both adrenergic and cholinergic innervation, whose importance varies according to species-specific characteristics and circumstances. While the heart is aneural in the hagfish or possesses only cholinergic nerves in lampreys and elasmobranchs, the heart of teleosts has an established sympathetic and parasympathetic innervation through the 'vagosympathetic' trunk (Laurent et al. 1983, Nilsson 1983, Taylor 1992). However, because of the lack of longitudinally connected sympathetic chains, the teleost sympathetic nervous system (SNS) is less developed in comparison with that of birds and mammals (Laurent et al. 1983, Taylor 1992). The peripheral nerves are replaced by aggregates of CA-containing chromaffin cells, which in many species become components of the diffuse neuroendocrine tissue (Burnstock 1969). These chromaffin cells, often associated with sympathetic nerves and/or cholinergic inputs, provide a zonal CA production, thereby contributing to the humoral cardiovascular regulation (Gannon and Burnstock 1969, Abrahamsson et al. 1979, Nilsson and Holmgren 1992, Tota et al. 2010).

Adrenergic effects

CA (adrenaline and noradrenaline) reach cardiac adrenoceptors (ARs) *via* the circulation and the SNS terminals. It is generally believed that in teleosts basal plasma CA levels are low, while the nervous activity plays a major role (Holmgren and Nilsson 1982, Axelsson 1988, Randall and Perry 1992). Circulating CA in most teleost species increased immediately in response to various physical and environmental stimuli (exhaustive exercise, hypoxia, hypercapnia, etc.) that require increase oxygen transport and are an indicator of acute stress, as best epitomized by salmonids (Randall and Perry 1992). In this respect, the genus *Anguilla* appears exceptional, since several of its members do not exhibit increased plasma CA nor cortisol, i.e., an indicator of sublethal chronic stress in response to stressful stimuli (Wendelar Bonga 1997). Thus, eels are considered stress-tolerant animals (McKenzie et al. 2003).

The cardiac response to CA stimulation is mediated by two main types of ARs, α–ARs and β–ARs; for each of them several subtypes have been identified in fish, similar to other vertebrates (Ask 1983). In the past, the adrenergic stimulation has been analysed in various teleost hearts (e.g., *Anguilla anguilla*: Forster 1976, Peyraud-Waitzenegger et al. 1980, Pennec and Peyraud 1983, Pennec and Le Bras 1988, *Anguilla dieffenbachti*: Forster 1981). For example, a basal adrenergic excitatory tone, which prevails over the cholinergic one (Pennec and Le Bras 1984), appears to be mediated by α- and β-ARs associated with both the pacemaker and the working myocardium (e.g., *Labrus mixtus*, *Zoarces viviparus*, *Myoxocephalus scorpius*: Axelsson et al. 1987, *Oncorhynchus mykiss*: Gamperl et al. 1994). Adrenergic-dependent increases of both HR (Graham and Farrell 1989) and the Frank-Starling response (Farrell et al. 1986) have also been proposed. Peyraud-Waitzenegger et al. (1980) suggested that seasonal changes in temperature may influence the cardiac adrenoceptor function; however, the specific mechanisms underpinning these, as well as other CA-elicited cardiac actions in teleosts, still remain to be clarified. By reinterpreting retrospectively these studies on the basis of the present knowledge, some theoretical and methodological considerations need to be made (Randall and Perry 1992). In fact, the cardiac response to ARs stimulation depends not only on the species-related differences and the organizational level under study (i.e., *in vivo* cardiovascular system *vs* isolated and denervated working heart), but also on the stress impact and the consequent compensatory responses (e.g., divergent stress coping styles). In the context of this rapidly growing area of fish physiology (Winberg and Nilsson 1993, Cossins et al. 2006, Johnson et al. 2006), it has become crucial to appreciate that the CA under study is, at the same time, strikingly implicated in the stressed condition of the laboratory animal. As discussed by Epple and Brinn (1987), another limitation has been the use of pharmacological but not physiological concentrations of CA. This may have delayed the identification of the today-acknowledged fine-tuned yin–yang modulation exerted by CA as components of the complex ARs–G-protein-coupled signal transduction pathways, including NO-cGMP (see below).

The recent identification of a novel type of cardiac β-ARs (β3) gave new insights into the adrenergic control of teleost heart function. As well as β1- and β2-ARs, β3-AR belongs to the G-protein coupled receptors characterized by seven transmembrane domains of 22–28 amino acids. In mammals, β3 activation elicits negative inotropism (Gauthier et al. 2000), involving a NO-cGMP signaling. Similarly, in the eel, β3-ARs activation decreases cardiac mechanical performance (Fig. 4) through a pertussis toxin (PTx)-sensitive Gi protein mechanism, consistent with a major β3-ARs myocardial localization, and requires the NO-cGMP-cGMP-activated protein kinase G (PKG) cascade (Imbrogno et al. 2006). Furthermore, in the *A. anguilla* heart, isoproterenol (ISO) stimulation, in addition to its classic positive inotropism, induces a negative inotropism (Fig. 4) abolished by β3-AR antagonist pretreatment (Imbrogno et al. 2006). The recent identification of β3-ARs and their involvement in counteracting ISO stimulation in the eel heart not only contributes to clarifying the adrenergic versatility that controls its function, but sheds new light on the role of the balanced regulation of β1/β2 and β3-AR systems in fish. Whether such an intrinsic β3-AR inhibition exerts cardio-

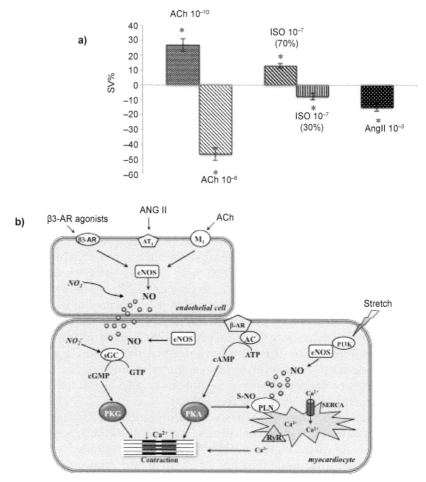

Figure 4. a) Effects of acetylcholine (ACh; 10^{-10}M and 10^{-6}M), isoproterenol (ISO; 10^{-7}M) and angiotensin II (Ang II; 10^{-8}M) on SV in isolated and perfused eel hearts. Percentage changes were evaluated as mean ± SEM (n=4–5 experiments for each group). Asterisks indicate significant differences (P<0.05) from rates in untreated controls normalized to 0% SV. Note that the negative inotropic effect induced by ACh (10^{-6}M), Ang II (10^{-8}M) and, in the 30% of preparations, ISO (10^{-7}M) involves a NOS/NO-dependent pathway. For details see Imbrogno et al. (2001, 2003, 2006). b) Schematic diagram showing the role of NO as a spatial paracrine/autocrine integrator. Chemical stimuli such as ACh, ANG II, and β3-AR agonists converge on the eNOS-NO signaling which requires the obligatory involvement of the EE (Imbrogno et al. 2001, 2003, 2006). Stretch conditions activate the release of autocrine NO which directly modulates the SR Ca2+ reuptake through PLN S-nitrosylation (Garofalo et al. 2009) (modified from Imbrogno et al. 2011).

protection against the stress-induced excessive excitatory stimulations (e.g., strenous exercise, exposure to systemic and/or intracardiac CA, angiotensin, endothelin-1, etc.) remains to be evaluated.

Cholinergic-induced effects

In teleosts, cholinergic fibers carried in the vagus are responsible for lowering HR, and several studies showed an abolition of bradycardia by vagotomy or atropin injection. Obviously, acetylcholine (ACh), released from the nerve terminals in the sinoatrial region hyperpolarizes pacemaker cells and slows the pacemaker rate (Saito 1973). Notably, differences in the cholinergic mechanisms that control HR in teleosts have been reported in relation to temperature. For example, the level of cholinergic inhibition of HR was greater in cold-acclimated trout compared with warm-acclimated rainbow trout (Wood et al. 1979). In contrast, the level of cholinergic inhibition of HR was lower in cold-acclimated *A. anguilla* (Seibert 1979). Negative ACh-induced contractility in teleosts has been reported in several studies (Randall 1970, Holmgren 1977, Cameron and Brown 1981).

Cardiac cholinergic stimuli are mediated by several ACh receptor (AChR) subtypes, whose relative amounts vary among species and tissues. Five different muscarinic receptor subtypes (M1–M5; Brodde and Michel 1999) have been described in vertebrates. The M2 and M4 subtypes, preferentially located on the myocardiocytes and principally coupled to adenylate cyclase inhibition, reduce intracellular cAMP levels and decrease the L-type Ca^{2+} current, thereby eliciting negative cholinergic chronotropic and inotropic effects (for references see Hove-Madsen et al. 1996, Gattuso et al. 1999). The M1, M3 and M5 subtypes, which are largely located on the vascular endothelial and EE cells, are functionally coupled to intracellular Ca^{2+} mobilization *via* phospholipase C, phospholipase A2 and phospholipase D and mediate positive cholinergic inotropism (Brodde and Michel 1999). In the *in vitro* perfused *A. anguilla* heart, it has been recently demonstrated that ACh elicits a biphasic concentration-dependent inotropism (Fig. 4); i.e., positive at nanomolar concentrations, mediated by M1 muscarinic receptors, and negative at micromolar concentrations, which involves M2 muscarinic receptors (Imbrogno et al. 2001). The ACh-mediated positive inotropism occurs through a NO-cGMP signal transduction mechanism and is abolished when the EE is functionally damaged (Imbrogno et al. 2001).

Angiotensin II stimulation

Angiotensin II (Ang II), the effector of the renin-angiotensin system (RAS), controls cardiac function and growth through plasma membrane AT1 and AT2 receptors and consequent activation of downstream pathways (see references in De Gasparo 2002). AT1 is responsible for most of the Ang II-mediated cardiac effects (i.e., chronotropism and inotropism) and rate of protein synthesis in isolated myocyte preparations (Schorb et al. 1993). The main post-receptor signal transduction pathways include slow Ca^{2+} channel activation, phosphoinositide hydrolysis acceleration (Baker et al. 1989) and stimulation of NOS activity (Paton et al. 2001). In contrast, cardiac AT2 receptors antagonize the AT1 growth-promoting effects *via* activation of a number of phosphatases. AT2 is also coupled with the NO-cGMP signaling, either directly or indirectly through enhanced bradykinin or eNOS expression (Dostal 2000). In teleosts, in which RAS is active in multiple effector systems, numerous examples indicate that Ang II-mediated responses parallel those reported in mammals (for references see Kobayashi and Takei 1996). Most of the Ang II-mediated effects documented in the cardiovascular system appear, however, species-specific and, notably, in synergy with the adrenergic system. For example, while Ang II injected in trout produces a hypertension-dependent reflex bradycardia with reduced CO (Olson 1998), the mechanism of Ang II action is different in the eel; the Ang II-induced hypertension is due to an increased CO while systemic resistence is unaffected in American eel *A. rostrata* (Butler and Oudit 1995, Oudit and Butler 1995). Both direct and indirect (*via* cardiac adrenoceptors) effects of Ang II have been reported in the heart of American eel and trout *Oncorhynchus mykiss* hearts (Oudit and Butler 1995, Bernier et al. 1999). In *A. anguilla* heart, an elevated sensitivity to the Ang II-dependent modulation has been documented (Imbrogno et al. 2003). In fact, Ang II exerts a cardiosuppressive influence, evaluated in terms of reduction of SV (Fig. 4), which implicates an EE interaction. In particular, endoluminal Ang II first interacts with the AT1-like receptors, thereby triggering a NO-cGMP-PKG signal transduction pathway (Imbrogno et al. 2003).

Natriuretic peptides (NPs)

The heart is the major site of synthesis and release of a number of peptide hormones belonging to the NP family, including atrial (ANP), brain (BNP), C (CNP) and ventricular (VNP) (Takei et al. 1991, Takei and Balment 1993, Evans 1995). First identified by de Bold and co-workers as the major constituents of rat atrial granules (de Bold et al. 1981), NPs were subsequently found in the hearts of a large number of non-mammalian vertebrates, including fish (Netchitailo et al. 1987, Takei et al. 1990, Bjenning et al. 1992, Larsen et al. 1994, Kawakoshi et al. 2003). In teleosts, both ANP and BNP are present in pufferfish and tilapia, but ANP was undetectable in medaka; in contrast, eel and trout contain VNP in addition to ANP and BNP (Takei et al. 1994a, 1994b, Inoue et al. 2005).

Application of homologous peptides for radioreceptor analysis identified four types of NP receptors (NPRs), named NPRA, NPRB, NPRC and NPRD, present from cyclostomes to mammals (for references see Takei et al. 2011). In particular, NPRC-like receptors are present in atrial and ventricular myocardium of the eel, while the ventricular EE appears to express an NPRA-type that is able to bind ANP and VNP with almost equal affinity (Cerra et al. 1996). A receptor with high CNP affinity, presumably NPRB, is also expressed in the endothelial and epicardial layers of the bulbus arteriosus of *A. anguilla* (Cerra et al. 1996), where it may mediate a CNP-dependent modulation of the bulbar hemodynamics (e.g., the Windkessel function) (see Icardo et al. 2000).

It is generally accepted that in all vertebrates the primary stimulus for acute atrial NPs release is the hypervolemia-induced stretch (Ruskoaho 1992). Once released, NPs activate complex signalling networks that control heart-vessel and volume-ion homeostasis at various levels (for references see Cerra and Pellegrino 2007). These networks coordinate directly chronotropic, inotropic and vasorelaxant responses, as well as indirectly to minimize cardiac hemodynamic loads. The hypothesis that NPs exert a cardioprotective action from excessive preloads and afterloads also in fish is supported by direct and indirect evidence obtained in various species (Olson et al. 1997, Johnson and Olson 2008, Tota et al. 2010). In addition, the finding that the transfer from freshwater to seawater of Japanese eel *A. japonica* transiently increases both ANP and VNP (Kaiya and Takei 1996a), suggests that an osmotic mechanism is also able to regulate NP release from the teleost heart. Although volemic stimuli are also stimulatory for NP release in the eel and trout, hyperosmolality may be more potent than hypervolemia in stimulating atrial ANP release (Kaiya and Takei 1996b, Cousins and Farrell 1996). Moreover, in seawater eels, ANP is antidipsogenic at the dose as low as non-hypotensive (Tsuchida and Takei 1998). These observations in fish indicate an opposite situation to that found in mammals, in which the blood volume decrease inhibits ANP secretion and the ANP-dependent hemodynamic actions appear dominant compared to the osmoregulatory effects.

NOS/NO-mediated Autocrine-paracrine Regulation

In the past two decades, the proliferation of relevant research on new cardiac autocrine-paracrine factors widened the scenario of heart autoregulation and neurovisceral integration (see Tota et al. 2010 for review). In particular, the concept of NO as a cardiac autocrine/paracrine integrator, largely consolidated in mammals, has been recently expanded also to the fish heart (Imbrogno et al. 2010). NOS isoenzymes [i.e., endothelial NOS (eNOS), neuronal NOS (nNOS) and inducible NOS (iNOS)], localized in almost every heart tissue, regulate, through their distinct spatial subcellular compartmentation, the production of NO close to its molecular targets (Seddon et al. 2007). NO, generated in one cell, can act on the adjacent cell (paracrine modulation) or on one or more processes of the cell itself (autocrine modulation) (Moncada et al. 1991). The fact that NO synthesized by the EE or the vascular endothelium diffuses and affects the function of the nearby cardiomyocytes well epitomizes the paracrine character of its signal (as depicted in Fig. 4b).

In the eel heart, Imbrogno et al. (2001) documented a tonic crosstalk between the EE-eNOS signaling and the subjacent myocardium, suggesting a major role of NO in the paracrine modulation (negative inotropism) of cardiac performance. This EE-eNOS signaling appears to be located at the cross-road of many extrinsic and intrinsic neuro-endocrine pathways, coordinating many chemically activated cascades (Imbrogno et al. 2001, 2003, 2004, 2006, 2010).

Since both constitutive and inducible NOSs are expressed in cardiomyocytes, NO-dependent autocrine modulation can directly influence their ion channels and pumps (Seddon et al. 2007). Thanks to their subtle spatial compartmentation and signalling networks coordination between cardiac NO and specific intracellular effectors (Hare 2003), NOSs can achieve local control of different cellular functions (Iwakiri et al. 2006). At the same time, a relatively high local concentration of NO equivalents may be generated, which in turns provides a favourable environment for protein S-nitrosylation (Lima et al. 2010). Interestingly, myocardial autocrine NO directly affects the Frank-Starling response in the *A. anguilla* heart through a non-classical cGMP-independent pathway which involves a "beat-to-beat" regulation of PLN S-nitrosylation-dependent calcium reuptake by SR-CA^{2+}ATPase (SERCA2a) pump and thus of myocardial relaxation (Garofalo et al. 2009) (Fig. 4b). This first evidence supports the idea that the NO ability to modulate ventricular performance through NOS isoforms compartmentation, as well as differences in their mode of stimulation and recruitment of distinct downstream pathways (S-nitrosylation or cGMP production), was indeed a crucial and early event during the vertebrate evolution.

Nitrite as NO sources

In various cells and tissues, nitrite (NO_2^-) represents an important physiological reservoir of NO, being a key intrinsic signaling molecule in many biological processes (Bryan et al. 2005). Its conversion into NO is achieved through both non-enzymatic and enzymatic pathways (Mayer et al. 1998, Modin et al. 2001, Cosby et al. 2003, Rassaf et al. 2007). The various biological responses mediated by nitrite include hypoxic vasodilation (Cosby et al. 2003, Crawford et al. 2006), inhibition of mitochondrial respiration (Huang et al. 2005), cytoprotection following ischemia/reperfusion (I/R) (Webb et al. 2004, Duranski et al. 2005), and regulation of protein and gene expression (Bryan et al. 2005). These actions are thought to be dependent on either the reduction of nitrite to NO or direct S-nitrosylation of thiol-containing proteins (Bryan et al. 2005, Perlman et al. 2009). Compared to terrestrial animals, water-breathing organisms such as fish have an additional direct uptake of exogenous nitrite from the environmental water across the respiratory surfaces (Jensen 2009). Therefore, the freshwater fish need to balance the advantageous access to an ambient pool of nitrite for internal NO production with the potentially dangerous effects of nitrite-polluted habitats (see Jensen and Hansen 2011).

In the eel, nitrite represents a significant source of bioactive NO that modulates heart function. In particular, it negatively affects basal mechanical performance through a NOS-dependent mechanism which involves a cGMP/PKG transduction signaling (Cerra et al. 2009). Moreover, nitrite profoundly influences the Frank-Starling response through a NO/cGMP/PKG pathway and S-nitrosylation of both membrane and cytosolic proteins (Angelone et al. 2012).

Other Factors Affecting Heart Function

Exercise

There is an extensive literature regarding the metabolic, respiratory and cardiovascular adjustments during swimming (Randall 1982, Randall and Perry 1992, McKenzie et al. 2004). In general, sustained swimming is aerobic, increasing both CO, through changes in HR and SV, and ventilation. However, eels may represent an exception. In fact, unlike most teleosts, the Australian short-finned eel *A. australis* does not increase CO but, because of branchial vessels constriction, increases ventral aortic blood pressure without changing dorsal aortic pressure (Davie and Foster 1980). The significance of this changes remains to be clarified. Contrary to active teleosts, like salmonids, in which aerobic swimming performance is hampered by elevated water CO_2 levels (Brauner et al. 1993) with consequent cardiovascular impairement (Gallaugher et al. 2001), chronic hypercapnia does not influence exercise performance in eel (McKenzie et al. 2003). It is likely that, in addition to other factors, this endurance capacity is related to the very high myocardial tolerance of the eel heart to hypercapnic acidosis (McKenzie et al. 2002) and hypoxia (Davie et al. 1992). In this regard, it would be of interest to verify in *A. anguilla* if the myocardial mitochondria exhibit the same resistance to hydrostatic pressure as those of the red myotomal muscle, as mentioned above (Sébert and Theron 2001). How and to which extent the eel myocardium is able to compensate its intracellular pH to preserve cardiac performance represents a challenge for future studies.

Temperature

Environmental temperature changes may have an impact on fish heart function and remodelling (see Klaiman et al. 2011). For example, in most teleosts, as temperature decrease, CO falls accordingly. Cold adaptation in many non-polar species is associated with an increased heart size, which, in turn, increases resting SV while lowering HR (Driedzic et al. 1996). It is well known that in fish, which transcend large thermoclines to forage or escape predation in lakes and oceans (salmon: Brett 1971, blue marlin: Block et al. 1992, tuna: Block et al. 1997, trout: Matthews and Berg 1997, Reid et al. 1997), the ability of cardiac muscle to maintain its functionality during temperature changes is essential for survival. The maintenance of a proper cardiac function at different temperatures requires profound compensatory changes in relative heart mass, energy metabolism and humoral control of cardiac contractility (Keen et al. 1993, Keen et al. 1994, Driedzic et al. 1996, Aho and Vornanen 1999, Vornanen et al. 2002). Such morpho-functional heart remodeling necessitates both qualitative and quantitative changes in thousands of macromolecules that constitute the cardiac phenotype (for eel see Cerra et al. 2004). For example, increased RVM in European eels acclimated at low temperature may be interpreted as a compensatory response for the depressing effect of low seasonal temperature. Similarly, in migrating eels that face acute temperature fluctuations, performing diel vertical migration, diving into colder waters (~ 6–8°C) during the day and ascending to shallow warmer waters (12–14°C) at night (Aarestrup et al. 2009), the consequent increases of ventricular power production can compensate for the cooling-induced upper limit placed on HR (Methling et al. 2012). The relationship between cholinergic tone and temperature has been metioned above.

Interestingly, a temperature sensitive paradigm of subtle heart regulation has been recently shown in *A. anguilla* by Amelio and coworkers (2013). The above mentioned NO-mediated modulation of the Frank-Starling response appears temperature dependent, since it is preserved during the physiological acclimation at different temperatures, but is abolished by acute thermal shock conditions. These results are paralleled by thermal-related decrease of both activated eNOS and pAkt expression, the latter being a major signaling protein kinase phosphorylating eNOS *in vitro* as well as *in vivo* (Dimmeler et al. 1999). This newly discovered thermal sensitivity of the NOS/NO-induced modulation of the Frank-Starling response provides another example of the fascinating plasticity of the eel heart.

Conclusions and Perspectives

Here we have highlighted the remarkable versatility and endurance of the eel heart, not only related to the plasticity of the basic teleost cardio-circulatory design but also to its particular biochemical–metabolic flexibility and elaborated neuro-endocrine traits. The high cardiac sensitivity appears fine-tuned by a number of intrinsic (e.g., hemodynamic forces) and extrinsic (season, temperature, etc.) factors that, downstream from their stimulation targets, can activate complex molecular signal-transduction networks. More recent evidence has shown in eel the major role played by the intracardiac endocrine and autocrine-paracrine system in orchestrating the whole ventricle function through spatially (EE sensory function) and temporally (beat-to-beat and short-and medium-term regulation) distinct pathways. This knowledge may contribute to better understand the extraordinary adaptation and evolutionary success of the genus Anguilla. It also emphasizes how the eel heart still represents since William Harvey's time a wonderful natural model for exploring novel physiological mechanisms.

Abbreviations

ACh	:	acetylcholine
AChR	:	ACh receptor
ARs	:	adrenoceptors
Ang II	:	angiotensin II
ANP	:	atrial natriuretic peptides
AV	:	atrio-ventricular
BNP	:	brain natriuretic peptides
CNP	:	C natriuretic peptides
CO	:	cardiac output
CA	:	catecholamines

EE	:	endocardial endothelium
EC	:	excitation/contraction
HR	:	heart rate
I/R	:	ischemia/reperfusion
ISO	:	isoproterenol
NPs	:	natriuretic peptides
NOS	:	nitric oxide synthases
NO	:	nitric oxide
NPRs	:	NP receptors
PO	:	output pressure
PTx	:	pertussis toxin
PLN	:	phospholamban
PKG	:	protein kinase G
RHW	:	relative heart weight
RVM	:	relative ventricular mass
RAS	:	renin-angiotensin system

SR-CA^{2+}ATPase (SERCA2a)

SV	:	stroke volume
SW	:	stroke work
SNS	:	sympathetic nervous system
VNP	:	ventricular natriuretic peptides

References

Aarestrup, K., F. Okland, M.M. Hansen, D. Righton, P. Gargan, M. Castonguay, L. Bernatchez, P. Howey, H. Sparholt, M.I. Pedersen and R.S. McKinley. 2009. Oceanic spawning migration of the European eel (*Anguilla anguilla*). Science 325(5948): 1660.

Abrahamsson, T., S. Holmgren, S. Nilsson and K. Pettersson. 1979. On the chromaffin system of the African lungfish, *Protopterus aethiopicus*. Acta Physiol. Scand. 107: 135–139.

Aho, E. and M. Vornanen. 1999. Contractile properties of atrial and ventricular myocardium of the heart of rainbow trout (*Oncorhynchus mykiss*): effects of thermal acclimation. J. Exp. Biol. 202: 2663–2677.

Amelio, D., F. Garofalo, D. Pellegrino, F. Giordano, B. Tota and M.C. Cerra. 2006. Cardiac expression and distribution of nitric oxide synthases in the ventricle of the cold-adaptedAntarcticteleosts, the hemoglobinless *Chionodraco hamatus* and the red-blooded *Trematomus bernacchii*. Nitric Oxide 15: 190–198.

Amelio, D., F. Garofalo, E. Brunelli, A.M. Loong, W.P. Wong, Y.K. Ip, B. Tota and M.C. Cerra. 2008. Differential NOS expression in freshwater and aestivating *Protopterus dolloi* (lungfish): heart *vs* kidney readjustments. Nitric Oxide 18: 1–10.

Amelio, D., F. Garofalo, C. Capria, B. Tota and S. Imbrogno. 2013. Effects of temperature on the nitric oxide-dependent modulation of the Frank–Starling mechanism: the fish heart as a case study. Comp. Biochem. Physiol. Part A 164: 356–362.

Angelone, T., A. Gattuso, S. Imbrogno, R. Mazza and B. Tota. 2012. Nitrite is a positive modulator of the Frank-Starling response in the vertebrate heart. Am. J. Physiol. Regul. Integr. Comp. Physiol. 302(11): R1271–81.

Aoyama, J., K. Hissmann, T. Yoshinaga, S. Sasai, T. Uto and H. Ueda. 1999. Swimming depth of migrating silver eels *Anguilla japonica* released at seamounts of the West Mariana Ridge, their estimated spawning sites. Mar. Ecol. Prog. Ser. 186: 265–269.

Ask, J.M-. 1983. Comparative aspects of adrenergic receptors in the heart of lower vertebrates. Comp. Biochem. Physiol. 76A: 543–552.

Axelsson, M., F. Ehrenstrom and S. Nilsson. 1987. Cholinergic and adrenergic influence on the teleost heart *in vivo*. Exp. Biol. 46: 179–186.

Axelsson, M. 1988. The importance of nervous and humoral mechanisms in the control of cardiac performance in the Atlantic cod *Gadus morhua* at rest and during non exhaustive exercise. J. Exp. Biol. 137: 287–301.

Bailey, J.R., J.L. West and W.R. Driedzic. 1997. Heart growth associated with sexual maturity in male rainbow trout (*Oncorhynchus mykiss*) is hyperplastic. Comp. Biochem. Physiol. 118B: 607–611.

Bailey, J.R., R. MacDougall, S. Clowe and W.R. Driedzic. 2000. Anoxic performance of the american eel (*Anguilla rostrata* L.) heart requires extracellular glucose. J. Exp. Zool. 286(7): 699–706.

Baker, K.M., H.A. Singer and J.F. Aceto. 1989. Angiotensin II receptor-mediated stimulation of cytosolic-free calcium and inositol phosphates in chick myocytes. J. Pharmacol Exp. Ther. 251: 578–585.

Barth, E., G. Stammler, B. Speiser and J. Schaper. 1992. Ultrastructural quantitation of mitochondria and myofilaments in cardiac muscle from 10 different animal species including man. J. Mol. Cell. Cardiol. 24: 669–681.

Bass, A., B. Ostadal, V. Pelouch and V. Vitek. 1973. Differences in weight parameters, myosin ATP-ase activity and the enzyme pattern of energy supplying metabolism between the compact and spongious cardiac musculature of carp and turtle. Pflug. Arch. 343: 65–77.

Bernier, N.J., J.E. McKendry and S.F. Perry. 1999. Blood pressure regulation during hypotension in two teleost species: differential involvement of the rennin-angiotensin and adrenergic systems. J. Exp. Biol. 202: 1677–1690.

Bjenning, C.E., Y. Takei, T.X. Watanabe, K. Nakajima, S. Sakakibara and N. Hazon. 1992. A C-type natriuretic peptide is a vasodilator *in vivo* and *in vitro* in the common dogfish. J. Endocrinol. 133: R1–R4.

Block, B.A., D. Booth and F.G. Carey. 1992. Direct measurements of swimming speed and depth of bluemarlin. J. Exp. Biol. 166: 267–284.

Block, B.A., J.E. Keen, B. Castillo, H. Dewar, E.V. Freund, D.J. Marcinek, R.W. Brill and C. Farwell. 1997. Environmental preferences of yellowfin tuna at the northern extent of their range. Mar. Biol. 130: 119–132.

Brauner, C.J., A.L. Val and D.J. Randall. 1993. The effect of graded methemoglobin levels on the swimming performance of chinook salmon (*Oncorhyncus tshawytscha*). J. Exp. Biol. 185: 121–135.

Brett, J.R. 1971. Energetic responses of salmon to temperature. A study of some thermal relations in the physiology and freshwater ecology of sockeye salmon (*Oncorhynchus nerka*). Am. Zool. 11: 99–113.

Brodde, O.E. and M.C. Michel. 1999. Adrenergic and muscarinic receptors in the human heart, Pharmac. Rev. 51: 651–689.

Bryan, N.S., B.O. Fernandez, S.M. Bauer, M.F. Garcia-Saura, A.B. Milsom, T. Rassaf, R.E. Maloney, A. Bharti, J. Rodriguez and M. Feelisch. 2005. Nitrite is a signaling molecule and regulator of gene expression in mammalian tissues. Nat. Chem. Biol. 1: 290–297.

Burnstock, G. 1969. Evolution of the autonomic innervation of visceral and cardiovascular systems in vertebrates. Pharmacol. Rev. 21: 247–324.

Butler, D.G. and G.Y. Oudit. 1995. Angiotensin I and III mediated cardiovascular responses in the freshwater North American eel, *Anguilla rostrata*: effect of Phe deletion. Gen. Comp. Endocrinol. 97: 259–269.

Calaghan, S.C., J. Colyer and E. White. 1999. Cyclic AMP but not phosphorylation of phospholamban contributes to the slow inotropic response to stretch in ferret papillary muscle. Pflugers Arch. 437(5): 780–782.

Cameron, J.S. and S.E. Brown. 1981. Adrenergic and cholinergic responses of the isolated heart of the goldfish *Carassius auratus*. Comp. Biochem. Physiol. C. 70(1): 109–15.

Casadei, B. and C.E. Sears. 2003. Nitric-oxide-mediated regulation of cardiac contractility and stretch responses. Prog. Biophys. Mol. Biol. 82: 67–80.

Cerra, M.C., M. Canonaco, Y. Takei and B. Tota. 1996. Characterization of natriuretic peptide binding sites in the heart of the eel *Anguilla anguilla*. J. Exp. Zool. 275: 27–35.

Cerra, M.C., S. Imbrogno, D. Amelio, F. Garofalo, E. Colvee, B. Tota and J.M. Icardo. 2004. Cardiac morphodynamic remodeling in the growing eel. J. Exp. Biol. 207: 2867–2875.

Cerra, M.C. and D. Pellegrino. 2007. Cardiovascular cGMP-generating systems in physiological and pathological conditions. Curr. Med. Chem. 14: 585–599.

Cerra, M.C., T. Angelone, M.L. Parisella, D. Pellegrino and B. Tota. 2009. Nitrite modulates contractility of teleost (*Anguilla anguilla* and *Chionodraco hamatus*, i.e., the Antarctic hemoglobinless icefish) and frog (*Rana esculenta*) hearts. Biochim. Biophys. Acta 1787: 849–855.

Clark, R.J. and K.J. Rodnick. 1998. Morphometric and biochemical characteristics of ventricular hypertrophy in male rainbow trout (*Oncorhynchus mykiss*). J. Exp. Biol. 201: 1541–1552.

Cosby, K., K.S. Partovi, J.H. Crawford, R.P. Patel, C.D. Reiter, S. Martyr, B.K. Yang, M.A. Waclawiw, G. Zalos, X. Xu, K.T. Huang, H. Shields, D.B. Kim-Shapiro, A.N. Schechter, R.O. Cannon and M.T. Gladwin. 2003. Nitrite reduction to nitric oxide by deoxyhemoglobin vasodilates the human circulation. Nat. Med. 9: 1498–1505.

Cossins, A.R., E.J. Fraser and A.Y. Gracey. 2006. Post-genomic approaches to understanding the mechanisms of environmentally-induced phenotypic plasticity. J. Exp. Biol. 209: 2328–2336.

Cousins, K.L. and A.P. Farrell. 1996. Stretch-induced release of atrial natriuretic factor from the heart of rainbow trout, *Oncorhynchus mykiss*. Can. J. Zool. 74(2): 380–387.

Crawford, J.H., T.S. Isbell, Z. Huang, S. Shiva, B.K. Chacko, A.N. Schechter, V.M. Darley-Usmar, J.D. Kerby, J.D. Lang, Jr., D. Kraus, C. Ho, M.T. Gladwin and R.P. Patel. 2006. Hypoxia, red bloodcells, and nitrite regulate NO-dependent hypoxic vasodilation. Blood. 107: 566–574.

Davie, P.S. and M.E. Foster. 1980. Cardiovascular responses to swimming in eels. Comp. Biochem. Physiol. 67A: 367–373.

Davie, P.S., A.P. Farrell and C.E. Franklin. 1992. Cardiac performance of an isolated eel heart: effects of hypoxia and responses to coronary artery perfusion. J. Exp. Zool. 262(2): 113–121.

de Bold, A.J., H.B. Borenstein, A.T. Veress and H. Sonnenberg. 1981. A rapid and potent natriuretic response to intravenous injection of atrial myocardial extract in rats. Life Sci. 28: 89–94.

De Gasparo, M. 2002. Angiotensin II and Nitric Oxide interaction, Heart Failure Rev. 7: 347–358.

Di Maio, A. and B.A. Block. 2008. Ultrastructure of the sarcoplasmic reticulum in cardiac myocytes from Pacific bluefin tuna. Cell Tissue Res. 334: 121–134.

Dimmeler, S., I. Fleming, B. Fisslthaler, C. Hermann, R. Busse and A.M. Zeiher. 1999. Activation of nitric oxide synthase in endothelial cells by Akt-dependent phosphorylation. Nature 399: 601–605.

Dostal, D. 2000. The cardiac renin–angiotensin system: novel signaling mechanisms related to cardiac growth and function. Reg. Pept. 91: 1–11.

Driedzic, W.R., J.R. Bailey and D.H. Sephton. 1996. Cardiac adaptations to low temperature in non-polar teleost fish. J. Exp. Biol. 275: 186–195.

Duranski, M.R., J.J. Greer, A. Dejam, S. Jaganmohan, N. Hogg, W. Langston, R.P. Patel, S.F. Yet, X. Wang, C.G. Kevil, M.T. Gladwin and D.J. Lefer. 2005. Cytoprotective effects of nitrite during *in vivo* ischemia-reperfusion of the heart and liver. J. Clin. Invest. 115: 1232–1240.

Ellerby, D.J., I.L.Y. Spiers and J.D. Altringham. 2001. Slow muscle power output of yellow- and silver-phase European eels (*Anguilla Anguilla* L.): changes in muscle performance prior to migration. J. Exp. Biol. 204: 1369–1379.

Epple, A. and J.E. Brinn. 1987. The comparative physiology of the pancreatic islet. *In*: Zoophysiology, Vol. 21, Springer-Verlag, Berlin, Germany. pp. 61–62.

Evans, D.H. 1995. The roles of natriuretic peptide hormones in fish osmoregulation and hemodynamics. Adv. Comp. Environ. Physiol. 22: 120–152.

Farrell, A.P. and P.L. Lutz. 1975. Apparent anion imbalance in the fresh water adapted eel. J. Comp. Physiol. 102: 159–166.

Farrell, A.P., A.M. Hammons, M.S. Graham and G.F. Tibbits. 1988. Cardiac growth in rainbow trout, *Salmo gairdneri*. Can. J. Zool. 66: 2368–2373.

Farrell, A.P. and D.R. Jones. 1992. The Heart. *In*: W.S. Hoar, D.J. Randall and A.P. Farrell [eds]. The Cardiovascular System. Academic Press, San Diego, USA. pp. 1–88.

Fiedler, K. 1991. Lehrbuch der Speziellen Zoologie. Band II: Wirbeltiere. 2. Teil: Fische D. Starck (ed.). Gustav Fischer Verlag, Jena, Germany.

Forster, M.E. 1976. Effects of catecholamines on the heart and on branchial and peripheral resistance of the eel, *Anguilla anguilla* (L.). Comp. Biochem. Physiol. C. Comp. Pharmacol. 55: 27–32.

Forster, M.E. 1981. Effects of catecholamines on the hearts and ventral aortas of the eel, *Anguilla australis schmidtii* and *Anguilla dieffenbachti*. Comp. Biochem. Physiol. 70C: 85–90.

Gallaugher, P.E., H. Thorarensen, A. Keissling and A.P. Farrell. 2001. Effects of high intensity exercise training on cardiovascular function, oxygen uptake, internal oxygen transport and osmotic balance in Chinook salmon (*Oncorhynchus tshawytscha*) during critical speed swimming. J. Exp. Biol. 204: 2861–2872.

Gamperl, A., M. Wilkinson and R. Boutilier. 1994. Beta-adrenoreceptors in the trout (*Oncorhynchus mykiss*) heart characterization, quantification, and effects of repeated catecholamine exposure. Gen. Comp. Endocrinol. 95: 259–272.

Gamperl, A.K. and A.P. Farrell. 2004. Cardiac plasticity in fishes: environmental influences and intraspecific differences. J. Exp. Biol. 207: 2539–2550.

Gannon, J.B. and G. Burnstock. 1969. Excitatory adrenergic innervation of the fish heart. Comp. Biochem. Physiol. 29: 765–773.

Garofalo, F., D. Amelio, M.C. Cerra, B. Tota, B.D. Sidell and D. Pellegrino. 2009. Morphological and physiological study of the cardiac NOS/NO system in the Antarctic (Hb⁻/Mb⁻) icefish *Chaenocephalus aceratus* and in the red-blooded *Trematomus bernacchii*. Nitric Oxide 20(2): 69–78.

Gattuso, A., R. Mazza, D. Pellegrino and B. Tota. 1999. Endocardial endothelium mediates luminal ACh-NO signaling in isolated frog heart. Am. J. Physiol. Heart Circ. Physiol. 276: H633–H641.

Gauthier, C., D. Langin and J.L. Balligand. 2000. Beta3-adrenoceptors in the cardiovascular system. Trends. Pharmacol. Sci. 21: 426–431.

Graham, M.S. and A.P. Farrell. 1989. Effect of temperature acclimation on cardiac performance in a perfused trout heart. Physiol. Zool. 62: 38–61.

Hare, J.M. 2003. Nitric oxide and excitation-contraction coupling. J. Mol. Cell. Cardiol. 35: 719–729.

Harwood, C.L., I.S. Young and J.D. Altringham. 2002. How the efficiency of rainbow trout (*Oncorhynchus mykiss*) ventricular muscle changes with cycle frequency. J. Exp. Biol. 205: 697–706.

Hochachka, P.W. and G.N. Somero. 2002. Biochemical Adaptation: Mechanism and Process in Physiological Evolution. Oxford University Press, New York, USA.

Holmgren, S. 1977. Regulation of the heart of a teleost, *Gadus morhua*, by autonomic nerves and circulating catecholamines. Acta Physiol. Scand. 99(1): 62–74.

Holmgren, S. and S. Nilsson. 1982. Neuropharmacology of adrenergicneurons in teleostfish. Comp. Biochem. Physiol. Part C: Comp. Pharmacol. 72(2): 289–302.

Hove-Madsen, L., P.F. Mery, J. Jurevicius, A.V. Skeberdis and R. Fischmeister. 1996. Regulation of myocardial calcium channels by cyclic AMP metabolism. Basic Res. Cardiol. 91 (Suppl 2): 1–8.

Huang, Z., S. Shiva, D.B. Kim-Shapiro, R.P. Patel, L.A. Ringwood, C.E. Irby, K.T. Huang, C. Ho, N. Hogg, A.N. Schechter and M.T. Gladwin. 2005. Enzymatic function of hemoglobin as a nitrite reductase that produces NO under allosteric control. J. Clin. Invest. 115: 2099–2107.

Hyde, D.A., T.W. Moon, S.F. and Perry. 1987. Physiological consequences of prolonged aerial exposure in the American eel, *Anguilla rostrata*: blood respiratory and acid-base status. J. Comp. Physiol. B. 157: 635–642.

Icardo, J.M., E. Colvee, M.C. Cerra and B. Tota. 2000. Light and electron microscopy of the bulbus arteriosus of the European eel (*Anguilla anguilla*). Cells Tiss. Organs 157: 184–198.

Icardo, J.M., S. Imbrogno, A. Gattuso, E. Colvee and B. Tota. 2005. The heart of *Sparus auratus*: a reappraisal of cardiac functional morphology in teleosts. J. Exp. Zoolog. A Comp. Exp. Biol. 303: 665–675.

Icardo, J.M. 2006. Conus arteriosus of the teleost heart: dismissed, but not missed. Anat. Rec. A Discov. Mol. Cell. Evol. Biol. 288: 900–908.

Icardo, J.M. and E. Colvee. 2011. The atrioventricular region of the teleost heart. A distinct heart segment. Anat. Rec. 294: 236–242.

Icardo, J.M. 2012. The teleost heart: A morphological approach. *In*: D. Sedmera and T. Wang [eds]. Ontogeny and Phylogeny of the Vertebrate Heart. Springer Science, New York, USA. pp. 35–53.

Imbrogno, S., L. De Iuri, R. Mazza and B. Tota. 2001. Nitric oxide modulates cardiac performance in the heart of *Anguilla anguilla*. J. Exp. Biol. 204: 1719–1727.

Imbrogno, S., M.C. Cerra and B. Tota. 2003. Angiotensin II-induced inotropism requires an endocardial endothelium-nitric oxide mechanism in the *in vitro* heart of *Anguilla anguilla*. J. Exp. Biol. 206: 2675–2684.

Imbrogno, S., T. Angelone, A. Corti, C. Adamo, K.B. Helle and B. Tota. 2004. Influence of vasostatins, the chromogranin A-derivedpeptides, on the working heart of the eel (*Anguilla anguilla*): negative inotropy and mechanism of action, Gen. Comp. Endocrinol. 139: 20–28.

Imbrogno, S., T. Angelone, C. Adamo, E. Pulerà, B. Tota and M.C. Cerra. 2006. Beta3-Adrenoceptor in the eel (*Anguilla anguilla*) heart: negative inotropy and NO-cGMP-dependent mechanism. J. Exp. Biol. 209: 4966–4973.

Imbrogno, S., F. Garofalo, M.C. Cerra, S.K. Mahata and B. Tota. 2010. The catecholamine release-inhibitory peptide catestatin (Chromogranin A344-364) modulates myocardial function in fish. J. Exp. Biol. 213: 3636–3643.

Imbrogno, S., B. Tota and A. Gattuso. 2011. The evolutionary functions of cardiac NOS/NO in vertebrates tracked by fish and amphibian paradigms. Nitric Oxide 25: 1–10.

Inoue, K., T. Sakamoto, S. Yuge, H. Iwatani, S. Yamagami, M. Tsutsumi, H. Hori, M.C. Cerra, B. Tota, N. Suzuki, N. Okamoto and Y. Takei. 2005. Structural and functional evolution of three cardiac natriuretic peptides. Mol. Biol. Evol. 22: 2428–2434.

Iwakiri, Y., A. Satoh, S. Chatterjee, D.K. Toomre, C.M. Chalouni, D. Fulton, R.J. Groszmann, V.H. Shah and W.C. Sessa. 2006. Nitric oxide synthase generates nitric oxide locally to regulate compartmentalized protein S-nitrosylation and protein trafficking. Proc. Natl. Acad. Sci. USA 103: 19777–19782.

Jensen, F.B. 2009. The role of nitrite in nitric oxide homeostasis: a comparative perspective. Biochim. Biophys. Acta. 1787(7): 841–848.

Jensen, F.B. and M.N. Hansen. 2011. Differential uptake and metabolism of nitrite in normoxic and hypoxic goldfish. Aquat. Toxicol. 101(2): 318–25.

Johansen, I.B., I.G. Lunde, H. Røsjø, G. Christensen, G.E. Nilsson, M. Bakken and O. Overli. 2011. Cortisol response to stress is associated with myocardial remodeling in salmonid fishes. J. Exp. Biol. 214(Pt 8): 1313–21.

Johnson, R.K., D. Hering, M.T. Furse and P.F.M. Verdonschot. 2006. Indicators of ecological change: comparison of the early response of four organism groups to stress gradients. Hydrobiologia. 566: 39–152.

Johnson, K.R. and K.R. Olson. 2008. Comparative physiology of the piscine natriuretic peptide system. Gen. Comp. Endocrinol. 157: 21–26.

Kaiya, H. and Y. Takei. 1996a. Changes in plasma atrial and ventricular natriuretic peptide concentrations after transfer of eels from fresh water to seawater or *vice versa*. Gen. Comp. Endocrinol. 104: 337–345.

Kaiya, H. and Y. Takei. 1996b. Osmotic and volaemic regulation of atrial and ventricular natriuretic peptide secretion in conscious eels. J. Endocrinol. 149: 441–447.

Katz, A.M. 2002. Ernest Henry Starling, his predecessors, and the 'Law of the Heart'. Circulation. 106: 2986–2992.

Kawakoshi, A., S. Hyodo, A. Yasudal and Y. Takei. 2003. A single and novel natriuretic peptide is expressed in the heart and brain of the most primitive vertebrate, the hagfish (*Eptatretus burgeri*). J. Mol. Endocrinol. 31: 209–222.

Kayar, S.R., K.E. Conley, H. Claassen and H. Hoppeler. 1986. Capillarity and mitochondrial distribution in rat myocardium following exercise training. J. Exp. Biol. 120: 189–199.

Keen, J.E., D.M. Vianzon, A.P. Farrell and G.F. Tibbits. 1993. Thermal acclimation alters both adrenergic sensitivity and adrenoceptor density in cardiac tissue of rainbow trout. J. Exp. Biol. 181: 27–47.

Keen, J.E., D.M. Vianzon, A.P. Farrell and G.F. Tibbits. 1994. Effect of temperature and temperature acclimation on the ryanodine sensitivity of the trout myocardium. J. Comp. Physiol. [B] 164: 438–443.

Klaiman, J.M., A.J. Fenna, H.A. Shiels, J. Macri and T.E. Gillis. 2011. Cardiac remodeling in fish: strategies to maintain heart function during temperature change. PLoS One 6: e24464.

Kobayashi, H. and Y. Takei. 1996. Biological actions of Ang II. *In*: S.D. Bradshaw, W. Burggren, H.C. Heller, S. Ishii, H. Langer, G. Neuweiler and D.J. Randall [eds]. The Renin-Angiotensin System: A Comparative Aspect Zoophysiology, vol. 35, Springer-Verlag, Berlin, Heidelberg, Germany. pp. 113–171.

Larsen, T.H., K.B. Helle and T. Saetersdal. 1994. Immunoreactive atrial natriuretic peptide and dopamine beta-hydroxylase in myocytes and chromaffin cells of the heart of the African lungfish, *Protopterus aethiopicus*. Gen. Comp. Endocrinol. 95: 1–12.

Larsen, B.K. and F.B. Jensen. 1997. Influence of ionic composition on acid-base regulation in rainbow trout (*Oncorhynchus mykiss*) exposed to environmental hypercapnia. Fish Physiol. Biochem. 16: 157–170.

Laurent, P., S. Holmgren and S. Nilsson. 1983. Nervous and humoral control of the fish heart: structure and function. Comp. Biochem. Physiol. 76A: 525–542.

Lepilina, A., A.N. Coon, K. Kikuchi, J.E. Holdway, R.W. Roberts, C.G. Burns and K.D. Poss. 2006. A dynamic epicardial injury response supports progenitor cell activity during zebrafish heart regeneration. Cell. 127(3): 607–19.

Lima, B., M.T. Forrester, D.T. Hess and J.S. Stamler. 2010. S-nitrosylation in cardiovascular signaling. Circ. Res. 106: 633–646.

Matthews, K.R. and N.H. Berg. 1997. Rainbow trout responses to water temperature and dissolved oxygen stress in two southern California streampools. J. Fish Biol. 50: 50–67.

Mayer, B., S. Pfeiffer, A. Schrammel, D. Koesling, K. Schmidt and F. Brunner. 1998. A new pathway of nitric oxide/cyclic GMP signaling involving S-nitrosoglutathione. J. Biol. Chem. 273: 3264–3270.

McCleave, J.D. and G.P. Arnold. 1999. Movements of yellowand silver-phase European eels (*Anguilla anguilla*) tracked in the western North Sea. J. Mar. Sci. 56: 510–536.

McKenzie, D.J., E.W. Taylor, A.Z. Dalla Valle and J.F. Steffensen. 2002. Tolerance of acute hypercapnic acidosis by the European eel (*Anguilla anguilla*). J. Comp. Physiol. B. 172: 339–346.

McKenzie, D.J., M. Piccolella, A.Z. Dalla Valle, E.W. Taylor, C.L. Bolis and J.F. Steffensen. 2003. Tolerance of chronic hypercapnia by the European eel *Anguilla anguilla*. J. Exp. Biol. 206(Pt 10): 1717–26.

McKenzie, D.J., S. Wong, D.J. Randall, S. Egginton, E.W. Taylor and A.P. Farrell. 2004. The effects of sustained exercise and hypoxia upon oxygen tensions in the red muscle of rainbow trout. J. Exp. Biol. 207: 3629–3637.

Methling, C., J.F. Steffensen and P.V. Skov. 2012. The temperature challenges on cardiac performance in winter-quiescent and migration-stage eels *Anguilla Anguilla*. Comp. Biochem. Physiol. A Mol. Integr. Physiol. 163(1): 66–73.

Modin, A., H. Bjorne, M. Herulf, K. Alving, E. Weitzberg and J.O. Lundberg. 2001. Nitrite-derived nitric oxide: a possible mediator of "acidic-metabolic" vasodilation. Acta Physiol. Scand. 171: 9–16.

Moncada, S., R.M. Palmer and E.A. Higgs. 1991. Nitric oxide: physiology, pathophysiology, and pharmacology. Pharmacol. Rev. 43: 109–142.

Netchitailo, P., M. Feuilloley, G. Pelletier, F. Leboulenge, M. Cantin, J. Gutkowska and H. Vaudry. 1987. Atrial natriuretic factor-like immunoreactivity in the central nervous system of the frog. Neuroscience 2: 341–359.

Nilsson, S. 1983. Autonomic Nerve Function in the Vertebrates. Springer-Verlag Berlin, Heidelberg, Germany and New York, USA.

Nilsson, S. and S. Holmgren. 1992. Cardiovascular control by purines, 5-hydroxytryptamine and neuropeptides. *In*: D. Randall and A.P. Farrell [eds]. Fish Physiology, vol. XII. Academic Press, New York, USA. pp. 301–341.

Olson, K.R., D.J. Conklin, A.P. Farrell, J.E. Keen, Y. Takei, L. Weaver Jr., M.P. Smith and Y. Zhang. 1997. Effects of natriuretic peptides and nitroprusside on venous function in trout. Am. J. Physiol. 273: 527–539.

Olson, R.K. 1998. The cardiovascular system. *In*: H.D. Evans [ed]. The Physiology of Fishes. CRC Press, Boca Raton, New York, USA. pp. 129–154.

Oudit, G.Y. and D.G. Butler. 1995. Angiotensin II and cardiovascular regulation in a freshwater teleost *Anguilla rostrata* Le Sueur, Am. J. Physiol. 269: R726–R735.

Paton, J.F.R., J. Deuchars, Z. Ahmad, L.F. Wong, D. Murphy and S. Kasparov. 2001. Adenoviral vector demonstrates that angiotensin II-induced depression of the cardiac baroreflex is mediated by endothelial nitric oxide synthase in the nucleus tractus solitarii of the rat. J. Physiol. 531: 445–458.

Pennec, J.P. and Y.M. Lebras. 1984. Storage and release of catecholamines by nervous endings in the isolated heart of the eel *Anguilla anguilla* L. Comp. Biochem. Physiol. C. 77: 167–172.

Pennec, J.P. and C. Peyraud. 1983. Effects of adrenaline on isolated heart of the eel (*Anguilla anguilla* L.) during winter. Comp. Biochem. Physiol. C. 74: 477–480.

Pennec, J.P. and Y.M. Le Bras. 1988. Diel and seasonal rhythms of the heart rate in the common eel (*Anguilla anguilla* L.): role of cardiac innervation. Exp. Biol. 47(3): 155–60.

Perlman, D.H., S.M. Bauer, H. Ashrafian, N.S. Bryan, M.F. Garcia-Saura, C.C. Lim, B.O. Fernandez, G. Infusini, M.E. McComb, C.E. Costello and M. Feelisch. 2009. Mechanistic insights into nitrite-induced cardio-protection using an integrated metabolomic/proteomic approach. Circ. Res. 104: 796–804.

Petroff, M.G., S.H. Kim, S. Pepe, C. Dessy, E. Marban, J.L. Balligand and S.J. Sollott. 2001. Endogenous nitric oxide mechanisms mediate the stretch-dependence of Ca^{2+} release in cardiomyocytes. Nat. Cell Biol. 3: 867–873.

Peyraud-Waitzenegger, M., L. Barthelemy and C. Peyraud. 1980. Cardiovascular and ventilatory effects of catecholamines in unrestrained eels (*Anguilla anguilla* L.). J. Comp. Physiol. 138: 367–375.

Peyraud-Waitzenegger, M. and P. Soulier. 1989. Ventilatory and circulatory adjustments in the European eel (*Anguilla Anguilla* L.) exposed to short term hypoxia. Exp. Biol. 48: 107–122.

Poupa, O., H. Gesser, S. Jonsson and L. Sullivan. 1974. Coronary supplied compact shell of ventricular myocardium in salmonids, growth and enzyme pattern. Comp. Biochem. Physiol. 48A: 85–95.

Poupa, O., L. Lindstrom, A. Maresca and B. Tota. 1981. Cardiac growth, myoglobin, proteins and DNA in developing tuna (*Thunnus thynnus thynnus*). Comp. Biochem. Physiol. 70A: 217–222.

Poupa, O. and L. Lindstrom. 1983. Comparative and scaling aspects of heart and body weights with reference to blood supply of cardiac fibres. Comp. Biochem. Physiol. 76A: 413–421.

Poss, K.D., L.G. Wilson and M.T. Keating. 2002. Heart regeneration in zebrafish. Science 298: 2188–2190.

Randall, D.J. 1970. The circulatory system. *In*: W.S. Hoar and D.J. Randall [eds]. Fish Physiology, 4. Academic Press, New York (USA) and London (UK). pp. 133–172.

Randall, D.J. 1982. The control of respiration and circulation in fish during exercise and hypoxia. J. Exp. Biol. 100: 275–288.

Randall, D.J. and S.F. Perry. 1992. Catecholamines. *In*: W.S. Hoar, D.J. Randall and A.P. Farrell [eds]. Fish Physiology vol.12b. Academic Press, San Diego, USA. pp. 255–300.

Rassaf, T., U. Flögel, C. Drexhage, U. Hendgen-Cotta, M. Kelm and J. Schrader. 2007. Nitrite reductase function of deoxymyoglobin: oxygen sensor and regulator of cardiac energetics and function. Circ. Res. 100: 1749–1754.

Reid, S.D., D.G. McDonald and C.M. Wood. 1997. Interactive effects of temperature and pollutant stress. *In*: C.M. Wood and D.G. McDonald [eds]. Global Warming: Implications for Freshwater and Marine Fish. University Press, Cambridge, UK. pp. 325–349.

Rodnick, K.J. and B.D. Sidell. 1997. Structural and biochemical analyses of cardiac ventricular enlargement in cold-acclimated striped bass. Am. J. Physiol. 273(1Pt 2): R252–258.

Ruskoaho, H. 1992. Atrial natriuretic peptide: Synthesis, release, and metabolism. Physiol. Rev. 44(4): 479–602.

Saito, T. 1973. Effects of vagal stimulation on the pacemaker action potentials of carp heart. Comp. Biochem. Physiol. 44A: 191–199.

Sancho, E., J.J. Cerón and M.D. Ferrando. 2000. Cholinesterase activity and hematological parameters as biomarkers of sublethalmolinate exposure in *Anguilla anguilla*. Ecotoxicol. Environ. Saf. 46(1): 81–6.

Santer, R.M. and J.L. Cobb. 1972. The fine structure of the heart of the teleost, *Pleuronectes platessa* L. Z. ZellforschMikrosk. Anat. 131: 1–14.

Santer, R.M. 1985. Morphology and innervation of the fish heart. Adv. Anat. Embryol. Cell Biol. 89: 1–102.

Schib, J.L., J.M. Icardo, A.C. Duran, A. Guerrero, D. Lopez, E. Colvee, A.V. de Andres and V. Sans Coma. 2002. The conus arteriosus of the adult gilthead seabream (*Sparus auratus*). J. Anat. 201: 395–404.

Schorb, W., G.W. Booz, D.E. Dostal, K.M. Conrad, K.G. Chang and K.M. Baker. 1993. Angiotensin II is mitogenic in neonatal rat cardiac fibroblasts. Circ. Res. 72: 1245–1255.

Sébert, P. 1997. Pressure effects on shallow-water fishes. *In*: D.J. Randall and A.P. Farrell (eds). Fish Physiology, Deep-Sea Fishes: vol. 16. Academic Press, New York, USA. pp. 279–323.

Sébert, P. 2001. Fish at high pressure: a hundred year history. Comp. Biochem. Physiol. Part A. 131: 575–585.

Sébert, P. and M. Theron. 2001. Why can the eel, unlike the trout, migrate under pressure. Mitochondrion 1: 79–85.

Seddon, M., A.M. Shah and B. Casadei. 2007. Cardiomyocytes as effectors of nitric oxide signalling. Cardiovasc. Res. 75: 315–326.

Seibert, H. 1979. Thermal adaptation of heart rate and its parasympathetic control in the European eel *Anguilla anguilla* (L.) Comp. Biochem. Physiol. C. 64: 275–278.

Shiels, H.A. and E. White. 2008. The Frank–Starling mechanism in vertebrate cardiacmyocytes. J. Exp. Biol. 211: 2005–2013.

Steffensen, J.F. and J.P. Lomholt. 1990. Accumulation of carbon dioxide in fish farms with recirculating water. *In*: R.C. Ryans [ed]. Fish Physiology, Fish Toxicology and Fisheries Management. Environmental Research Laboratories—US Environmental Protection Agency, Athens, Georgia, USA. pp. 157–161.

Takei, Y., A. Takahashi, T.X. Watanabe, K. Nakajima, S. Sakakibara, T. Takao and Y. Shimonishi. 1990. Amino acid sequence and relative biological activity of a natriuretic peptide isolated from eel brain. Biochem. Biophys. Res. Commun. 170: 883–891.

Takei, Y., A. Takahashi, T.X. Watanabe, K. Nakajima and S. Sakakibara. 1991. A novel natriuretic peptide isolated from eel cardiac ventricles. FEBS Lett. 282: 317–320.

Takei, Y. and R.J. Balment. 1993. Biochemistry and physiology of a family of eel natriuretic peptides. Fish Physiol. Biochem. 11: 183–188.

Takei, Y., A. Takahashi, T.X. Watanabe, K. Nakajima and K. Ando. 1994a. Eel ventricular natriuretic peptide: isolation of a low molecular size form and characterization of plasma form by homologous radioimmunoassay. J. Endocrinol. 141: 81–89.

Takei, Y., M. Takano, Y. Itahara, T.X. Watanabe, K. Nakajima, D.J. Conklin, D.W. Duff and K.R. Olson. 1994b. Rainbow trout ventricular natriuretic peptide: isolation, sequencing and determination of biological activity. Gen. Comp. Endocrinol. 96: 420–426.

Takei, Y., K. Inoue, S. Trajanovska and J.A. Donald. 2011. B-type natriuretic peptide (BNP), not ANP, is the principal cardiac natriuretic peptide in vertebrates as revealed by comparative studies. Gen. Comp. Endocrinol. 171: 258–266.

Taylor, E.W. 1992. Nervous control of the heart and cardiorespiratory interactions. *In*: W.S. Hoar, D.J. Randall and A.P. Farrell [eds]. Fish Physiology XIIB. Academic Press, New York, USA. pp. 343–387.

Tota, B. 1978. Functional cardiac morphology and biochemistry in Atlantic Bluefin tuna. *In*: G. Sharp and A. Dizon [eds]. The Physiological Ecology of Tuna. Academic Press, New York, USA. pp. 89–112.

Tota, B., V. Cimini, G. Salvatore and G. Zummo. 1983. Comparative study of the arterial and lacunary systems of the ventricular myocardium of the elasmobranch and teleost fishes. Am. J. Anat. 167: 15–32.

Tota, B. 1989. Vascular and metabolic zonation in the ventricular myocardium of mammals and fishes. Comp. Biochem. Physiol. 76: 423–427.

Tota, B., R. Acierno and C. Agnisola. 1991. Mechanical performance ofthe isolated and perfused heart of the hemoglobineless Antarctic icefish *Chionodraco hamatus* (Lonnberg), effects of loading conditions and temperature. Phil. Trans. R. Soc. B. 332: 191–198.

Tota, B. and A. Gattuso. 1996. Heart ventricle pumps in teleosts and elasmobranchs: a morphodynamic approach. J. Exp. Zool. 275: 162–171.

Tota, B., D. Amelio, D. Pellegrino, Y.K. Ip and M.C. Cerra. 2005. NO modulation of myocardial performance in fish hearts. Comp. Biochem. Physiol. 142: 164–177.

Tota, B., M.C. Cerra and A. Gattuso. 2010. Catecholamines, cardiac natriuretic peptides and chromogranin A: evolution and physiopathology of a 'whip-brake' system of the endocrine heart. J. Exp. Biol. 213: 3081–3103.

Tsuchida, T. and Y. Takei. 1998. Effects of homologous atrial natriuretic peptide on drinking and plasma Ang II level in eels. Am. J. Physiol. Regul. Integr. Comp. Physiol. 275: 1605–1610.

van Ginneken, V. and G. van den Thillart. 2000. Eel fat stores are enoughto reach the Sargasso. Nature 403: 156–157.

van Ginneken, V., E. Antonissen, U.K. Müller, R. Booms, E. Eding, J. Verreth and G. van den Thillart. 2005. Eel migration to the Sargasso: remarkably high swimming efficiency and low energy costs. J. Exp. Biol. 208 (Pt 7): 1329–1335.

Vornanen, M., A. Ryokkynen and A. Murmi. 2002. Temperature dependent expression of sarcolemmal $K^{(+)}$ currents in rainbow trout atrial and ventricular myocytes. Am. J. Physiol. 282: 1191–1199.

Webb, A., R. Bond, P. McLean, R. Uppal, N. Benjamin and A. Ahluwalia. 2004. Reduction of nitrite to nitric oxide during ischemia protects against myocardial ischemia-reperfusion damage. Proc. Natl. Acad. Sci. USA 101: 13683–13688.

Wendelar Bonga, S.E. 1997. The stress response in fish. Physiol. Rev. 77: 591–625.

Winberg, S. and G.E. Nilsson. 1993. Roles of brain monoamine neurotransmitters in agonistic behaviour and stress reactions with particular reference to fish. Comp. Biochem. Physiol. 106C: 597–614.

Wood, C.M., P. Pieprzak, and J.N. Trott. 1979. The influence of temperature and anaemia on the adrenergic and cholinergic mechanisms controlling heart rate in the rainbow trout. Can. J. Zool. 57: 2440–2447.

Yamauchi, A. 1980. Fine structure of the fish heart. *In*: G. Bourne [ed]. Heart and Heart-like Organs, vol. 1. Academic Press, New York, USA. pp. 119–148.

Sex Differences in Energy Metabolism

Philippe Sébert

Introduction

Papers and books about the Eel and its migration are numerous. However, when reading this literature one could think that the eel is unisex because most studies concern only females. This is probably because females are bigger than males, and thus easier to study: this difference is the basis of this chapter. Male and female European eels, *Anguilla anguilla*, exhibit important morphological differences which, to our knowledge, have never been pointed out though these differences have a bearing on physiological consequences on their migratory activity (Sébert et al. 2009b, Amérand et al. 2010). One of the main and easily "visible" differences is the size with its immediate consequence on migration: males being smaller must swim faster than females to attain the Sargasso Sea at the same time in order to reproduce! How do they ensure this performance?

ORPHY-EA4324, Université de Brest, 6, Avenue Le Gorgeu, CS 93837 29238 Brest Cedex3 France.
E-mail: Philippe.sebert@univ-brest.fr

Main Differences in Morphology and Way of Life

In 2003, Kloppmann devoted a chapter on eel body structure and function (Tesch 2003) but without really pointing out the sex differences. In a very well-documented and extensive study, Durif introduced some tools which helped in differentiating between yellow and silver eel stages and for the silver stage, it is possible to classify the maturation level according to sex (Durif 2003, Durif et al. 2005 and Table 1). In this section, we will only report the facts which could be involved in eel migratory performance: this restriction explains the limited number of citations. The reader interested in eel physiology can consult different topics in this book and in Van den Thillart et al. (2009).

Table 1. Determination of silvering stage and sex differences. Ranks for ocular index, pectoral index and body length in males and females eels at yellow (Y) or silver (S) stages. From Durif (2003).

	Males	Females
Body length, BL, cm	S ≤ 45	S > 45
Ocular Index, OI	Y < 6.8 < S	Y < 8 < S
Pectoral Index, PI, %	Y < 4 < S	Y < 4 < S

For silver eels, in the migratory stage, males generally have a body length BL ≤ 40 cm (females can have BL > 100 cm) and a body mass BM ≤ 100 g (females can have BM > 1200g). These two parameters allow an estimation of the body surface area BS (Sébert et al. 2004b) which is important to consider because it could be used to standardize physiological and metabolic parameters but it could also be useful in fish farming activity. Interestingly, two morphological indexes could also be determined (see Table 1). Firstly, the ocular index OI which is calculated from vertical and horizontal eye diameters (Pankhurst 1982). The OI increases from about 4 to 15 during the silvering and maturation process. This means that during maturation the eye size increases, thus participating supposedly in adapting to low light intensity during migration at depth. Such exophtalmia gives the male a characteristic aspect due to its small size (Ide et al. 2011). Secondly, the pectoral index which corresponds to the relative length of the pectoral fin to the total BL (in the range of 3–7%). An increase in the pectoral fin length could help neutralize buoyancy as the fins play the role of hydrofoils during swimming, thus saving energy (Pelster 1997). Another parameter easy to estimate but rarely used is body density which can provide some information about lipid stores (low density) and thus energy stores with a role in neutral buoyancy (see review by Pelster 1997). Male body density is generally higher than in females and can attain 1.15 depending on the site (see Vettier 2005): this means that they can dive more easily (see Sébert et al. 2009b). Sexual differentiation occurs when 30 cm length is attained

(Palstra 2006) at the same time as the silvering process, whereas female gonads differentiate well before (Colombo et al. 1984, Durif et al. 2005). The differences in male and female growth rates and strategies have often been discussed but it is not clear why some eels choose to maximize size while others opt for time minimizing strategies. However it is clear that eel size increases from downstream to upstream (Durif et al. 2009). In fact, as the males are smaller than females, they are more numerous downstream which is important for migration (Palstra et al. 2009 and see after). However, the male or female density (relative number) depends on the site (Palstra 2006). The feeding stage before migration is as long as 5 to 8 years for males and up to 8 to 20 years for females (Tesch 2003).

The general information given above could be summed up by saying that males are smaller but generally live downstream which decreases their swimming distance but it accounts only for a small fraction of the total migratory distance to the Sargasso Sea. To illustrate this, Table 2 gives theoretical examples about migration duration possibilities. If eels swim at the average speed suggested by Van Ginneken and Van den Thillart (2000), i.e., 0.5 BL/sec, it appears that in the best case males will arrive around 5 months after the females, but "only" 3.5 months if they start their migration 1.5 months before the females, making reproduction impossible. In the case described above, reproduction would be possible only if the migrating males of year 1 waited *in situ* for the females of year 2. The same problem remains at a higher swimming speed but the waiting time would be shorter. In contrast, if males swim faster than females (Table 2) for example 0.5 BL/sec for females and 0.8 BL/sec for males, male migration would only take 20 extra days. This hypothesis could be pertinent if males started 20 days earlier. The earlier departure of males is well documented (Todd 1981, Palstra et al. 2009) but the question remains whether they have the

Table 2. Consequences (example) of size and eel river position on duration of migratory activity. Considering that males need to cope with a shorter distance, values in italics show that if males start their migration sooner (20 days) than females and swim faster, they can meet at the Sargasso Sea.

	Males	Females
Body length, cm	40	80
Swimming speed, cm/s		
at 0.5 BL/s	20	40
at 0.8 BL/s	32	64
Distance, km	5500	6000
Distance/day, km		
at 0.5 BL/s	17	34
at 0.8 BL/s	28	55
Migration duration, days		
at 0.5 BL/s	323	*176*
at 0.8 BL/s	*196*	110

physiological and metabolic capacities to swim faster: the swimming cost for male eels would be much higher than for females because they are smaller and have to swim at higher speeds (Van den Thillart et al. 2009).

Differences in Energy Metabolism

Aerobic Metabolism

Clearly, the long journey to the Sargasso sea is a typical endurance exercise involving mainly red muscles and consequently aerobic metabolism fed from energy stores (lipids). In Table 3 the values of oxygen consumption of red muscle fibres and whole animal (males and females) are reported. For a given temperature, the respiration rates of muscle and whole body are higher in males than in females when standardized to mass of tissue or body mass. It is not the case when the values are standardized to body surface (see after). The calculations of Q_{10} (Table 3) taken as an index of temperature sensitivity show that males are more sensitive to high temperature whereas females are more sensitive ($Q_{10} = 4$) to low temperature (Scaion and Sébert 2008) which supposes that males could well prefer cold waters (depth) and females warmer waters corresponding to their preferedum (Haro 1991). When completed with results obtained under pressure (see p.106), we hypothesize that males could migrate more deeply than females. By using the measured maximal oxygen consumption of red and white fibres, it is possible to estimate the maximal aerobic power in males and females considering that red muscle represents 3% of body mass and 6% of muscle mass (Goolish 1991) bearing in mind that these percentages may be different during the yellow and then silver stages. Such calculations indicate that maximal aerobic power is about 25%–40% higher in males than in females:

Table 3. Oxygen consumption of muscle fibres and overall eel. The values inside parenthesis are Q_{10} calculated between the temperature indicated on the same horizontal line and the temperature immediately lower (Fibres: 15/5 and 25/15; whole animal: 15/9 and 22/15). Values are mean ± SEM and data from Scaion et al. (2008).

	Males	Females
Muscle fibres, µmol/mn/g		
5°C	ND	0.15 ± 0.015
15°C	0.48 ± 0.05	0.26 ± 0.031 *(1.7)*
25°C	ND	0.49 ± 0.045 *(1.9)*
Whole animal, mmol/h/kg		
9°C	0.57 ± 0.09	0.29 ± 0.07
15°C	0.85 ± 0.14 *(1.9)*	0.76 ± 0.13 *(4.9)*
22°C	1.63 ± 0.20 *(2.5)*	1.07 ± 0.14 *(1.6)*

this depends on the origin and/or the size of the eels (Vettier 2005, Amérand et al. 2010; Table 4). Consequently, males really have the aerobic capacities to swim faster than females. As the metabolic aerobic rate appears higher in males, it is supposed that they are consequently submitted to a higher production of deleterious reactive oxygen species (ROS). In fact, any rise in oxygen consumption is accompanied by a higher rate in electron transfer through the mitochondrial respiratory chain, which is liable to enhance electron leak and thus ROS production (Chen et al. 2003, Dröge 2002, Indo et al. 2007). This hypothesis was verified by Mortelette et al. (2010a) but only at atmospheric pressure as in mammals (Ku et al. 1993).

Table 4. Metabolic (aerobic energy) data for males and females eels from Loire river. Values for maximal estimated aerobic capacity (including red and white muscle) are from Vettier (2005); relative cost of swimming is estimated from Van Ginneken and Van den Thillart (2000) at a speed of 0.5 body length/s. The relationship MO_2 vs Swimming speed (S) is: $MO_2 = aS + b$ with S expressed in m/s. For this relationship, data for males are from Sébert et al. (2009a), data for females are from Palstra et al. (2008). Body surface areas are calculated from Sébert et al. (2004b).

		Males	Females
Maximal estimated aerobic capacity, mmol/h/kg		7.4	5.2
Relative cost of swimming, %		30	42
Relationship MO_2 vs Speed			
(mmol/h/kg)	a	6.23	4.58
	b	1.08	0.54
(mmol/h/m²)	a	22.5	33.9
	b	3.89	4.03

Anaerobic Metabolism

Migration of *Anguilla anguilla* is a very long journey from Europe to the Sargasso Sea. Consequently, the metabolic process used to ensure energy requirements is the aerobic one, as the performance is mainly achieved by red muscles (Van Ginneken and Van den Thillart 2000) at least at atmospheric pressure. In this context, white muscle and/or the anaerobic process are rarely considered although several experimental factors underline their important role. White muscle can help red muscle during the swimming activity in particular conditions (Pritchard et al. 1971). Lactate from white fibres can be used by red muscle fibres due to the adjustments in inter-tissue coupling of anaerobic and aerobic metabolisms (Sullivan and Somero 1980). It must not be forgotten that white muscle contains most of the whole body aerobic capacity and together with viscera (e.g., liver) may be responsible for a substantial part of whole body oxygen consumption, with the involvement of red muscle appearing as minimal (Goolish 1991,

Sébert et al. 2011). Likewise, it has been shown that under pressure, there are numerous modifications (physiology and muscle morphology) in white muscle (Simon et al. 1991, Sébert et al. 1998). Keeping in mind that migration is performed at depth and at low water temperature, the cold probably has inhibitor effects on red muscle functions (Dalla Via et al. 1989, Block 1991, Coughlin 2003). In these conditions, it would be very useful for the eel if fast fibres (white muscle) at low temperature use ATP as rapidly as slow fibres (red muscle) at high temperature (Rome 1998). Finally, the recent study by Sébert and co-workers (2011) showing that the anaerobic process is probably involved in a relatively low swimming speed (0.3 m/sec) indicates that the anaerobic process and white muscle must not be neglected when considering migratory activity. Here again, sex differences have been observed that is to say, males are more aerobic and females more anaerobic as shown in the trout (Battiprolu et al. 2007). The ratio J_B/J_A, that is to say, the ratio of anaerobic flux to aerobic flux is about 60% higher in females than in males whatever the temperature from 5°C to 25°C. Consequently, the metabolic "reprise", ρ, is twice higher in females (52 ± 11 s^{-1} and 27 ± 7 s^{-1} for females and males respectively). This means that females can increase glycolytic flux twice faster than males but with a very high temperature sensitivity: $Q_{10} \sim 10$ for males and $Q_{10} \sim 20$ for females (Scaion and Sébert 2008).

To summarize, the above sections clearly show that males have higher aerobic capacities than females which in turn have higher anaerobic capacities than males. It must be pointed out that in this context, the term capacity signifies ability and is not used in the sense of energy quantity (energy stores) as opposed to energy power.

Pressure Sensitivity in Energy Metabolism

This aspect has been previously reviewed in detail (Sébert 2002, Sébert et al. 2004a, 2009b). Briefly, high hydrostatic pressure (101 ATA ~ 1000 m depth) has deleterious effects on aerobic energy production. It probably acts on the respiratory chain and oxidative phosphorylation via a pressure effect on cell and/or mitochondrial membranes. This is true for yellow eels but the silvering process, which prepares the eel for migration, sets up the required pressure adaptation thus allowing the eel to migrate at depth without further energy use (Vettier et al. 2005). An essential part of this adaptation consists in increasing polyunsaturated fatty acids (PUFA) in order to ensure optimal membrane fluidity (Vettier et al. 2006). By studying glycolytic fluxes in European eels after 1 month at 100 atmospheres, Sébert et al. (1998) have shown that high hydrostatic pressure changes the eel's metabolic design by increasing significantly the glycolytic fluxes and the transition speed from the aerobic to the anaerobic pathways (see above).

This gives the eel the possibility of increasing delivered glycolytic power 1.6 times more rapidly than at atmospheric pressure, the condition under which the eel is generally studied. The reader who is interested in the general effects of high hydrostatic pressure on muscle function can refer to recent reviews considering cellular mechanisms and their implications in taxonomic environmental locomotion or dedicated to fish (Sébert 2008, Friedrich 2010).

Temperature Sensitivity of Energy Metabolism; Interactions with Pressure

The temperature effects on fish metabolism are well known and the eel does not differ from other fishes. Due to the Arrhenius law, a rise in temperature increases the aerobic and anaerobic rates: this thermal sensitivity is generally quantified by using the Q_{10}. However, interactions between temperature and pressure are less known due to technical constraints and difficulties in working under pressure (Sébert 1993). These interactions are highly dependent on the protocol used: they are maximal when the temperature and pressure are concomitantly modified but temperature has little effect on eel acclimatized to high pressure (Sébert et al. 1995). At the cell level (respiratory chain and oxidative phosphorylation), pressure and temperature interactions have been reviewed in detail (Sébert et al. 2004a). As temperature always increases the reaction rates, pressure effects vary considerably from inhibition to stimulation of the reaction rates due to differences in volume changes during the different reactions. This means that the targets for temperature and pressure could be different with an important role for uncoupling proteins (Sébert et al. 2004a). However, *in vitro* results must be considered with care, as pressure resistance and sensitivity depend on complexity: isolated tissue is less sensitive than an organ, invertebrates are less sensitive than vertebrates etc.…. Are there any sex differences in European eels which are concomitantly exposed to high pressure and low temperature, conditions for migration? The answer is clearly yes!

The glycolytic pathway is few sensitive to pressure exposure because this process is located in the cytoplasm and has a low dependence on the membrane state of the mitochondria (see above). In contrast, this pathway is very sensitive to temperature as shown by Scaion and Sébert (2008). Briefly, the glycolytic flux increases when temperature increases which responds to thermodynamic laws. However, anaerobic glycolysis is more sensitive to cold water ($Q_{10} \geq 6.5$), females being doubly more affected by cold temperatures than males, whereas the aerobic pathway is more affected by warm temperatures. The authors suggest that male and female eels could migrate at different depths in order to optimize their energy use by aerobic and/or anaerobic pathways. This hypothesis proposed from *in vitro* results, has been confirmed by a study on pressure-temperature interactions on aerobic metabolism of migrating silver eels. Scaion et al. (2008) studied oxygen consumption, M_{O2}, of male and female silver eels at different pressures (from 1 to 121 ATA i.e., from atmospheric pressure to 1200 m depth) and 3 temperatures (9°C, 15°C, 22°C). The main results show that, 1) the rise in hydrostatic pressure increases M_{O2} at a given temperature in both sexes, 2) at atmospheric pressure and in both sexes, a temperature rise induces an increase in M_{O2}. However, temperature sensitivity in males does not depend on the temperature zone (Q_{10} ~ 2.5 in the cold zone or warm zone) in contrast with females whose thermal sensitivity is high in the cold zone (Q_{10} ~ 4.2 between 9°C and 15°C) but low in the warm zone (Q_{10} ~ 1.4 between 15 and 22°C). The third and most important result is that, in males, pressure sensitivity increases with a rise in temperature but the reverse is observed in females. In other words, males are less sensitive to high pressure at low temperature (migration at depth) whereas females are less sensitive to "high" temperature (surface). Due to conditions required for complete maturation, it is probable that in the last days of migration and before reproduction, eels move to shallower waters (Haenen et al. 2009). It is surprising that, although females respond to thermodynamic laws (increasing temperature reduces pressure effects) this is not the case in males where temperature and pressure increases act in synergy. At the present time, we do not have a convincing explanation for this phenomenon, except to suggest that males have more biochemical regulations than females. In fact, *in vivo* biological processes involve coupling between uncatalyzed potentially unregulated processes (physical, some chemical) and catalyzed regulated processes (biochemical), thus allowing integrated physiological functions. The uncatalyzed processes are set by thermodynamics whereas the biochemical catalyzed processes are the only ones which can be regulated in terms of temperature (and/or pressure?) dependence (Hochachka 1991). The above results led authors Scaion et al. (2008) to suggest that males migrate at depth where they have lower energy expenditure. It must be pointed out that males and females have the same pressure sensitivity at 17°C which is close to the Sargasso Sea temperature but is also their thermal preferendum (Haro 1991).

To summarize, when compared to females, males have: 1) higher aerobic capacities, 2) lower anaerobic capacities, 3) lower pressure sensitivity at low temperatures. Do these sex differences in energy metabolism at rest have any consequences on swimming activity?

Swimming Activity

Swimming physiology of European eel, *Anguilla anguilla*, has been recently reviewed (Palstra and Van den Thillart 2010, Righton et al. 2012). Together with these papers, the study by Palstra and Planas (2011) analysing the contents of the Barcelona congress (FITFISH) and the book on the swimming physiology of fish, edited by Palstra and Planas (2012) will give a complete and recent overview of eel swimming. Therefore there is no point giving an introduction to this section devoted only to male-female comparison which, in contrast, has very rarely been considered.

In the same way as running for humans, swimming is a natural activity for fish. Whatever the species, there are subjects which are pathological, many are sedentary but some are athletes. The silver eel which swims over 6000 kms could thus be compared to human marathon runners as high level athletes. This comparison between humans and eels could sound surprising but, in fact, it is pertinent. Firstly as fish physiology can help us understand human physiology (Rome 1998), the reverse could also be true when some routine human tests are adapted. Secondly, as in humans, female eels have lower aerobic capacities than males (see above) which could be explained by taking into account the higher fat content which decreases the "active tissue mass". Thirdly, training improves performances (Palstra et al. 2008, Mortelette et al. 2010b) as it does in humans. Other similarities could also be pointed out when comparing human exercise physiology and eel swimming physiology (Fig. 1).

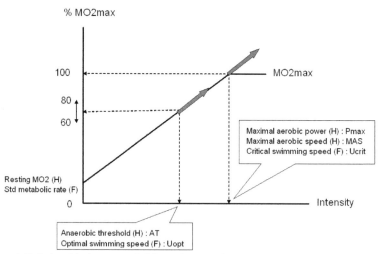

Figure 1. Relationship between % $M_{O_2}^{max}$ (maximal oxygen consumption) and exercise intensity. Grey arrows illustrate the training effects. Training improves the levels of oxygen consumption and intensity but also, by increasing the limit time, increases the working capacity. H: human; F: fish.

In fact, both human and eel species exhibit maximal oxygen consumption corresponding to critical swimming speed (Ucrit in fish) or maximal aerobic power or maximal aerobic speed (aerobic P^{max} in humans). Another important point is the anaerobic threshold in humans. It was originally described as exercise intensity where there is an imbalance between lactate production by white muscles and lactate consumption by red muscles: at this moment, lactates appear more intensively in the blood and lactatemia sharply increases. The anaerobic threshold ranges from 40%–50% of maximal oxygen consumption, $M_{O_2}^{max}$, in sedentary humans up to 70%–80% in athletes. Consequently, athletes running for a long time (aerobic exercise) must stay under this level in order to restrict blood lactate increase and its consequences. It must be pointed out that the anaerobic threshold does not correspond to the activation of anaerobic glycolysis (which is largely anterior) but rather to the imbalance between aerobic and anaerobic processes. In fish and thus in eels, the anaerobic threshold AT has not been precisely described even though, in a recent paper (2010) Marras et al. identified a gait transition speed (Ugait) at about 60% of maximal speed (Ucat) in sea bass performing a constant acceleration test (see also Hammer 1995). However, as Ucat is about 1.6 Ucrit (obtained with a conventional protocol), this means that Ugait corresponds to Ucrit which is not surprising because both correspond to a transition in the swimming mode (anaerobic?). Maximal aerobic power is evaluated from the direct measurement of oxygen consumption in relation to maximal swimming speed (Ucrit). Optimal speed (Uopt, see Palstra and Van den Thillart 2010) is often used to determine the swimming speed at which the energy transport cost is minimal. For silver eels, optimal speed is in the range of 60%–80% of Ucrit (Palstra and Van den Thillart 2010, Methling et al. 2011) and the maximal sustained swimming speed is usually around 80% Ucrit: these values are in accordance with the anaerobic threshold of human athletes, although the interpretation is somewhat different (see above). However, can we reasonably rule out the hypothesis that Uopt may correspond to AT? In a recent paper, Sébert et al. (2011) measured glycolytic and aerobic fluxes in white and red muscles after swimming. They calculated that to be exclusively aerobic, the eel swimming session must be performed at a speed below 0.27 m/sec (0.5 BL/sec for the used eels). Such a speed fits in well with the average speed estimated for eel migration (Van Ginneken and Van den Thillart 2000; see also Palstra and Van den Thillart 2010) or the optimal speed of 0.6 BL/sec reported by Methling et al. (2011). Consequently, it could be hypothesized that Uopt in the eel may well match the anaerobic threshold in humans, Uopt corresponding to a strategy where the eel uses maximal aerobic power without activating the anaerobic process too much. If not, glycolysis produces more protons which can decrease free energy from hydrolysis of ATP and consequently can decrease efficiency. Such a pH effect probably has little importance in the case of a short exercise but it can become deleterious in the case of a long swimming activity. Although human males and females are comparable in terms of body mass and size (10% to 30% lower in females which have a higher fat content) it is not the case for migrating eels where females could have a body size 3 times longer than males and body mass more than 10 times greater which restricts the human/fish comparison. This has some consequences on migratory activity as previously shown. To meet the females in the Sargasso Sea (the putative breeding area) males must swim faster! It is for this reason that they have higher aerobic abilities. The calculated values for individuals from the Loire river (France) are 7.4 mmol/h/kg and 5.2 mmol/h/kg for males and females respectively (Table 4). Considering the total swimming cost reported by Van Ginneken and Van den Thillart (2000), it represents about 42% of aerobic capacities in females but only 30% in males if both are assumed to swim at a speed of 0.5 BL/sec. With the assumption that the energy cost is the same in both sexes and that this cost is in linear relationship with the swimming speed, we can calculate at what speed males could swim if they used the same percentage of aerobic capacities as females. The obtained value is 0.67 BL/sec which allows the males to arrive at the Sargasso Sea at the same time as females if they start sooner (Sébert et al. 2009b). However, the sentence "males swim faster than females" is ambiguous. Quintella et al. (2010) measuring Ucrit obtained 1.70 BL/sec and 1.20 BL/sec for males and females respectively. However, when expressed in ground speed (m/sec) corresponding to real displacement relative to the bottom, the Ucrit are very similar in both sexes (0.65 m/sec and 0.70 m/sec for males and females respectively). It must be remembered that males start sooner and that females are heavier and bigger, thus potentially increasing the drag and affecting thrust (Quintella et al. 2010). In a preceding section (Table 3) we reported a common observation that males have higher oxygen consumption than females, considering isolated muscle cells or whole animal. This is true at rest but also during swimming activity (Fig. 2).

However, when oxygen consumption during swimming activity is standardized to body surface (important for swimming and drag consequences) rather than body mass the reverse is observed: males have lower oxygen consumption (see Table 4). The reader may or may not be convinced by the argument that the standardization to body mass rather than to body surface could introduce a bias. Nevertheless, the question must be raised bearing in mind that energy metabolism (evaluated from oxygen consumption) corresponds to energy/time ratio that is to say, the dimension of a flow or a surface (see Sébert et al. 2004b). Basically, the above data clearly show the sex difference in terms of eel energy metabolism: males can migrate at the same ground speed as females because they have higher metabolic abilities which compensate for their small size. However

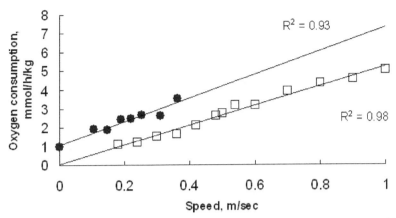

Figure 2. Oxygen consumption versus swimming speed. Data for males (closed symbols) are from Sébert et al. (2009a). Data for females (open squares) are from Methling et al. (2011) for the range 0.18 to 0.54 m/s and from Palstra et al. (2008) for the range 0.5 to 1.0 m/s.

these abilities have been evaluated in terms of maximal aerobic power for muscle fibres or in terms of Ucrit for whole animals. These are measurements corresponding to maximal capacities exhibited by the fish, or its muscle fibres, on the day of experiment, i.e., for a short period of time. Is it true for a period lasting several months in the case of a migratory journey? We must remember that this migration is performed while fasting, i.e., only using energy stores, mainly fat stores. It is known that these fat stores are sufficient for females and must represent about 20% of body mass (Van Ginneken and Van den Thillart 2000). But what about males? Assuming that males migrate over 250 days, this represents 6000 hours. For a male 40 cm in length and swimming at 0.67 BL/sec, Table 4 indicates an oxygen consumption of about 2.8 mmol/h/kg. Consequently, a 100g male would consume about 1.7 moles of oxygen for the total swimming activity during migration. This represents about 38LSTPD of oxygen or 760 kJ: by using 37 kJ/g of fat, the males need about 20g of fat just for the energy requirements of locomotion. This means that such a male must have at least 20% of its body mass as lipid stores corresponding to what is expected in females (Van Ginneken and Van den Thillart 2000). However, on examining silver migrating males, it is hard to believe that fat represents 20% of their body mass and females are also known to have a higher fat content (Garcia-Gallego and Akharbach 1998). Moreover, eel fat content has been decreasing over the last 20 years and depends on the site where the eels grow (Belpaire et al. 2009). It is improbable that males perform their migration having just what they need as energy stores. Belpaire et al. (2009) have published some calculations based on fat content and transport cost in males from different origins: they consider that currently, many male eels are not able to reach their spawning grounds but Clevestam et al. (2011) draw the same conclusion about females. However we can only assume that some strategies are used by males to reduce their energy requirements, while maintaining a higher relative swimming speed than females.

Possible Strategies Reducing Male Energy Requirements

Swimming in Groups

In a recent study Burgerhout and co-workers (Burgerhout et al. 2010) have performed swimming trials with male silver eels: they reported greater efficiency while swimming in groups. In fact, the cost of swimming individually is about 25 mgO_2/kg/km but only 9 mgO_2/kg/km when swimming in groups: the difference is attributed, at least in part, to a decrease in drag and thus the cost of transport. This is an interesting observation which has been previously reported in other fish species (Schuett 1933, Shlaifer 1938, Geyer and Mann 1939): unfortunately these authors have also shown that the "group effect" disappears when the size of the container is increased in proportion to the increase in the number of fish. The study by Burgerhout et al. (2010) thus confirms the group effect using eels (in a restricted volume) but what are the consequences from an ecophysiological point of view? Males could well swim in groups in order to reduce energy cost during migration. Fish shoals are known among other species but not for eels even though they start their migration in great number (making their capture easier!). In contrast if males really swam in shoals during their migration, they would undoubtedly be fished in the open ocean which is not the case (Van Ginneken and Maes 2005). To be efficient (mainly regarding the drag effect) such a strategy implies a dense and relatively compact group. Experiments with juvenile mullet have shown the complexity of such schools, depending on aerobic scope and standard metabolic rate: the optimal position for a given fish is 1–2 body length diagonally behind two anterior individuals (Killen et al. 2011). It can thus be concluded that if group effect exists in experimental conditions of restricted volumes, it certainly does not exist in the open ocean for the eel.

Better Mechanical Efficiency

Long distance migration is probably easier thanks to a greater body size which is consistent with the size of swimming migrants (Alerstam et al. 2003). Consequently, using the equations proposed by Hedenström (2003), a 1kg female eel would migrate at a speed about 50% greater than a 100g male. Although some studies indicate that conversion efficiency increases with a greater body size (Casey 1992) it is generally assumed that efficiency remains a constant regarding body size. There is thus an apparent contradiction between the negative impact of small size and the necessity for males to swim faster. Van Ginneken et al. (2005) has analysed the factors which could explain the particularity of swimming efficiency in the eel. There is no reference to sex difference but it is reported that swimming speed depends linearly on tail beat frequency, and tail beat frequency corresponds to muscle contraction frequency. The relationship between tail-beat frequency (TB) and swimming speed (S) is linear for both sexes: TB = 1.36 S + 0.26 for males (Farhat 2011) and TB = 1.41 S + 0.31 for females (Methling et al. 2011). Consequently, for a given swimming speed at 0.5 BL/sec, the tail beat frequency is 0.94 for males and 1.01 for females. The difference is not significant, meaning that contraction frequency and thus mechanical efficiency are similar in both sexes. The conclusion is that the greater swimming performance and energy spares in males cannot probably be attributed to biomechanics.

High Pressure-low Temperature Environment

Due to their higher density, males tend to sink more than females. Moreover, by using their fat stores all along the migratory journey, their density progressively increases, as it does in females, thus decreasing buoyancy. It is probable that males and females continuously sink from the European coasts to the Sargasso Sea although the differences in water densities (temperature and/or salinity) could compensate for this supposed decrease in buoyancy. In a recent book, Righton et al. (2012) have detailed the different aspects of extreme swimming conditions for eel migration. Briefly, whatever their sex, eels migrate at depth, that is to say, high pressure and low temperature. High pressure increases oxidative phosphorylation efficiency (Theron et al. 2000) together with an optimization of membrane fluidity during the silvering process. Although the experiments were performed on females, there is no serious reason to doubt it would be similar in males. However, the study by Scaion et al. (2008) clearly shows that resting males are significantly less sensitive to the combined effects of high pressure and low temperature in terms of aerobic metabolism. This means that males consume less oxygen when the pressure is high and the temperature is low. Consequently, Scaion et al. (2008) and Scaion and Sébert (2008) concluded that males and females probably migrate at different depths. Considering that energy stores are limited (no refuelling), it could be reasonably hypothesized that eels migrate in the most favourable environment capitalizing on conditions for where their oxygen consumption is minimum in order to optimize their energy budget. We are thus convinced that males migrate more deeply than females which is in agreement with their higher density. Such a behavior could explain why tracking experiments (Stasko and Rommel 1974, Westerberg 1975, Tesch 2003, McCleave and Arnold 1999, Jellyman and Tsukamoto 2002, Aarestrup et al. 2009), which have been performed using only females, estimate the depth of eel migration to be a few hundred meters although the maximal depth for eel migration could be estimated at about 2000 m (Sébert 2008). However results from such tracking experiments could be questioned considering the negative impact induced by pop-up tags which can increase swimming cost twofold (Methling et al. 2011, Burgerhout et al. 2011) and the high energy cost induced by vertical migrations, if they really exist, in eels swimming with very limited budget (Sébert 2008). Besides the supplementary energy cost of transport, such tags produce a positive buoyancy (Burgerhout et al. 2011) which can force the eel to go up when they are tired. Are the described eel vertical migrations the results of an unnatural combination of the tag effect and trying to orient itself after release into the ocean (Tesch 2003)?

Besides the energetic advantages of swimming at depth, the required increase in swimming speed corresponds to an increase (limited, see after) in oxygen consumption. This is accompanied by a higher rate of electron transfer through the mitochondrial respiratory chain and thus ROS production with its deleterious effects (Dröge 2002). However, hydrostatic pressure seems to reveal differences in the regulation of ROS production between the two genders. The usual positive and linear relationship between ROS production and oxygen consumption, also observed in female eels at atmospheric pressure is reversed at depth but only in males: the more male muscle mitochondria consume oxygen, the less ROS are produced (Amérand et al. 2010). Thus, by swimming deeper, males can increase their swimming speed (more than females) and consequently their oxygen consumption without increasing ROS production, even decreasing it. In contrast, females maintain the positive relationship between oxygen consumption and ROS production even at depth but they could use (among other possibilities) sexual hormones as antioxidant systems (Vina et al. 2006, Amérand et al. 2010).

By migrating deeper than females, males decrease their energy requirements thus acting in synergy with the direct consequences of hydrostatic pressure on swimming activity. In 2009, Sébert and colleagues showed that swimming under pressure (101 ATA ~ 1000 m depth) significantly decreases male oxygen consumption when compared to eels swimming at atmospheric pressure (Sébert et al. 2009a). For example, for a swimming speed at 0.67 BL/sec M_{O2} = 2.8 mmol/h/kg at 1 ATA but only 1.5 mmol/h/kg at 101 ATA: swimming at depth decreases oxygen consumption by 40% compared to surface swimming. This pressure effect is surely less important for females swimming in shallower waters. Such a decrease in transport cost at depth explains how eels are able to perform such a long migration using only their energy stores which are very limited. But how could such a gain in metabolic efficiency be explained? It is known that during the silvering process, membrane fluidity significantly increases through higher PUFA recruitment (Vettier et al. 2006). This means that when the pressure effect (decrease in membrane fluidity) is applied on eels entering the ocean, membrane fluidity is optimized by a return to normal value from a hyperfluid state (Sébert et al. 2009a). Such an explanation has several consequences. Bearing in mind that membrane fluidity is optimal at 101 ATA (1000 m depth) , this means that at surface, the membrane is too fluid to ensure optimal functioning of the respiratory chain and oxidative phosphorylation: this has been shown *in vitro* by Theron et al. (2000). This reasoning implies that we must accept that silvering is not an optimal transient state at atmospheric pressure and consequently the normal environmental condition for silver eels is high pressure and low temperature which allow them to minimize their energy expenditure. This also shows that the study of energy metabolism and its efficiency at surface atmospheric pressure induces a bias because such an environmental condition does not correspond to the optimal environment.

Conclusion

Sex dimorphism in the eel has important consequences on its ability to migrate over 6000 km. Due to their smaller sizes, males must swim faster in order to meet the females at the appropriate time and at the right location. To ensure such a performance, males have greater aerobic capacities. However, even the most powerful vehicle cannot work over a long period with an empty or small-sized fuel tank. Likewise for the eel which must reduce its fuel consumption (fat and thus oxygen consumption) due to its limited energy stores. Among the possible strategies to reduce energy consumption, it appears that swimming at depth is the most efficient for males which probably migrate more deeply than females. If high pressure and low temperature are considered as environmental conditions which optimize energy budgets during migration, we must conclude that life at atmospheric pressure is an inefficient state for silver eels and can only be a transient state. The fact that silver eels which end up by not migrating return to the yellow stage (Durif et al. 2009) is in agreement with this hypothesis. This also means that the results from experiments on energy metabolism efficiency at atmospheric pressure must be considered with caution when extrapolation is performed involving migration. Finally, the stimulating topic of sex differences in the eel must lead to further research, not only on sexual maturation but also on energy metabolism by studying combined environmental factors: temperature-pressure-salinity-light-swimming. This is a huge yet tremendous challenge.

Acknowledgments

I thank Pr Christine MOISAN and Pr Francesca TRISCHITTA for careful reading of the manuscript and helpful suggestions.

References

Aarestrup, K., F. Okland, M. Hansen, D. Righton, P. Gargan, M. Castonguay, L. Bernatchez, P. Howey, H. Sparholt, M.I. Pedersen and R.S. McKinley. 2009. Oceanic Spawning Migration of the European Eel (*Anguilla anguilla*). Science 325: 1660.

Alerstam, T., A. Hedenström and S. Åkesson. 2003. Long-distance migration : evolution and determinants. OIKOS 103: 247–260.

Amérand, A., A. Vettier, C. Moisan, M. Belhomme and P. Sébert. 2010. Sex-related differences in aerobic capacities and reactive oxygen species metabolism in the silver eel. Fish Physiol. Biochem. 36: 741–747.

Battiprolu, P.K., K.J. Harmon and K.J. Rodnick. 2007. Sex differences in energy metabolism and performance of teleost cardiac tissue. Am. J. Physiol. 292: R827–R836.

Belpaire, C., G. Goemans, C. Geeraerts, P. Quataert, K. Parmentier, P. Hagel and J. de Boer. 2009. Decreasing eel stocks: survival of the fattest ? Ecol. Freshwater Fish 18: 197–214.

Block, B.A. 1991. Endothermy in fish: thermogenesis, ecology and evolution. *In*: P.W. Hochachka and T.P. Mommsen [eds]. Phylogenetic and Biochemical Perspectives, Elsevier, Amsterdam, Netherlands. pp. 269–311.

Burgerhout, E., R. Manabe, S.A. Brittijn, J. Aoyama, K. Tsukamoto and G.E.E.J.M. Van den Thillart. 2011. Dramatic effect of pop-up satellite tags on eel swimming. Naturwissenschaften 98: 631–634.

Burgerhout, E., S.A. Brittijn, A. Palstra and G.E.E.J.M. Van den Thillart. 2010. Swimming trials with male silver eels indicating a higher efficiency by swimming in groups. The proceeding of the Fitfish-workshop on the swimming physiology of fish. Barcelona. http://ub.edu/fitfish2010/.

Casey, T.M. 1992. Energetics of locomotion. *In*: R. McN. Alexander [ed]. Advances in Comparative and Environmental Physiology. Springer-Verlag, Berlin, Germany. 11: 251–275.

Chen, Q., E.J. Vazquez, S. Moghadas, C.L. Hoppel and E.J. Lesnefsky. 2003. Production of reactive oxygen species by mitochondria: central role of complex III. J. Biol. Chem. 278: 36027–36031.

Clevestam, P.D., M. Ogonowski, N.B. Sjoberg and H. Wickström. 2011. Too short to spawn? Implications of small body size and swimming distance on successful migration and maturation of the European eel Anguilla Anguilla. J. Fish Biol. 78: 1073–1089.

Colombo, G., G. Grandi and R. Rossi. 1984. Gonad differentiation and body growth in *Anguilla anguilla*. J. Fish Biol. 24: 215–228.

Coughlin, D.J. 2003. Steady swimming by fishes: kinetic properties and power production by the aerobic musculature. *In*: A.L. Val and B.G. Kapoor [eds]. Fish Adaptations. Science Publishers, Enfield (NH), USA. pp. 55–72.

Dalla Via, J., M. Huber, W. Wieser and R. Lackner. 1989. Temperature-related responses of intermediary metabolism to forced exercises and recovery in juvenile *Rutilus rutilis* (L.) (Cyprinidae: Teleostei). Physiol. Zool. 62: 964–976.

Dröge, W. 2002. Free radicals in the physiological control of cell function. Physiol. Rev. 82: 47–95.

Durif, C. 2003. La migration d'avalaison de l'anguille européenne, *Anguilla anguilla*. PhD Thesis, Université Toulouse III, Toulouse, France.

Durif, C., S. Dufour and P. Elie. 2005. The Silvering Process of *Anguilla anguilla*: A new classification from the yellow resident to the silver migrating stage. J. Fish Biol. 66: 1025–1043.

Durif, C., V. Van Ginneken, S. Dufour, T. Müller and P. Elie. 2009. Seasonal evolution and individual differences in silvering eels from different locations. *In*: G. Van den Thillart, S. Dufour and J.C. Rankin [eds]. Spawning Migration of the European Eel. Springer. pp. 13–38.

Farhat, F. 2011. Effets de l'entraînement à la nage sur la relation métabolisme et production radicalaire chez l'anguille argentée mâle. Master 2 report, Université de Brest, France.

Friedrich, O. 2010. Muscle function and high hydrostatic pressure. *In*: P. Sébert [ed]. Comparative High Pressure Biology. Science Publisher, Enfield (NH), USA. pp. 211–249.

Garcia-Gallego, M. and H. Akharbach. 1998. Evolution of body composition of European eels during their growth phase on a fish farm, with special emphasis on the lipid component. Aquacult. Int. 6: 345–356.

Geyer, F. and H. Mann. 1939. Beitrage zur atmung der fische. III. Der Saueratoffverbrauch im gruppenversuch. Zeitschr. Vergl. Physiol. 27: 429–433.

Goolish, E.M. 1991. Aerobic and Anaerobic Scaling in Fish. Biol. Rev. 66: 33–56.

Haenen, O., V. Van Ginneken, M. Engelsma and G. Van den Thillart. 2009. Impact of eel viruses on the recruitmenr of european eel. *In*: G. Van den Thillart, S. Dufour and J.C. Rankin [eds]. Spawning Migration of the European Eel. Springer. pp. 387–400.

Hammer, C. 1995. Fatigue and exercise tests with fish. Comp. Biochem. Physiol. 112A: 1–20.

Haro, A.J. 1991. Thermal *preferenda* and behavior of Atlantic eels (genus *Anguilla*) in relation to their spawning migration. Environ. Biol. Fish 31: 171–184.

Hedenström, A. 2003. Scaling migration speed in animals that run, swim and fly. J. Zool. Lond. 259: 155–160.

Hochachka, P.W. 1991. Temperature: the ectothermy option. *In*: P.W. Hochachka and T.P. Mommsen [eds]. Phylogenetic and Biochemical Perspectives. Elsevier, Amsterdam, Netherlands. pp. 313–322.

Ide, C., N. de Schepper, J. Christiaens, C. van Liefferinge, A. Herrel, G. Goemans, P. Meire, C. Belpaire, C. Geeraerts and D. Adriaens. 2011. Bimodality in head shape in European eel. J. Zool. DOI: 10.1111/j.1469-7998.2011.00834.x.

Indo, H.P., M. Davidson, H.C. Yen, S. Suenaga, K. Tomita, T. Nishii, M. Higuchi, Y. Koga, T. Ozawa and H.J. Majima. 2007. Evidence of ROS generation by mitochondria in cells with impaired electron transport chain and mitochondrial DNA damage. Mitochondrion 7: 106–118.

Jellyman, D. and K. Tsukamoto. 2002. First Use of Archival Transmitters to Track Migrating Freshwater Eels Anguilla Dieffenbachii at Sea. Mar. Ecol. Prog. Ser. 233: 207–215.

Killen, S.S., S. Marras, J.F. Steffensen and D.J. McKenzie. 2011. Aerobic capacity influences the spatial position of individuals within fish schools. Proc. R. Soc. B. DOI: 10.1098/rspb.2011.1006.

Kloppmann, M. 2003. Body structure and functions. *In*: F.W. Tesch [ed]. The Eel. Blackwell Publishing. Oxford, UK. pp. 1–71.

Ku, H.H., U.T. Brunk and R.S. Shoal. 1993. Relationship between mitochondrial superoxide and hydrogen peroxide production and longevity of mammalian species. Free Radic. Biol. Med. 15: 621–627.

Marras, S., G. Claireaux, D.J. McKenzie and J.A. Nelson. 2010. Individual variation and repeatability in aerobic and anaerobic swimming performance of European sea bass, *Dicentrarchus labrax*. J. Exp. Biol. 213: 26–32.

McCleave, J.D. and G.P. Arnold. 1999. Movements of yellow- and silver-phase European eels (*Anguilla anguilla*) tracked in the western North Sea. J. Mar. Sci. 56: 510–536.

Methling, C., C. Tudorache, P.V. Skov and J.F. Steffensen. 2011. Pop up satellite tags impair swimming performance and energetics of the European eel (*Anguilla anguilla*). Plos One 6: 1–7.

Mortelette, H., C. Moisan, P. Sébert, M. Belhomme and A. Amérand. 2010a. Fish as a model in investigations about the relationship between oxygen consumption and hydroxyl radical production in permeabilized muscle fibers. Mitochondrion 10: 555–558.

Mortelette, H., A. Amérand, P. Sébert, M. Belhomme, P. Calves and C. Moisan. 2010b. Effect of exercise training on respiration and reactive oxygen species metabolism in eel red muscle. Resp. Physiol. Neurobiol. 172: 201–205.

Palstra, A. 2006. Energetic Requirements and Environmental Constraints of Reproductive Migration and Maturation of European Silver Eel (*Anguilla anguilla* L.). Ph.D thesis, University of Leiden, Leiden, Netherlands.

Palstra, A., V. Van Ginneken and G. Van den Thillart. 2009. Effects of swimming on silvering and maturation of the European eel, *Anguilla anguilla* L. *In*: G. Van den Thillart, S. Dufour and J.C. Rankin [eds]. Spawning Migration of the European Eel. Springer. pp. 229–251.

Palstra, A.P. and J.V. Planas. 2011. Fish under exercise. Fish Physiol. Biochem. 37: 259–272.

Palstra, A.P. and J.V. Planas. 2012. Swimming Physiology of Fish. Springer.

Palstra, A., V. Van Ginneken and G. Van den Thillart. 2008. Cost of transport and optimal swimming speed in farmed and wild European silver eels (*Anguilla anguilla*). Comp. Biochem. Physiol. 151A: 37–44.

Palstra, A.P. and G.E.E.J.M. Van den Thillart. 2010. Swimming physiology of European silver eels (*Anguilla anguilla* L.): energetic costs and effects on sexual maturation and reproduction. Fish Physiol. Biochem. 36: 297–322.

Pankhurst, N.W. 1982. Relation of visual changes to the onset of sexual maturation in the European eel *Anguilla anguilla* (L.) J. Fish Biol. 21: 127–140.

Pelster, B. 1997. Buoyancy at depth. Chap. 5. *In*: D.J. Randall and A.P. Farrell [eds]. Fish Physiology, vol. 16: Deep-sea fishes. Academic Press, San Diego, USA. pp. 195–238.

Pritchard, A.W., J.R. Hunter and R. Lasker. 1971. The relation between exercise and biochemical changes in red and white muscle and liver in Jack mackerel, *Trachurus symmetricus*. Fish Bull. Nat. Ocean. Atm. Adm. 69: 379–386.

Quintella, B.R., C.S. Mateus, J.L. Costa, I. Domingos and P.R. Almeida. 2010. Critical swimming speed of yellow- and silver-phase European eel (*Anguilla anguilla*, L.). J. Appl. Ichthyol. 26: 432–435.

Righton, D., K. Aarestrup, K. Tsukamoto, D. Jellyman, P. Sébert and G. Van den Thillart. 2012. Extreme swimming : the oceanic migrations of anguillide eels. *In*: A.P. Palstra and J.V. Planas. Swimming Physiology of Fish. Springer, Berlin, Germany. pp. 19–44.

Rome, S. 1998. Some advances in integrative muscle physiology. Comp. Biochem. Pysiol. 120B: 51–72.

Scaion, D., M. Belhomme and P. Sébert. 2008. Pressure and temperature interactions on aerobic metabolism of europeansilver eel. Resp. Physiol. Neurobiol. 164: 319–322.

Scaion, D. and P. Sébert. 2008. Glycolytic fluxes in european silver eel, *Anguilla anguilla*: sex differences and temperature sensitivity. Comp. Biochem. Physiol. 151A: 687–690.

Schuett, F. 1933. Studies in mass physiology: the effect of numbers upon oxygen consumption of fishes. Ecol. J. 4: 10–122.

Sébert, P. 1993. Energy metabolism of fish under hydrostatic pressure: a review. Trends Comp. Biochem. Physiol. 1: 289–317.

Sébert, P. 2002. Fish metabolism at pressure: a hundred years history. Comp. Biochem. Physiol. 131A: 575–585.

Sébert, P. 2008. Fish muscle function and pressure- Chap. 9. *In*: P. Sébert, D. Onyango and B.G. Kapoor [eds]. Fish Life in Special Environments: Physiological Functions. Science Publishers, Enfield (NH), USA. pp. 234–255.

Sébert, P., M. Theron and A. Vettier. 2004a. Pressure and temperature interactions on cellular respiration: a review. Cell. Mol. Biol. 50: 491–500.

Sébert, P., A. Vettier and M. Belhomme. 2004b. A simple relationship to calculate eel surface area : interest in studying energy metabolism. An. Biol. 54: 131–136.

Sébert, P., J. Peragon, J.B. Barroso, B. Simon and E. Melendez Hevia. 1998. High hydrostatic pressure (101 Ata) changes the metabolic design of yellow freshwater eel muscle. Comp. Biochem. Physiol. 121B: 195–200.

Sébert, P., D. Scaion and M. Belhomme. 2009a. High hydrostatic pressure improves the swimming efficiency of European migrating silver eel. Resp. Physiol. Neurobiol. 165: 112–114.

Sébert, P., B. Simon and L. Barthélémy. 1995. Effects of a temperature increase on oxygen consumption of yellow fresh water eel exposed to high hydrostatic pressure. Exp. Physiol. 80: 1039–1046.

Sébert, P., A. Vettier, A. Amérand and C. Moisan. 2009b. High pressure resistance and adaptation of european eels. *In*: G. Van den Thillart, S. Dufour and J.C. Rankin [eds]. Spawning Migration of the European Eel. Springer, Berlin, Germany. pp. 99–127.

Sébert, P., H. Mortelette, J. Nicolas, A. Amérand, M. Belhomme and C. Moisan. 2011.
.   *In vitro* aerobic and anaerobic muscle capacities in the european eel, *Anguilla anguilla* : effects of a swimming session. Resp. Physiol. Neurobiol. 176: 118–122.

Shlaifer, A. 1938. Studies in mass physiology : effect of numbers upon the oxygen consumption and locomotor activity of *Carassius auratus*. Physiol. Zool. 1: 408–424.

Simon, B., P. Sébert and L. Barthélémy. 1991. Eel, *Anguilla Anguilla* (L.), muscle modifications induced by long-term exposure to 101 Ata hydrostatic pressure. J. Fish Biol. 38: 89–94.

Stasko, A.B. and S.A. Rommel. 1974. Swimming depth of adult American eels (*Anguilla rostrata*) in a saltwater bay as determined by ultrasonic tracking. J. Fish. Res. Bd. Can. 31: 1148–1150.

Sullivan, K.M. and G.M. Somero. 1980. Enzyme activities of fish skeletal muscle and brain as influenced by depth of occurence and habits of feeding and locomotion. Mar. Biol. 60: 91–99.

Tesch, F.W. 2003. The Eel. Blackwell Science, Oxford, UK.

Theron, M., F. Guerrero and P. Sébert. 2000. Improvement in the efficiency of oxidative phosphorylation in the freshwater eel acclimated to 10.1 Mpa hydrostatic pressure. J. Exp. Biol. 203: 3019–3023.

Todd, P.R. 1981. Timing and periodicity of migrating New Zealand freshwater eels (*Anguilla* spp.). New Zeal. J. Mar. Fresh. Res. 15: 225–235.

Van den Thillart, G., S. Dufour and J.C. Rankin. 2009. Spawning Migration of the European Eel. Springer, Berlin, Germany.

Van Ginneken, V.J.T. and G.E.E.J.M. Van den Thillart. 2000. Eel fat stores are enough to reach the Sargasso. Nature 403: 156–157.

Van Ginneken, V., E. Antonissen, U.K. Müller, R. Booms, E. Eding, J. Verreth and G. Van den Thillart. 2005. Eel migration to the Sargasso: remarkably high swimming efficiency and low energy costs. J. Exp. Biol. 208: 1329–1335.

Van Ginneken, V.J.T. and G.E. Maes. 2005. The European eel (*Anguilla Anguilla* L.), its lifecycle, evolution and reproductioin: a literature review. Rev. Fish Biol. Fisheries 15: 367–398.

Vettier, A. 2005. La migration de reproduction de l'anguille Européenne (*Anguilla anguillai* L.) : effets de la pression hydrostatique et de la métamorphose. PhD Thesis, Université de Brest, Brest, France.

Vettier, A., A. Amérand, C. Cann-Moisan and P. Sébert. 2005. Is the Silvering Process Similar to the Effects of Pressure Acclimatization on Yellow Eels? Respir. Physiol. Neurobiol. 145: 243–250.

Vettier, A., C. Labbé, A. Amérand, G. Da Costa, E. Le Rumeur, C. Moisan and P. Sébert. 2006. Hydrostatic pressure effects on eel mitochondrial functioning and membrane fluidity. Und. Hyperb. Med. 33: 149–156.

Vina, J., J. Sastre, F.V. Pallardo, J. Gambini and C. Borrás. 2006. Role of mitochondrial oxidative stress to explain the different longevity between genders: protective effect of estrogens. Free Rad. Res. 40: 1359–1365.

Westerberg, H. 1975. Counter current orientation in the migration of the European eel (*Anguilla anguilla* L.). Göteborgs Univ. Oceanogr. Inst. Rep. 9: 1–18.

Sensory Systems

David M. Hunt,[a],* Nathan S. Hart[b] and Shaun P. Collin[c]

Introduction

Eels have the same five senses of sight, hearing, touch, smell and taste that most animals possess. In addition and as in many teleost fishes, eels are capable of detecting sound and water movements using mechanoreceptive structures within the acoustic and lateral line systems. Eels also appear to possess the ability to detect the earth's magnetic field. This is critically important for navigation during the migratory phases of the life cycle seen in certain species and it is this migratory behavior that impacts directly on another sense, that of vision, with adaptations in visual sensitivity driven by changes in the quality and quantity of the available light during migration. Chemoreception also plays an important role in migration and feeding with many species possessing high sensitivity to a range of water-borne molecules.

School of Animal Biology, Lions Eye Institute and UWA Oceans Institute, University of Western Australia, 35 Stirling Highway, Crawley, Perth, Western Australia, 6009, Australia.
[a]E-mail: david.hunt@uwa.edu.au
[b]E-mail: nathan.hart@uwa.edu.au
[c]E-mail: shaun.collin@uwa.edu.au
*Corresponding author

Magnetoreception

Catadromous freshwater eels (Anguillidae) such as the European (*Anguilla anguilla*), American (*A. rostrata*) and Japanese (*A. japonica*) eels are famous for their oceanic migration to spawning grounds, in the Sargasso Sea region of the North Atlantic Ocean for European (Schmidt 1923, Aarestrup et al. 2009) and American eels (Schmidt 1923), and to an area west of the Mariana Islands in the North-West Pacific Ocean for Japanese eels (Chow et al. 2009, Kurogi et al. 2011). However, as with other vertebrates that undergo long-distance migrations (Walker et al. 2002), the sensory mechanisms used by the eels for navigation and homing are still largely a matter of speculation.

Beyond the European continental shelf, migrating European eels swim at depths between 196 and 344 m (Tesch 1989); similarly, migrating Japanese eels released close to their spawning grounds in the Western Pacific Ocean swim at depths between 81 and 172 m (Aoyama et al. 1999) and American eels migrate and spawn in the upper few hundred meters of the water column (Kleckner et al. 1983, McCleave and Kleckner 1985). At such depths, there are no visual landmarks or cues that can be used for navigation. Sound cues may be used for homing by some fish (Montgomery et al. 2006), but these have an effective maximum detection range of just a few kilometres and it is unlikely that hearing subserves navigation over the large distances (hundreds or thousands of kilometres) travelled by migratory eels. Olfactory cues (including the olfactory detection of changes in water salinity) have also been implicated in the navigation and homing abilities of migratory eels (Westin 1990), as they have in salmonids (Dittman and Quinn 1996), but these cues are likely to be effective over relatively short distances once the animals are close to their destination, such as their natal stream (Lohmann et al. 2008).

Consequently, for trans-oceanic migrants such as eels, the earth's magnetic field is probably the only geophysical signal that provides consistent directional and positional information at all times of the day and in all habitats (Kirschvink et al. 2001). Although the ability to detect and use the geomagnetic field for navigation is thought to be widespread amongst both vertebrates and invertebrates, the exact way the nervous system interprets this information remains unclear and most likely differs between species (Johnsen and Lohmann 2005). For example, marine turtles are able to orient themselves using magnetic field information in complete darkness (Irwin and Lohmann 2003), suggesting a light-independent mechanism, whereas the magnetic compass of migratory birds is perturbed by drastic changes in the spectral composition and/or intensity of the ambient illumination (Wiltschko et al. 1993), suggesting a light-dependent mechanism.

Evidence for a light-dependent magnetoreceptor mechanism comes mainly from experiments in fruitflies (Gegear et al. 2010), salamanders (Phillips and Borland 1992) and birds (Wiltschko and Wiltschko 2001) and implicates either the retinal photoreceptor visual pigments (Leask 1977, Ritz et al. 2000) or a class of blue-light-sensitive proteins called cryptochromes found, at least in birds, in specific populations of retinal ganglion cells (Mouritsen et al. 2004).

Light-independent magnetoreception may occur by two very different mechanisms. Firstly, endogenous magnetic materials, such as the ferrimagnetic mineral magnetite, may allow direct detection of the magnetic field when coupled with a mechanotransducer that converts the motion of (or torque acting upon) the magnetite crystals into a signal that is detectable by the nervous system (Kirschvink and Gould 1981, Diebel et al. 2000). Secondly, the Earth's magnetic field can be detected indirectly by the process of electromagnetic induction, either in the form of electrical currents generated within the body of an animal moving through the magnetic field, or, in the case of marine animals, the detection of electrical currents generated within oceanic water currents as they move through the earth's magnetic field (Kalmijn 2000).

A. anguilla, A. rostrata and *A. japonica* all show behavioral responses to changes in the ambient magnetic field. In *A. anguilla*, changes in the position of magnetic north, the angle of inclination and the magnetic field intensity all result in shifts in the preferred swimming direction of both migratory ('silver') and stationary ('yellow') eels (Tesch et al. 1992, van Ginneken et al. 2005). Although earlier studies failed to elicit behavioral responses to altered magnetic fields (Rommel and McCleave 1973, Zimmerman and McCleave 1975, McCleave and Power 1978), *A. rostrata* has subsequently been shown to display altered directional preferences when the intensity and direction of the ambient magnetic field is changed (Souza et al. 1988). Using a classical cardiac conditioning paradigm, *A. japonica* was shown to respond to changes in the ambient magnetic field with a sensitivity sufficient to detect a change equivalent to a 21° shift in the horizontal component of the field (Nishi et al. 2004).

The mechanism used by eels to detect the earth's magnetic field is unknown. Eels lack the highly sensitive electroreceptive organs (ampullae of Lorenzini) possessed by sharks and rays that form the physiological basis of the electromagnetic induction hypothesis of magnetoreception in elasmobranchs (Brown and Ilyinsky 1978). Nevertheless, *A. rostrata* can be conditioned to respond to weak electric fields of strength comparable to that produced by oceanic water currents (Rommel and McCleave 1973). Prominent skin papillae with anatomical features resembling electroreceptive structures in other fishes have been described in the deep-sea bobtail eel, *Cyema atrum* (Meyer-Rochow 1978), but the actual function of these structures is unknown and no such papillae were reported in a detailed study of the integument of *A. rostrata* (Leonard and Summers 1976).

Magnetic material has been isolated from the skull and backbone of *A. anguilla* (Hanson and Westerberg 1986), but although this was thought to be magnetite, the authors of a subsequent study (Hanson and Walker 1987) concluded that the observed magnetic behavior of the isolated material was not consistent with the presence of single-domain magnetite crystals that would be optimal for magnetoreception. However, magnetically active particles resembling the single-domain magnetite crystals located in the putative magnetoreceptor organs of the rainbow trout (Walker et al. 1997) have since been found in the mandibular canals of the lateral line in *A. anguilla* (Moore and Riley 2009), leaving open the idea that magnetoreception in eels is mediated by a magnetite-based transducer. Candidate magnetite-based magnetoreceptors in both birds (Williams and Wild 2001) and fishes (Walker et al. 1997) are located close to the olfactory tissues; interestingly, *A. japonica* rendered temporarily anosmic cannot be conditioned to respond to changes in the ambient magnetic field (Nishi et al. 2005), perhaps suggesting a common structural organisation of the magnetoreceptive organs among vertebrates.

Vision

The Eel Eye

The eel eye conforms to the typical camera design of most other teleosts (Williamson and Castle 1975). The cornea is split into dermal and scleral components separated by loose lamellar tissue, thereby permitting some degree of movement of the globe under the dermal cornea (Hein 1913, Walls 1942, Duke-Elder 1958). The dermal cornea or secondary spectacle in eels is used to protect the eye from dessication while the animal is out of water (Collin and Collin 2001). Light enters the transparent cornea and is focussed on to the retina by a spherical lens. Accommodation is provided by the posterior movement of the lens following the contraction of the small *retractor lentis* muscle. The eye of *Anguilla* sp. is unique amongst bony fishes in that it lacks a choroidal gland and a choroid, which, in other fishes, provides oxygenated arterial blood to the back of the eye (Duke-Elder 1958). However, there is an extensive pattern of vitreal vascularisation, which uniquely pierces the inner limiting membrane and forms two capillary strata within the inner and outer nuclear layers of the retina, converging into a hyaloid vein (Hanyu 1959, Ali et al. 1968). The eel retina of *A. anguilla*, which lacks a falciform process, appears to be the only teleost retina to possess intraretinal circulation (Walls 1942).

Owing to the lateral placement of the eyes on the head, the binocular visual field of many eels is small, although the protrusion of the eyes from the slender body provides small binocular fields both anteriorly and posteriorly, for example as in the deep-sea snipe eel, *Borodinula infans* (Duke-Elder 1958). The pupil is round and contractile, thereby regulating the amount of light entering the eye (Walls 1942).

Structure of the eel retina

The vertebrate retina comprises distinct layers of cells and processes (Fig. 1). The outermost layer comprises a monolayer of non-neural pigment

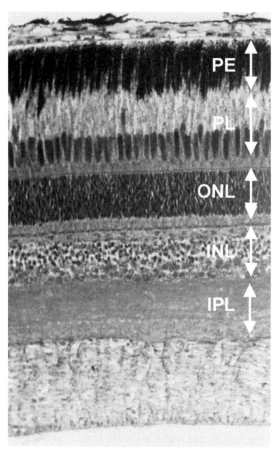

Figure 1. Transverse section of a typical teleost retina. PE, pigmented epithelium; PL, photoreceptor layer; ONL, outer nuclear layer, INL, inner nuclear layer; IPL, inner plexiform layer.

Color image of this figure appears in the color plate section at the end of the book.

epithelial cells (PE). This is adjacent to the photoreceptor layer (PL), which contains the photosensitive outer segments of the photoreceptor cells. Photoreceptor cell bodies form the outer nuclear layer (ONL), while the cell bodies of interneurons (bipolar, amacrine and horizontal cells) form the inner nuclear layer (INL). The ganglion cell layer containing both ganglion cells and amacrine cells is situated close to the inner margin of the retina. These features are labeled in Fig. 1 which shows a transverse section through a typical teleost retina. The eel retina conforms to this basic pattern (Ali and Anctil 1976). Two distinct types of photoreceptors are present in the majority of teleosts so far examined (Gordon et al. 1978, Pankhurst 1984, Bozzano 2003, Wang et al. 2011), rods that are functional in dim light and cones that are functional in daylight. These photoreceptors can also be distinguished on the basis of their spectral sensitivity. For rods, there is generally a single class with a peak spectral sensitivity (λ_{max}) at around 500 nm whereas cones are considerably more variable and fall into distinct spectral classes. This variability in spectral sensitivity provides the basis for colour vision, whereby the photon capture of the different cone classes is compared via a process called opponency to give the perception of colour. For colour vision to be present, at least two spectrally distinct types of cones must the present in the retina. In many vertebrate species including most species of teleost fishes, a subset of cones is composed of two adjoining cells to give double cones.

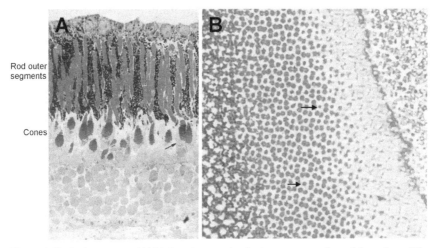

Figure 2. Retinal structure. (A) Light micrograph of a transverse section of the retina of the eel, *Anguilla rostrata* showing the single row of long thin rod photoreceptors and double cones (arrow). (B) Tangential section of the retina of the moray eel, *Gymnothorax* showing the regular arrangement of cone photoreceptors within a square mosaic (two examples arrowed). Reprinted with permission from Ali and Anctil (1976).

Structural Adaptations of the Eel Retina

The overall structure of the teleost retina shows significant differences in diurnal and nocturnal species (Walls 1942, Munz and McFarland 1973, Ali and Anctil 1976, Pankhurst 1989). Amongst the eels, the moray eel is generally considered to be nocturnal, although the spectral sensitivities of the photoreceptors indicate that this may be an over-simplification. In nocturnal teleosts, the retinal pigment epithelium (PE) and the inner nuclear layer (INL) are generally thinner and the photoreceptor layer and outer nuclear layer (ONL) are thicker, with this pattern reversed in diurnal teleosts. Amongst the moray eels studied (Wang et al. 2011), *Gymnothorax favagineus*, *G. reticularis* and *Strophidon sathete* show the characteristics of a nocturnal species, whereas *Rhinomuraena quaesita* possesses retinal layer dimensions typical of a diurnal species, and this is replicated in the length ratios of the ONL to the INL where *R. quaesita* has a value at 0.47, typical of a diurnal species whereas values >2 are found for *G. favagineus* and *G. reticularis*, indicating nocturnality. *S. sathete* is intermediate at a value of 1.43, which suggests a more crepuscular life style (Munz and McFarland 1973).

Amongst the different eel species studied (Garten 1907, Wunder 1925, Verrier 1928, Vilter 1951, Mirzaliev and Koloss 1964, Ali et al. 1968, Beatty 1975), the retina is well differentiated with the retinal pigment epithelial cells filled with large numbers of melanin granules. Rods are long and slender and generally occur in very high numbers compared to cones. Cones are small and either single or double with their nuclei lying within the sclerad or outer-most region of the outer nuclear layer (Beatty 1975, Ali and Anctil 1976) (Fig. 2A). In the moray eel, *Gymnothorax richardsonii*, the cone photoreceptors are arranged into a regular square mosaic with four double cones bordering a central single cone (Ali and Anctil 1976) (Fig. 2B) but in the European eel, *A. anguilla*, the cone mosaic is lost (Braekevelt 1984). The number of ganglion cells is generally low, although higher densities occur in moray eels (Ali and Anctil 1976). Rods of *A. anguilla* appear to be capable of photomechanical movements (Braekevelt 1985, Braekevelt 1988c, Braekevelt 1988b, Braekevelt 1988a) although retinomotor movements of cones has not been confirmed (Walls 1942, Duke-Elder 1958).

The retinae of deep-sea eels such as the conger eels, *Coloconger raniceps* and *Conger conger*, are comprised purely of rods ((Franz 1910, Verrier 1928, Ali and Anctil 1976). In contrast to most shallow water species which possess a single layer of long slender rods, some deep-sea species possess multiple banks or rows of rod photoreceptors (Hewss et al. 1998). Kaup's arrowtooth eel, *Synaphobranchus kaupi*, possesses three banks of rods at an average density of 64 x 10^3 rods per mm^2 (Wagner et al. 1998). The advantage of possessing multiple banks of rods is not clear especially since most of the light will be absorbed by the most proximal banks (Land 1990). However, these species possess a number of other adaptations to enhance light capture that include a longer overall photoreceptor thickness, a mirror-like tapetum behind the retina for reflecting light back on to the rod photoreceptors and an increased density of visual pigment (Partridge et al. 1989), so the benefit of the extra banks of rods must have a selective advantage. All these adaptations optimise sensitivity in a deep-sea environment, where the only source of illumination is the intermittent flashes of bioluminescent light within an otherwise dark background.

As in most other teleost species examined, the density of retinal sampling elements in eels is not homogeneously distributed across the retina. This means that not all parts of the visual field of the eyes of these fishes are sampled equally, with retinal acute zones subtending specific regions of the visual world, whether these are for feeding or for predator avoidance. The topography of retinal ganglion cells has been examined in four species of eel, *Ophichthus rufus*, *Serrivomer beani*, *S. kaupi* and *Nemichthys* sp. Between one and three localised regions of increased density have been revealed (Collin and Partridge 1996, Wagner et al. 1998, Bozzano 2003), with peak ganglion cell density ranging from 3.9×10^3 to 19.0×10^3 cells per mm^2, indicating differences in the peak spatial resolving power of the eye based on the visual demands of each species. The total number of ganglion cells in the retina (indicative of the number of ganglion cell axons or fibres entering the brain from the eye) ranges from 1.13×10^5 to 5.27×10^5 (Wagner et al. 1998). The terminal fields of these visual projections in the brain are also considered a reflection of the importance of vision; in a study of eight species of eels, Wagner (2002) showed that the optic tectum occupies the largest volume of all the sensory brain areas examined, although the size of the olfactory bulbs is also large. This suggests that both vision and olfaction play a large part in the lifestyles of eels in both the mesopelagic and demersal zones.

Spectral Sensitivity Adaptations in Anguilla spp.

Amongst the eels, members of the genus *Anguilla* have a complex life cycle (Berry et al. 1972, Tesch 1977) that exposes them at different stages of their lifecycle to very different photic environments. The European eel, *A. anguilla*, the American eel, *A. rostrata*, and the Japanese eel, *A. japonica*, are catadromous species that spend most of their lives in freshwater but migrate to the sea to breed. For the European and American eels, this occurs in the Sargasso Sea (Schmidt 1923, van Ginneken and Maes 2005) and for the Japanese eel, in an area west of the Mariana Islands in the North-West Pacific Ocean (Tsukamoto 1992), with spawning at depths of around 200 m. After hatching, the leptocephalus larvae of the European eel spend 1–2 years drifting with the Gulf Stream into the North Atlantic, travelling around 6000 km to the European continental shelf. Here they metamorphose into glass eels, initially as elvers or pigmented glass eels but then as the larger yellow eels, before travelling up European rivers to spend 6–20 years in a fresh water environment where they grow and mature as a freshwater species. Mature eels must then cross the Atlantic Ocean to return to the Sargasso Sea to spawn. Just prior to and during this migration they undergo 'silvering' (Aroua et al. 2005), a pubertal change (Rousseau et al. 2009), to become sexually mature adult fish. During this part of the life cycle, the photopic environment of the eel changes markedly from the yellow/brown of shallow freshwater rivers to green coastal waters and then to the blue of the deep ocean. A notable feature of the visual system of eels from the genus *Anguilla* is the change in sensitivity to different regions of the light spectrum at different stages of the life cycle.

Spectral sensitivity is dependent on the types of photoreceptors present in the retina and the photosensitive visual pigments that they contain in their outer segments. There is generally a single class of rod photoreceptors expressing a rod or *Rh1* opsin gene, whereas up to four classes of cone photoreceptors are found in jawed vertebrates. Each cone class expresses a different cone opsin gene that encodes visual pigments that are maximally sensitive to different regions of the spectrum. The four classes of cone pigments can be summarised as follows: long wavelength-sensitive (LWS) with λ_{max} values ranging from 500 nm to 570 nm, middle wavelength-sensitive (MWS or Rh2) with λ_{max} values ranging from 480 nm to 530 nm, and two short wavelength-sensitive pigments, SWS2 with λ_{max} values ranging from 400 nm to 470 nm and SWS1 with λ_{max} values ranging from 355 nm to 445 nm (Yokoyama 2000, Bowmaker and Hunt 2006, Bowmaker 2008) (Fig. 3). The actual number of cone classes present in individual species varies substantially; cyprinids like the goldfish and zebrafish possess all four classes (Johnson et al. 1993, Chinen et al. 2003) whereas many species that exist in more light limited environments have dispensed with some or all cones, retaining only the more light sensitive rod photoreceptors in a rod-dominant or rod-only retina (Partridge et al. 1989, Partridge et al. 1992, Douglas and Partridge 1997). Visual pigments are members of an extensive family of G-protein-linked membrane receptors. Each pigment comprises a unique opsin protein of around 350 amino acids encoded by a corresponding opsin gene, linked to a chromophore. In all opsin classes, the spectral sensitivity of the pigment arises from interactions between the chromophore and the amino acid residues that form a binding pocket in the opsin for the chromophore. The chromophore is bound to the opsin via a Schiff base linkage and may be either retinal, a derivative of vitamin A1, or 3,4-dehydroretinal, a derivative of vitamin A2, with the respective pigments described as either rhodopsins or porphyropsins.

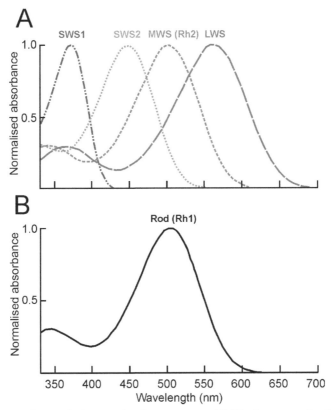

Figure 3. Example visual pigment absorbance spectra. (A) The four cone visual pigment classes. (B) The rod (Rh1) class.

During the downstream migration and subsequent entry of eels into the deep-sea and the associated transition from the yellow to silver stage, the λ_{max} of the rod photoreceptors of the European and Japanese eels are known to change from around 523 nm to around 482 nm. Photoreceptors and the visual pigments that they contain are generally tuned to the wavelengths of available light. This is achieved in two ways, either by a change in chromophore, with the A2-derived form causing a shift in spectral sensitivity to longer wavelengths (Harosi 1994, Parry and Bowmaker 2000), or by changes at key amino acid sites in the opsin protein that alter the interaction of the opsin protein with the chromophore (Bowmaker and Hunt 2006, Bowmaker 2008). Both these mechanisms are used by the eel to achieve a short wavelength shift in rod photoreceptor spectral sensitivity at maturation in preparation for migration from the river to the deep ocean. Firstly, the A2 chromophore is replaced by the A1 form (Carlisle and Denton 1959, Beatty 1975) to change the pigment from porphyropsin to rhodopsin;

the pigments of freshwater fish tend to be porphyropsins, whereas marine species generally have rhodopsins (Bridges 1972), so this switch from A2 to A1 pigments in eels as they migrate from freshwater to the ocean is consistent with such interspecific differences.

This short wavelength (SW) shift in rod λ_{max} accompanying this natural maturation in European eels has been experimentally induced by gonadotropin administration (Wood and Partridge 1993), thereby enabling the monitoring of the λ_{max} of individual rod photoreceptors during maturation. In this way, Wood and Partridge (1993) showed that not only is there a change in chromophore but the rod opsin form (Rh1a) expressed during the freshwater phase of the life cycle is progressively replaced by a SW-shifted "deep-sea" or Rh1b form (Fig. 4A). During this transition, both forms are present in the outer segments of the rod photoreceptors, with the "freshwater" form at the distal end of the outer segment progressively replaced by the "deep-sea" form which enters the outer segment at the proximal end as newly synthesized opsin (Fig. 4B).

The rod opsin genes that underlie this SW shift were identified for the European eel by Archer et al. (1995) and confirmed in the Japanese eel by Zhang et al. (2000). Both genes lack introns, in line with the observation that the four introns normally present in the rod opsin gene were lost at the base of the teleost lineage (Fitzgibbon et al. 1995). The two genes encode opsins that show a 14% amino acid non-identity in the European eel and 12% in the Japanese eel. These amino acid differences include substitutions at two sites, 83 and 292, that are known to be important for the spectral tuning of visual pigments. The replacement of Asp83 in the "freshwater" form with Asn in the "deep-sea" form has been linked to a 6–10 nm SW-shift in fish rod pigments (Hunt et al. 1996, Hunt et al. 2001) and changes in the conformation of the opsin protein (Rath et al. 1993, Lehmann et al. 2007). Moreover, recent work (Sugawara et al. 2010) on the kinetics of rhodopsin activation has shown that the presence of Asn83 results in an accelerated formation of metarhodopsin II, an intermediate in the activation of the G-protein transducin in the photo transduction cascade, thereby increasing sensitivity in dim light conditions. The replacement of Ala292 in the "freshwater" form by Ser in the "deep-sea" form will also generate a 20 to 30 nm SW-shift, as demonstrated in studies of rod opsin pigments in a number of species that include the elephant shark (Davies et al. 2009), many marine mammals (Fasick et al. 1998, Newman and Robinson 2005) and the mouse (Sun et al. 1997, Davies et al. 2012). If the effects of these changes in the eel pigments are additive, then they would be sufficient to generate the SW-shift in the "deep-sea" pigment.

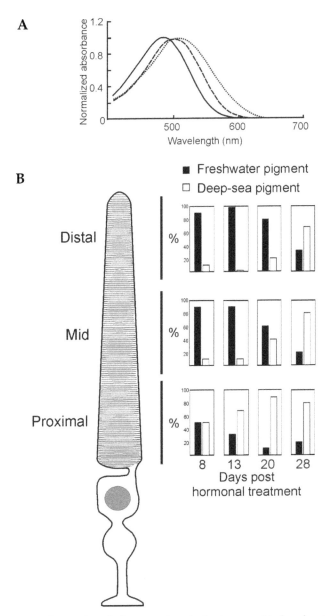

Figure 4. Pigment changes in rod photoreceptors during hormonal-induced metamorphosis. (A) Normalized absorbance spectra showing spectral shifts in the rod pigment obtained from three different eels. The proportions of A1 and A2 chromophore and deep-sea (DS) and freshwater (FW) opsin have been determined by best-fitting of visual pigment template mixtures. Re-drawn from Wood and Partridge (1993). (B) Progressive changes in the proportions of deep-sea and freshwater opsins along rod outer segments at four intervals post hormonal treatment. Data obtained from Wood and Partridge (1993).

Developmental Changes in Cone Photoreceptors in *Anguilla* spp.

The retina of leptocephalus larvae of the European eel is similar to the retinae of most larval teleosts which undergo metamorphosis, in the pre-metamorphic presence of cones (Omura et al. 2003). Post-metamorphosis, the retina remains duplex with a mixed population of rod and cone photoreceptors (Fig. 5).

A single class of cones sensitive in the green region of the spectrum is found in glass eels, increasing to two classes in eels undergoing the transition from freshwater yellow to marine silver eels (Bowmaker et al. 2008). Consistent with this, molecular studies by Cottrill et al. (2009) have shown that only two cone opsin genes are expressed in the eel retina, belonging respectively to the *SWS2* and *Rh2* opsin gene classes. Colour vision is therefore at best dichromatic with only two cone channels; indirect evidence

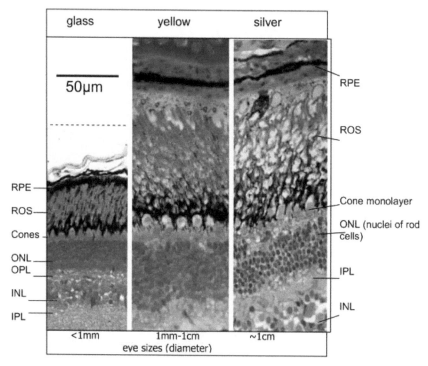

Figure 5. Changes in the thickness of the layers of the retina at the glass, yellow and silver stages of the life cycle. This is especially pronounced in the ONL of the glass eel compared to the later stages. The size of the eye also changes between the different stages, increasing tenfold in diameter. RPE, retinal pigment epithelium; ROS, rod outer segments; ONL, outer nuclear layer; INL, inner nuclear layer; OPL, outer plexiform layer; IPL, inner plexiform layer. From Cottrill et al. (2009).

Color image of this figure appears in the color plate section at the end of the book.

for colour opponency was previously obtained in the study of rod and cone inputs on to retinal neurons (Damjanovic et al. 2005) in the European eel. Spectral analysis of these cone pigments synthesized *in vitro* have shown that the corresponding rhodopsin (A1) pigments have λ_{max} values at 446 nm and 525 nm, respectively. These values are consistent with the expression of these pigments in the two cone classes identified in yellow eels by Bowmaker et al. (2008) with respective λ_{max} values at around 436 nm and between 520–546 nm; the range in values for the latter class is attributable to varying proportions of A1/A2 chromophore. During the early stages of the second metamorphosis, the cone pigments switch from porphyropsins to rhodopsins (Bowmaker et al. 2008), with the maximum sensitivity of the "green"-sensitive cone shifting to around 525 nm, paralleling but preceding the change in rod photoreceptors. *In situ* hybridisation with *Rh2* and *SWS2* gene probes revealed that glass eels express only the *Rh2* opsin, while larger yellow eels continue to express *Rh2* in the majority of their cones, but also have a low number (<5%) of cones which express the *SWS2* opsin (Fig. 6). Silver eels showed the same expression pattern as the larger yellow eels.

Prior to migration to the deep-sea, there is a decrease in the number of ganglion cells and the cells of the INL (Pankhurst and Lythgoe 1983), together with a marked depletion in the number of cones; in glass and yellow stage eels, rod cells are more numerous than cone cells and this increases to a rod to cone ratio of 100:1 in pre-maturation silver eels and 200:1 post-maturation (Braekevelt 1984, Braekevelt 1985, Braekevelt 1988c, Braekevelt 1988a). It remains unclear however whether all cones are subsequently lost. Coincident with this, it has been shown in the Japanese eel, that there is a substantial addition of new rod photoreceptors arising from rod progenitor cells, and this is thought to continue throughout its life cycle (Omura et al. 2003).

Spectral Shifts in the Pigments of Other Eel Species

Conger Eels. Coding sequence for two forms of the *Rh1* rod pigment gene in the white spotted Conger eel, *Conger myriaster*, have been deposited under accession numbers AB043817 and AB043818, but only a single gene has been reported in the European Conger eel, *C. conger* (Archer and Hirano 1996). *C. myriaster* inhabits shallow water whereas *C. conger* moves between shallow coastal water and deep water (to a depth of 1000 m). The retina of *C. conger* is described as rod-only (Lythgoe and Lythgoe 1991) with peak sensitivities in the blue region of the spectrum (Munz and McFarland 1973, Ali and Anctil 1976). It is similar therefore to many deep-sea teleosts that have dispensed with cone photoreceptors and have SW-shifted rod pigments (Partridge et al. 1988, Partridge et al. 1992, Douglas et al. 1995, Douglas and Partridge 1997, Douglas et al. 2003). *In situ* spectral analysis of

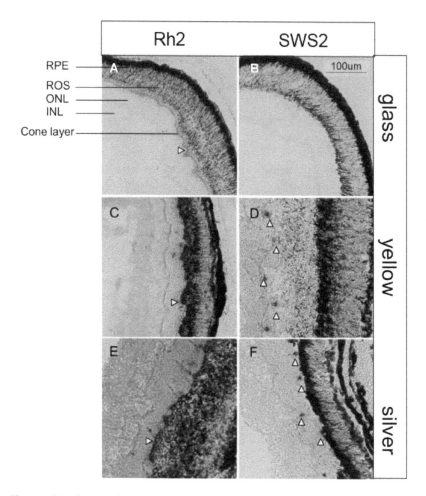

Figure 6. Distribution of opsin-expressing cone cells in eel retina. *In situ* hybridisation of *Rh2* or *SWS2* opsin probes of transverse sections of the eel retina from different developmental stages. Cone cell bodies in the eel are arranged in a single layer between the inner nuclear layer (INL) and the rod outer segments (ROS). Glass eel: (A) *Rh2* opsin expression (horizontal arrowhead), (B) no *SWS2* opsin expression. Yellow eel: (C) *Rh2* opsin (horizontal arrowhead), (D) occasional cone cells expressing *SWS2* opsin (vertical arrowhead). Silver eel: (E) monolayer of cone cells expressing *Rh2* opsin (horizontal arrowhead), (F) occasional cone cells expressing *SWS2* opsin (vertical arrowheads). RPE, retinal pigment epithelium; ROS, rod outer segments; ONL, outer nuclear layer; INL, inner nuclear layer. From Cottrill et al. (2009).

Color image of this figure appears in the color plate section at the end of the book.

the pigment revealed a λ_{max} at 487 nm and opsin gene sequencing showed that like the "deep-sea" opsin of the European eel, the pigment in the Conger eel possesses Ser292 although it has retained Asp83 (Archer and Hirano 1996). The presence of Ser292 would be sufficient to SW-shift the λ_{max} of the pigment into the blue region of the spectrum and the retention of Asp83 may account for the slightly smaller SW-shift in λ_{max} of the Conger eel pigment at 487 nm compared to European eel "deep-sea" pigment at 482 nm.

Deepwater Arrowtooth Eel. The Rh1 rod pigment of the deepwater arrowtooth eel, *Histiobranchus bathybius*, a member of the Synaphobranchidae family of eels that are recovered at depths ranging from 1,790 to 4,790 m, shows a similar SW-shift to that seen in the "deep-sea" form of the rod pigment in the European eel (Hope et al. 1997). The retina in this extreme deep-sea teleost contains only rods with λ_{max} at 477 nm and the molecular basis for this SW-shift is almost certainly again due to the presence of two residues, Asn83 and Ser292 (Hope et al. 1997). Indeed, this would appear to be a common mechanism for achieving a SW-shift in the peak sensitivity of rod pigments in deep-sea teleosts where the main source of light comes from bioluminescence (Hope et al. 1997, Hunt et al. 2001).

Moray Eels. SW-shifts in pigment sensitivity are also seen in moray eels. Wang et al. (2011) studied four species, *R. quaesita* and *G. favagineus*, which are shallow-water reef crevice-dwellers, and *G. reticularis* and *S. sathete*, which live in sand-muddy sediments at depths between 200 and 300 m (Table 1). Rods and a single class of middle wavelength-sensitive (green) single cones are present within the photoreceptor layer, so it is unlikely that these species possess colour vision. *In situ* spectral analysis has shown that the λ_{max} values for rods in *G. favagineus*, *G. reticularis* and *S. sathete* are shifted to around 487 nm compared to *R. quaesita* at 498 nm. These shifts are consistent with the deeper habitats of *G. reticularis* and *S. sathete*, but this does not apply to *G. favagineus* which lives in shallower water. Alternative explanations are that the SW-shift in the latter species may be associated with a diurnal vertical migration as seen in the conger eel (Shapley and Gordon 1980, Archer and Hirano 1996), or it may serve to increase sensitivity of rods to the available light at twilight, when it is known that the spectrum is weighted towards the blue by atmospheric absorption at low solar angles

Table 1. Mean spectral sensitivities of the photoreceptors of moray eels measured by microspectrophotometry. Data from Wang et al. (2011).

Species	Peak sensitivities ± SD		Habitat
	Rods	**Cones**	
Rhinomuraena quaesita	498 ± 4.8 nm	493 ± 7.0 nm	Shallow water reef crevices
Gymnothorax favagineus	487 ± 5.4 nm	501 ± 7.7 nm	Shallow water reef crevices
Gymnothorax reticularis	486 ± 4.0 nm	494 ± 5.8 nm	Sand-muddy sediments at depths of 200 m
Strophidon sathete	487 ± 4.8 nm	509 ± 6.6 nm	Sand-muddy sediments at depths of 300 m.

(McFarland and Munz 1975, McFarland 1986, Sandstrom 1999). Sequencing of the *Rh1* opsin genes in the four species revealed that *R. quaesita* possesses Ala292, whereas all the SW-shifted pigments have Ser292. As discussed above for the deep-sea rod pigments of the European and Japanese eels (Archer et al. 1995, Zhang et al. 2000), this substitution would be sufficient to SW-shift the λ_{max} of the pigments.

The λ_{max} values for the single green cones also differ between species, with *R. quaesita* and *G. reticularis* around 494 nm and *G. favagineus* and *S. sathete* at 501 and 509 nm, respectively (Wang et al. 2011). The gene encoding the pigment in these cones was identified as belonging to the *Rh2* class, with no evidence for SWS1, SWS2 or LWS pigments. However, as found for many teleost species (Johnson et al. 1993, Chinen et al. 2003, Parry et al. 2005, Shand et al. 2008), two copies of the *Rh2* gene (*Rh2A* and *Rh2B*) are present and expressed in the retinae of *R. quaesita*, *G. favagineus* and *G. reticularis*, although only one gene, *Rh2A*, is present in *S. sathete*. Significantly, these pigments differ at site 122 with Glu in Rh2A and Gln in Rh2B; this substitution has been shown in many fish species to generate a SW shift (Yokoyama et al. 1999, Chinen et al. 2005, Wang et al. 2008, Yokoyama 2008). If the Rh2B opsin is the predominant cone pigment in the photoreceptors of *R. quaesita* and *G. reticularis*, whereas *G. favagineus* expresses both pigments in the same cone photoreceptors to give a mixture of pigments, then this would account for the spectral shifts from 494 nm in *R. quaesita* and *G. reticularis* to 501 nm in *G. favagineus*. Likewise, the presence of only a single pigment in *S. sathete*, but in this case, Rh2A rather than Rh2B, would account for the LW-shift to 509 nm. These changes in gene expression may reflect ontogenetic changes, with *Rh2A* expressed at earlier developmental stages in *R. quaesita* and *G. reticularis*; such developmental changes are not uncommon amongst teleosts (Carleton and Kocher 2001, Parry et al. 2005, Shand et al. 2008, Carleton 2009, Cottrill et al. 2009, Carleton et al. 2010). Unlike the tuning of rod pigments, the spectral shifts of the cone pigments of moray eels do not correlate with the different depth habitats of the four species. If the *Rh2* genes show differing expression levels at different stages of the life cycle, then the spectral sensitivity shifts may have a developmental dimension, although this cannot be the case for *S. sathete* where the *Rh2B* gene appears to have been lost. This species does however move from depths around 300 m to shallow brackish waters and may sometimes venture into rivers (Randall et al. 1990, Myers 1999), where it will encounter a very different photic environment of turbid estuarine water (Munz 1958); the LW-shift to 509 nm may be therefore an adaptation to these changing light habitats.

Origin of Rod Opsin Gene Duplication

As detailed above, the rod opsin gene duplication has been studied in detail in *Anguilla* but is also reported as present in white spotted conger eel, *C. myriaster*, but not in the European conger eel, *C. conger*. Phylogenetic analysis (Fig. 7) indicates that the *Rh1* gene duplication most likely occurred within the Anguilliformes; the freshwater form was subsequently lost in *H. bathybius* and the deep-sea form in moray eels. When the key substitution at site 292 responsible for SW spectral shifts is placed on to the tree (Fig. 7), the most parsimonious explanation is that the ancestral pigment possessed Ser292 (S292) and that Ser292Ala (S292A) substitutions have occurred twice within the lineage, in the duplicated copy that gave the freshwater variants in *Anguilla* sps and in one species of moray eel, *R. quaesita*. The ancestral pigment would therefore have been SW-shifted and would imply a deep-sea origin for all anguilliform teleosts. This is consistent with a recent study (Inoue et al. 2010) that indicated that even freshwater eels have a deep-sea origin. Site 83 is also important for spectral tuning at these wavelengths; the ancestral gene encoded Asp83 (D83) and its replacement by Asn83 (N83) in the deep-sea Rh1b forms of *A. anguilla* and *A. japonica* may contribute to the SW-shifts of these pigments. Asn83 is also found in *H. bathybius* so the combination of Asn83 and Ser292 would account for the SW shift in this deep-sea species to 477 nm (Hope et al. 1997).

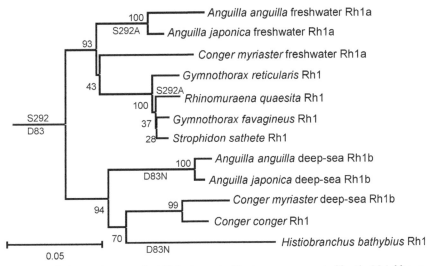

Figure 7. Phylogenetic trees of eel Rh1 pigments. The tree was generated by the Neighbour-joining method (Saitou and Nei 1987) using aligned amino acid sequences. The scale bar shows distance equivalent to 0.05 amino acid substitutions.

Visual Adaptations in other Species of Teleosts

As described above, deep-sea species of eel have evolved a rod-only retina. However, even in the shallower *Anguilla* spp., only two of the four classes of cone pigments, SWS2 and Rh2, have been retained and even then, the relative number of cone photoreceptors remains low compared to the number of rods. In other species like the moray eels, some of which live in shallow water, only a single class of cone visual pigment, *Rh2*, has been retained, although this gene appears to have undergone a duplication to give two spectrally distinct pigments which may provide a very limited form of dichromatic colour vision if expressed in separate cone cells. In contrast, many species of shallow dwelling fish with access to a wide visual spectrum usually express a full complement of a rod and four cone pigments, and this may be supplemented by the expression of duplicate copies of some cone opsins (Chinen et al. 2003, Parry et al. 2005, Weadick and Chang 2007). Deeper dwelling benthic species living at depth in excess of 1000m which have limited or no access to down-welling sunlight have largely dispensed with cone photoreceptors, retaining only rod photoreceptors (Partridge et al. 1989, Douglas and Partridge 1997). The ontogenetic changes seen in the eel retina parallel these evolutionary changes with the apparent loss of cone photoreceptors in migrants as they head into the deep ocean.

Many fish alter their complement of visual pigments during development and, as found for the eel, this can often be attributed to environmental changes during the life cycle, which result in altered light levels (Partridge and Cummings 1999, Douglas 2001, Bowmaker and Loew 2008). A notable example is the loss of ultraviolet-sensitive cones expressing the SWS1 pigment in a wide range of teleost fishes (Lyall 1957). In salmonids, this occurs during the first year of development (Bowmaker and Kunz 1987, Hawryshyn et al. 1989). However, the switching in the expression of rod pigments within intact photoreceptors as found in the eel has not been reported in any other species. In fact, an *Rh1* rod opsin gene duplication has only been reported in one other species of teleost, the pearl eye, *Scopelarchus analis*, a deep-sea fish belonging to the Order Aulopiformes. Both forms, *Rh1A* and *Rh1B*, are expressed in the retina and both show a SW-shift in λ_{max}, typical of a deep-sea fish (Pointer et al. 2007). However, only *Rh1A* is expressed in both young and more adult fish with the expression of *Rh1B* confined to adult fish, although it is not known whether the appearance of *Rh1B* expression involves a switch-over from *Rh1A* within existing rod photoreceptors as found in *Anguilla*, or requires newly added cells. Microspectrophotometry of the retina of the pearl-eye has shown that two other pigments are present with λ_{max} values at 505 and 444 nm. Both are present in the same outer segments of photoreceptors located in the main and accessory retinae, with the 505 nm pigment always located at the distal end and the 444 nm pigment at the proximal end of individual photoreceptors (Partridge et al. 1992), with an abrupt transition between the two. This is reminiscent of the distribution of the "freshwater" and "deep-sea" variants of the eel rod pigment during maturation, but whether this represents a similar switch-over in opsin gene expression from the 505 nm to the 444 nm pigment, or a fixed spatial separation of pigments along the outer segment, remains uncertain.

The Octavolateralis System

The octavolateralis system comprises the sensory structures that fishes use to detect sound and water motion: the ears and the lateral line, respectively (Popper and Higgs 2008). Although anatomically, developmentally and, to some extent, functionally distinct, both of these sensory systems use sensory hair cells to transduce mechanical energy into an electrical signal that can be processed by the nervous system.

Fishes possess a pair of internal ears that resemble the inner ears of terrestrial vertebrates (Retzius 1881). Although extensive anatomical diversity exists, inner ear structure in fishes is broadly similar: each comprises three orthogonally arranged semi-circular canals and three otolithic organs (utricle, saccule and lagena) (Fig. 8). Although the internal ears of fishes were described in the early part of the 17th century, their use as organs of hearing, rather than just balance, was not confirmed for almost 300 years (Parker 1918). In fact, the European eel was one of the first fish species for which positive evidence was obtained for the use of the ears in

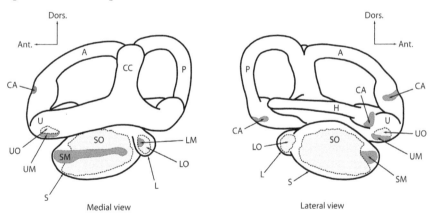

Medial view   Lateral view

Figure 8. Schematic diagram of the inner ear of the European eel *Anguilla anguilla* (Retzius 1881; redrawn from Popper 1979). A, anterior semicircular canal; Ant., anterior; CA, crista ampullaris; CC, crus commune; Dors., dorsal; H, horizontal semicircular canal; L, lagena; LM, lagenar macula; LO, lagenar otolith; P, posterior semicircular canal; S, saccule; SM, saccular macula; SO, saccular otolith; U, utricle; UM, utricular macula; UO, utricular otolith.

the detection of sounds, in this case sound-evoked action currents recorded from the 8th cranial nerve that innervates them (Piper 1906).

Sensory hair cells are found on a ridge (crista ampullaris) in a swelling (ampulla) at the base of each canal and in a sensory epithelium (macula) on the wall of each otolithic organ (Popper et al. 2003). On the apical surface of each sensory hair cell is a bundle of hair-like projections, comprising a single eccentrically placed cilium (kinocilium) and numerous (40–70 depending on species) stereocilia (Popper and Coombs 1980). Stereocilia are arranged in parallel rows that get progressively shorter in length with increasing distance from the kinocilium (Flock 1964). This asymmetric organisation reflects a strong directional sensitivity: mechanical deflection of the hair bundle in the direction of the kinocilium causes cation channels in the apical membrane to open and the cell to become depolarised; deflection in the opposite direction causes the ion channels to close and the cell to become hyperpolarised (Hudspeth and Corey 1977). These graded changes in membrane potential are transmitted via chemical synapses to afferent fibres of the 8th cranial nerve that project to the brain (Meredith et al. 1987).

The hair bundles of sensory cells in the ampullae and the otolithic organs are embedded within a gelatinous matrix (called the cupula and otolith membrane, respectively). In the case of the otolithic organs, the matrix contains a dense aggregation of calcareous material that forms a rock-like otolith (Popper and Coombs 1982). The sizes and shapes of the saccular, utricular and lagenar otoliths vary and are species-specific, but the functional significance of these variations is poorly understood (Paxton 2000).

The sensory structures of the inner ear function as accelerometers. A fish exposed to a sound wave moves in concert with the water surrounding it, because the density of its body is similar to that of the water and because it is physically small compared to the wavelength of the sound wave (Bretschneider et al. 2001). However, because the otolith is approximately 2–3 times denser than water (Hunt 1992), it accelerates more slowly than the rest of the fish. This inertial lag is transmitted to the macula via the otolith membrane and causes the hair bundles to bend. Moreover, because hair cells within the maculae are organised into discrete populations with opposing directional preferences, information can be obtained about the direction of the sound source, in addition to its intensity (Popper et al. 2003). The inner ear is also sensitive to accelerations caused by the movement of the fish itself and the action of gravity. When a fish moves its head, the inertial lag of the endolymph filling the semicircular canals deflects the cupula of the crista ampullaris and, consequently, the hair bundles of the underlying sensory epithelium (Popper et al. 2003). Similarly, the otolith's inertial mass provides a stimulus for the macular hair cells under the effects of gravity or head movements. Thus, the inner ear is primarily and evolutionarily a motion detector, and is an essential part of the vestibular system that controls posture and balance (Ladich and Popper 2004).

Although it is likely that all of the otolithic organs in teleosts have both a vestibular and an auditory role, the saccule is thought to be most specialised for hearing (Popper et al. 2003). The saccule shows the widest inter-specific variability in otolith size and shape, and in sensory hair cell orientation patterns, and in both cases, the presence of convergence across taxonomically diverse species implies a strong correlation (albeit poorly understood) between structure and ecological function (Popper and Schilt 2008). The saccular hair cell patterns found in *A. anguilla* (Mathiesen 1984) and the moray eel *Gymnothorax* sp. (Popper 1979) are of the 'alternating' type with anteriorly and posteriorly oriented cells, which is usually characteristic of species that also have additional structures to enhance hearing ability, such as specialised connections between the ear and the swimbladder or other gas-filled vesicles in the skull (Popper and Coombs 1982). However, neither of these types of auditory specialisation appears to be present in eels. The orientation patterns of hair cells in the three otolithic maculae of *A. anguilla* do not change throughout development, although minor changes in the shape of the saccular macula are observed (Mathiesen 1984).

The swimbladder is thought to significantly reduce hearing thresholds by transducing the (normally undetectable) pressure component of a sound wave into particle motion that can be detected by the otolithic organs (Sand and Enger 1973, Fay and Popper 1974, Finneran and Hastings 2000). Eels possess a swim-bladder, but it is located some distance from the ears (approximately 10 cm away in a 50 cm long fish, Jerkø et al. 1989); they also lack a Weberian apparatus, a chain of interconnected ossicles that connects the swimbladder with the ear in ostariophysian fishes such as the catfishes and minnows (Weber 1820, Poggendorf 1952). Nevertheless, there is evidence that the swimbladder in *A. anguilla* provides an auditory gain that extends the upper frequency limit of its hearing to at least 300–600 Hz (Diesselhorst 1938, Jerkø et al. 1989), although the impact on these processes of changes in the volume of the swimbladder which accompany vertical migrations is unknown. This upper limit of 300–600 Hz is modest however compared to the limit of 5 kHz measured in the goldfish *Carassius auratus* and the limit of 140 kHz ('ultrasound') measured in the blueback herring *Alosa aestivalis* (Higgs 2004). Taken together with a relatively high absolute threshold of sensitivity compared to other fishes (Jerkø et al. 1989) and the lack of any obvious anatomical specialisations designed to enhance auditory range, the eel is perhaps best described as a hearing 'generalist' with 'medium' hearing ability (Kastelein et al. 2008).

Many fish are able to detect sounds of very low frequency ('infrasound', i.e., <20Hz) down to below 1Hz using their otolithic organs (Karlsen 1992, Sand et al. 2001). Captive European eels exposed to an intense source of infrasound (11.8 Hz) display startle responses and other prolonged stress reactions, and the same frequency of sound can be used to create an "acoustic fence" that deflects migrating eels from their intended path (Sand et al. 2000). The hydrodynamic sounds that fish make when they swim are mostly of low frequency and the detection of infrasound may be adaptive for avoiding predators (Karlsen et al. 2004). It has also been suggested that some fish may use ultrasound for navigation and/or homing (Sand and Karlsen 1986).

Although the detection of low frequency sounds can be mediated by the otolithic organs alone, the mechanosensory lateral line is also sensitive to the mechanical disturbances produced by sound sources, in addition to non-auditory hydrodynamic stimuli (Dijkgraaf 1963). The sensory units of the lateral line (neuromasts) closely resemble those of the vestibular-auditory system. Each neuromast contains two populations of sensory hair cells with opposing directional selectivity (Flock and Wersäll 1962); the two populations are innervated separately by afferent nerve fibres, rendering the neuromasts bidirectionally sensitive (Mogdans et al. 2004). The hair bundles are embedded within a gelatinous mass called a cupula, and displacement of the cupula relative to the underlying sensory epithelium by fluid movement causes bending of the hair cells.

There are two types of neuromast in teleosts, superficial neuromasts which are located on the surface of the skin (sometimes occurring in pits or on protuberances), and canal neuromasts which are located in fluid-filled canals just beneath the skin surface (Dijkgraaf 1963). Superficial neuromasts are directly exposed to the water around the fish and respond in proportion to the velocity of water flow over the cupula; they respond best to very low frequency (20 Hz) unidirectional water currents (Braun et al. 2002). Lateral line canals are connected to the surrounding water by tubes/pores and canal neuromasts respond to water flow along the canals caused by differences in pressure at adjacent pores; they respond best to higher frequency water motion (up to 150 Hz). In addition to the detection of low frequency sounds, the lateral line is used to detect the hydrodynamic movements of other animals, the presence of objects in the vicinity of the fish and the detection of water current flow (rheotaxis) (Mogdans and Bleckmann 2001).

The functional overlap in frequency response of the otolithic organs and the lateral line covers the frequency range of approximately 50–150 Hz (Popper and Higgs 2009). However, the lateral line is only able to detect water vibrations at short distances (a few body lengths) where large local variations in water flow relative to the body of the fish exist (Popper et al. 2003). Thus, although both the lateral line and ears function to detect sound in the so-called 'near field', only the ears are sensitive to sound in the 'far field'.

The structure of the lateral line system varies widely between species. For example, fishes that live in still or slow moving water tend to have numerous superficial neuromasts and less elaborate canal systems, whereas fishes exposed to turbulent and/or fast moving water tend to have fewer superficial neuromasts and extended and/or highly branched canals (Dijkgraaf 1963, Bleckmann and Münz 1990). There is also extensive interspecific variation in neuromast shape and orientation, hair cell density, and canal/pore number, length and diameter, although the functional significance of many of the observed differences is unclear (Coombs and Braun 2003).

The canal system of *A. anguilla* has been described in detail by Zacchei and Tavolaro (1988). There are three clearly defined canals in the head region: the supra-orbital canal running above the eye, the infra-orbital canal running beneath the eye and the mandibular canal running along the edge of the lower jaw (Fig. 9). The supra-orbital canal joins with the infra-orbital canal behind the eye at an enlarged chamber called the post-orbital commissure.

Glass Eel (6 cm)

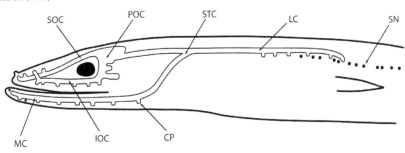

Silver Eel (78 cm)

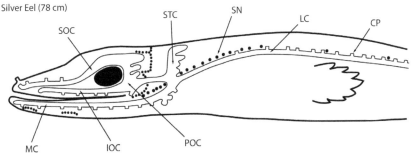

Figure 9. Schematic diagram of the mechanosensory lateral line systems of juvenile (glass eel) and adult (silver eel) stage European eels *Anguilla anguilla* (redrawn from Zacchei and Tavolaro 1988). CP, canal pore; IOC, infra-orbital canal; LC, lateral canal; MC, mandibular canal; POC, post-orbital commissure; SN, superficial neuromast; SOC, supra-orbital canal; STC, supra-temporal commissure. Note the elaboration of the canal system and the increased number of superficial neuromasts in the silver eel.

This combined canal joins with the mandibular canal (and the canals from the other side of the body) at the supra-temporal commissure. The lateral canal arises from the supra-temporal commissure and runs along the flank. As the eel matures, the supra-orbital canal, post-orbital commissure and supra-temporal commissure all become wider and more elaborate, and the lateral canal extends all the way along the body. Superficial neuromasts (pit organs) are also found in rows on the head and trunk and these become more numerous during development. These ontogenetic changes in lateral line morphology are thought to reflect the increase in motility of the mature silver eels compared to the more sedentary yellow eel stage.

The sensory hair cells of *A. japonica* and the white spotted eel *C. myriaster* each possess a single kinocilium and 40–60 stereocilia (Yamada and Hama 1972, Hama 1978), whereas another sea eel *Lyncozymba nystromi* possesses fewer (20–40) stereocilia per hair bundle (Hama 1965). Differences in hair cell morphology may underlie differences in sensitivity and/or frequency response characteristics of the neuromast (Mogdans et al. 2004). Neighbouring hair cells in the neuromast tend to have opposite directional preferences. Neuromasts in the lateral canals are arranged so as to respond to water motion along the canal in a rostral or caudal direction; superficial neuromasts arranged in a row dorsal to the lateral canal are oriented to respond to water motion along a dorso-ventral axis (Hama 1978).

Afferent lateral line nerve fibres send signals from the sensory hair cells to the brain (crista cerebellaris). Two morphological subtypes of afferent fibre have been identified in *A. anguilla* (Alnæs 1973), which likely correspond to the two different physiological subtypes identified in other fishes, one of which innervates superficial neuromasts and the other innervates the canal neuromasts (Mogdans et al. 2004). In addition, octavolateralis efferent neurons with cell bodies in the brainstem synapse directly on to hair cells (Hama 1965) and modulate both the spontaneous and mechanically evoked discharges of the afferent fibres in response to a variety of other sensory inputs (Tricas and Highstein 1991), emphasising the truly multimodal nature of the octavolateralis system (Braun et al. 2002).

Chemoreception

Teleost fishes typically possess five types of chemosensory systems as follows. Olfaction or smell is important for social interactions, homing, alerting to the presence of food and avoiding predators. The sense of smell is mediated through an olfactory organ found within the paired nostrils. Gustation or taste is used in feeding and the localisation, acceptance or rejection of food. The gustatory system comprises taste buds or aggregations of sensory receptors situated often on raised papillae within the oral cavity, pharynx or over the head and body. Isolated chemosensory receptors are used for feeding and predator avoidance. These solitary chemoreceptors exist as small, encapsulated nerve endings over the epidermis or, more commonly, on the pectoral fin rays. A common chemical sense from free nerve endings is important for irritant detection (Finger 1997). The common chemical sense is closely associated with the somatosensory system. Chemoreceptors are involved in cardiovascular responses to changes in oxygen and carbon dioxide levels. Branchial and extrabranchial chemoreceptors play a role in controlling cardiorespiratory responses to changes in oxygen and carbon dioxide levels in addition to pH. These receptors are sensitive to dissolved gases and are innervated by cranial nerves V, VII, IX and X. Only the olfactory and gustatory systems will be discussed in detail.

Olfaction: Sensitivity and Odour Discrimination

The olfactory epithelium (rosette) in eels lies within a water-filled chamber that sits between anterior tubular inlet nares and slit-like posterior outlet openings (Fishelson 1995). The olfactory rosettes are round in shape in members of the *Siderea* and *Echidna* genera, whereas they are elongated in *A. anguilla* (Hansen and Zielinski 2005) (Fig. 10). Between 16 (*Siderea* sp.) and 168 (*Gymnothorax* sp.) lamellae, divided by a central raphe, are present (Fishelson 1995, Chakrabarti and Guin 2011). Anguilliformes lack both the ethmoid and lacrimal nasal sacs used by sedentary, bottom-dwelling species to regulate the flow of water over the olfactory epithelium (Burne 1909, Kapoor and Ojha 1973). Eels are considered therefore to be isosmates, relying on ciliary beating for distributing odorants over the olfactory epithelium (Doving et al. 1977). Each lamella is adorned with three types of ciliated sensory receptors, each with significant differences in the number, density and coverage of stereocilia, which is thought to be a good predictor of the relative importance of olfaction (compared to vision), at least in moray eels (Fishelson 1995).

Four major classes of chemicals (amino acids, sex steroids, bile acids/salts and prostaglandins) have been identified as specific olfactory stimuli in teleosts. Most studies have used the electro-olfactogram (EOG) technique to assess the sensitivity thresholds to the amino acids L-glycine, L-alanine, L-valine, L-leucine, L-asparagine, L-glutamine and L-methionine; in *A. anguilla* sensitivities have been found to be as low as 10^{-9} M (Silver 1982, Crnjar et al. 1992). With the exception of D-asparagine and D-alanine, most other non-protein amino acids were found to be strong attractants to glass eels (Sola and Tongiorgi 1998), although mixed responses (attraction, neutral behavior or repulsion) to a variety of amino acids mixed with different media and at different concentrations may also occur (Sola et al. 1993). The same level of variability was found by Sorensen (1986) after presentation of

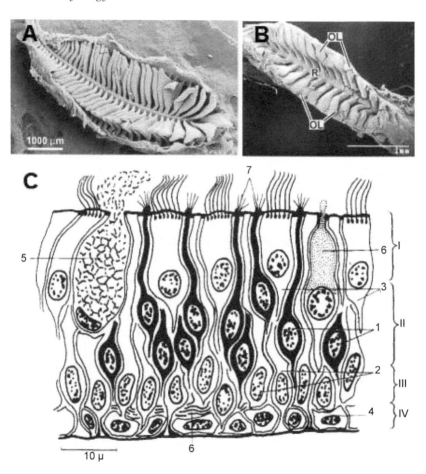

Figure 10. Scanning electron micrographs of the olfactory rosette of the European eel, *Anguilla anguilla* (A) and the freshwater lesser spiny eel, *Macrognathus aculeatus* (B) showing its elongated form and large number of thin lamellae (OL) emanating from the central raphe (R) within the narrow olfactory cavity. (C) Schematic diagram of the olfactory epithelium in the eel, *Anguilla anguilla*. 1, receptor cells; 2, supporting cells; 3, ciliated ccells; 4, basal cells; 5, goblet cells; 6, club-shaped secretory cells; 7, olfactory knob with cilia. Reprinted with permission from Hansen and Zielinski (2005) (A), Chakrabarti and Guin (2011) (B) and Hara (1975) (C).

more natural substances to *A. anguilla* elvers (Fig. 11). Huertas et al. (2007) have recently shown that the odour of bile from *A. anguilla* is different for the two sexes and changes with sexual maturity, suggesting a possible role for bile salts as sex pheromones. Given the large distances travelled in order to spawn, a high olfactory sensitivity to bile salts may aid in finding reproductive partners (Huertas et al. 2010). As part of their catadromous life history, yellow phase *A. anguilla* migrate up river where they become sexually mature. Silver phase eels then migrate from freshwater to the

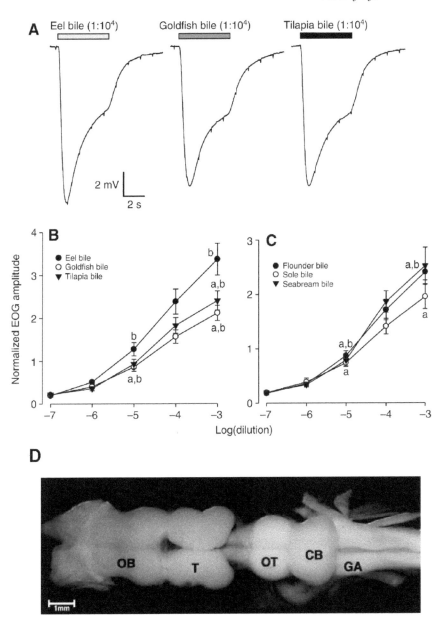

Figure 11. Olfactory sensitivity to conspecifc and heterospecific bile fluid in the eel, *Anguilla anguilla* (A) Semi-logarithmic plot of pooled normalized electro-olfactogram (EOG) trace amplitudes recorded in immature male eels in response to dilution of bile fluid from eel, goldfish and tilapia. Data are mean ± s.e.m. (N=6). (B) Dorsal view of the brain of the deep-sea eel, *Histobranchus bathybius,* showing the relatively large size of the olfactory bulbs (OB) and gustatory region (GA). CB, cerebellum; OT, optic tectum; T, telencephalon. Redrawn (A) from Huertas et al. (2010) and (B) reprinted with permission from Wagner (2001b).

Sargasso Sea to spawn. Both yellow and silver phase eels use selective tidal stream transport to migrate in and out of freshwater (McCleave and Kleckner 1982, Barbin 1998, Barbin et al. 1998, DeBose and Nevitt 2008). Olfaction seems to be important for the discrimination of the appropriate tide for transport and location of a home site for these eels but is not the only orientational mechanism used in estuaries. Mechanisms used to detect rates of change of water mass characteristics, including salinity (Hain 1975) and rheotaxis (Carton and Montgomery 2003), may also be important for guidance of estuarine migrations (Barbin 1998), but olfaction is still considered to play an important role. Strong attractive behavior to bile salts and taurine is exhibited by the glass eels of *A. anguilla* (Sola and Tosi 1993), providing further evidence that these fish employ chemical cues emanating from conspecifics to recognize siblings, to locate spawning grounds (Foster 1985), and to orientate to native streams. A range of other substances (e.g., 2-methyl-3-methoxypyrazine and 1-ethyl-2,2,26-trimethylcyclohexanol) have also been found to act as potent chemoattractants (kairomones) at very low concentrations (10^{-13} mg l^{-1}, Sola (1995)) to European eels during their migration.

Primary and Secondary Olfactory Input to the CNS

The relative importance of olfaction in eels can also be assessed quantitatively by the size of primary and/or secondary inputs to the CNS. There are no studies to date that have examined the surface area of the olfactory epithelium in eels (although the number of lamellae have been studied, see above). The size of the olfactory bulbs has been assessed however in a number of deep-sea eels (Wagner 2001a, Wagner 2001b), where all species examined (*Eurypharynx pelecanoides*, *S. beani*, *Avocettina infans*, *Nemichthys curvirostris* and *Cyema atrum*) possess above average olfactory bulbs, often combined with a larger optic tectum (Fig. 11D). This suggests that in the mesopelagic zone, deep-sea eels rely heavily on both vision and olfaction for survival. The most convincing evidence of olfactory specialization however is found in the synaphobranchid eels, *H. bathybius* and *Synaphobranchus kaupi*, two deep-sea demersal species with olfactory bulbs that exceed the average by more than three times (Wagner 2001a, Wagner 2001b, Wagner 2002). These two species are large mobile piscivorous scavengers that appear quickly and in high abundance at food falls (Jannasch 1978, Merrett and Domanski 1985). They are considered to be 'swimming noses', a feature that enables them to effectively locate their prey in an otherwise depauperate sensory environment.

In *A. rostrata* and *A. japonica*, the terminal nerve system (*nervus terminalis*) is also considered to mediate reproductive responses to sex pheromones. This system comprises a group of ganglion cells situated at the rostral and caudal extremities of the olfactory bulbs, which possess fibres projecting from the nasal olfactory epithelium and terminate in various regions of the CNS including the retina (Von Bartheld 2004). These fibres are immunoreactive to luteinising hormone releasing hormone (LHRH) along with a host of neurotransmitters and neuropeptides and might play a role in integrating or coordinating sex-related pheromonal or visual activity (Nozaki et al. 1985, Grober et al. 1987).

Gustation

Taste buds are aggregations of sensory receptors that are often found as raised papillae within the mouth and oropharyngeal cavity. Unlike olfactory neurons, taste receptors lack an axon, and connect to afferent nerve fibers via chemical synapses. The receptor cells of taste buds are considered secondary sensory cells (Reutter and Witt 1993) and are always stimulated by nutritionally important substances. Although there is considerable variation in the morphology of taste buds, four types of cells have been characterised in teleosts: dark cells (type I), light cells (type II), basal cells and stem cells (Reutter and Witt 1993, Finger 1997). The taste buds over the epithelial region within the mouth and oropharyngeal cavity of *A. anguilla* are comprised of cells with large and small microvilli (Reutter and Hansen 2005), which allows the detection of chemical substances in the immediate environment and thereby the ability to "test" any potential food source before ingestion. At the level of the scanning electron microscope, it appears that the larger microvilli receptors correspond to the "light" cells observed using transmission electron microscopy (and used by many authors to differentiate sensory cell types) whereas the smaller microvilli receptors belong to the "dark" cells (Pevzner 1978). However, more work is needed to ascertain morphologically the number of cell types in eel taste buds and whether there are any functional differences between them.

Taste buds on bony fishes are used to assess the palatability of food and may be "tuned" to specific prey items or nutritional needs or both. In teleost fishes, taste buds are sensitive to different amino acids such as arginine, proline, and alanine (Caprio et al. 1993) and various organic acids, nucleotides and bile salts (Hara 1994). Electrophysiological work specifically on gustatory responses of the Japanese eel (measured at the palatine nerve) has revealed a high sensitivity to a range of amino acids (L-Arginine, L-Glycine, L-Alanine, L-Proline, L-Lysine and L-Serine with thresholds between 10^{-8} and 10^{-9}) and carboxylic acids (with thresholds between 10^{-4} and 10^{-7}) (Yoshii et al. 1979). It has also been shown that the binding of cations to the gustatory membrane plays an important role in holding the stimulant receptor complex in an active state, thereby increasing the affinity of the receptor site for water-borne stimulants (Yoshii and Kurihara 1983).

In an aquaculture setting, elvers and fingerlings of *A. anguilla* appear to use gustatory rather than visual cues to ingest food, a bias which is reversed in larger eels (Knights 1983). The importance of gustation can also be assessed quantitatively by the size of the terminal fields within the CNS. Volumetric measurements of the relative size of the sensory regions of the brain in the deep-sea bobtail snipe eel, *Cyema atrum,* and the pelican eel, *Eurypharynx pelecanoides*, reveals that both species are generalist feeders with an above average relative size in the sensory brain areas that receive gustatory inputs (Wagner 2005).

Summary

It is evident from the foregoing that eels have a range of complex and highly specialised senses, with sight, hearing, touch, smell and taste all represented. There is however considerable degree of ontogenetic and interspecific variation.

The visual system has been the most extensively studied sensory system in the eel, with many species possessing either a rod-only retina or, at best, a retina with a very limited complement of cones. A feature that appears to be unique to eels is found in members of the genus *Anguilla*. In these species, a duplication of the *Rh1* rod visual pigment gene has occurred and the sequential expression of the duplicated copies in the retina during ontogeny and migration is responsible for changes in the spectral sensitivity of the eye that can be directly correlated with the visual environment that pertains to the different stages of the life cycle. In other species, multiple visual pigment genes may again be present but whether they show differential expression during ontogeny is presently unknown. Other adaptations to the visual environment are seen amongst the moray eels, where changes in spectral sensitivities of the eye can be related to specific changes in the amino acid composition of the visual pigments and to their differential expression across closely related species.

As part of their life cycle, many eels migrate over large distances so an ability to navigate is critically important. For trans-oceanic migrants such as eels, the earth's magnetic field is probably the only geophysical signal that provides consistent directional and positional information at all times of the day and in all habitats. The ability to detect and use the geomagnetic field for navigation is thought to underlie migration in many vertebrates and would certainly appear to be the most likely mechanism in eels. However, although there is some evidence that eels can orientate in relation to magnetic north, the precise mechanism used by eels to detect the earth's magnetic field is unknown.

The detection of sound and water movement is the role of the octavolateralis system. Both these senses utilise sensory hair cells to convert mechanical energy into an electrical signal that can be processed by the nervous system. Internal ears are used to detect sounds with bundles of sensory hairs in the ampullae and the otolithic organs embedded in a gelatinous matrix. This matrix in otoliths contains a dense aggregation of calcareous material that forms a rock-like structure. The otoliths of the inner ear also function as an accelerometer and motion detector. The detection of very low frequency sound is thought to be important for navigation and/or homing and this is largely achieved by the mechanosensory lateral line that is sensitive to the mechanical disturbances produced by sound sources, in addition to non-auditory hydrodynamic stimuli.

Teleost fishes typically possess five types of chemosensory systems: olfaction, gustation or taste, chemoreception for feeding and predator avoidance, free nerve endings for the detection of irritants, and a somatosensory system involved in cardiovascular responses to changes in oxygen and carbon dioxide levels. Olfactory rosettes are found in water-filled chambers in eels that sit between anterior tubular inlet nares and slit-like posterior outlet openings. The shape and size of these rosettes varies amongst eels. Anguilliformes lack the nasal sacs used by sedentary, bottom-dwelling species and are considered to be isosmates. Four major classes of chemicals (amino acids, sex steroids, bile acids/salts and prostaglandins) have been identified as specific olfactory stimuli in teleosts; in *A. anguilla*, sensitivities to a number of L-amino acids is as low as 10^{-9} M, and most D-amino acids have been shown to act as strong attractants to glass eels. Evidence from *A. anguilla* indicates a possible role for bile salts as sex pheromones and it is thought that a high olfactory sensitivity to bile salts may aid in finding reproductive partners, which may be especially important given the large distances travelled in order to spawn. Olfaction would also appear to be important for the discrimination of the appropriate tide for transport and location of a home site for these eels. Taste buds that are important in gustation are found within the mouth and oropharyngeal cavity. In *A. anguilla*, these cells possess either large and small microvilli but how cell morphology correlates with function has yet to be determined.

In conclusion, there is a remarkable variety in eel sensory systems and this has contributed not only to our understanding of the regulation and evolution of sensory mechanisms but also to the realisation of the critically important role of sensory adaptation in the evolution of teleost fishes and vertebrates in general.

References

Aarestrup, K., F. Økland, M.M. Hansen, D. Righton, P. Gargan, M. Castonguay, L. Bernatchez, P. Howey, H. Sparholt, M.I. Pedersen and R.S. McKinley. 2009. Oceanic Spawning Migration of the European Eel (*Anguilla anguilla*). Science 325: 1660.

Ali, M.A. and M. Anctil. 1976. Retinas of Fishes. Springer-Verlag, Berlin, Germany.

Ali, M.A., M. Anctil and H.M. Mohideen. 1968. Structure rétinienne et la vascularisation intraoculaire chez qulque poisons marins de la region de Gaspé. Can J. Zool. 46: 729–745.

Alnæs, E. 1973. Two types of lateral line afferents in the eel (*Anguilla anguilla*). Acta Physiol. Scand. 87: 535–548.

Aoyama, J., K. Hissmann, T. Yoshinaga, S. Sasai, T. Uto and H. Ueda. 1999. Swimming depth of migrating silver eels *Anguilla japonica* released at seamounts of the West Mariana Ridge, their estimated spawning sites. Mar. Ecol. Prog. Ser. 186: 265–269.

Archer, S. and J. Hirano. 1996. Absorbance spectra and molecular structure of the blue-sensitive rod visual pigment in the conger eel (*Conger conger*). Proc. Biol. Sci. 263: 761–767.

Archer, S., A. Hope and J.C. Partridge. 1995. The molecular basis for the green-blue sensitivity shift in the rod visual pigments of the European eel. Proc. R. Soc. Lond. B. Biol. Sci. 262: 289–95.

Aroua, S., M. Schmitz, S. Baloche, B. Vidal, K. Rousseau and S. Dufour. 2005. Endocrine evidence that silvering, a secondary metamorphosis in the eel, is a pubertal rather than a metamorphic event. Neuroendocrinology 82: 221–232.

Barbin, G.P. 1998. The role of olfaction in homing and estuarine migratory behavior of yellow-phase American eels. Can J. Fish Aquat. Sci. 55: 564–575.

Barbin, G.P., S.J. Parker and J.D. McCleave. 1998. Olfactory clues play a critical role in the estuarine migration of silver-phase American eels. Environ Biol Fishes 53: 283–291.

Beatty, D.D. 1975. Visual pigments of the american eel *Anguilla rostrata*. Vision Res. 15: 771–776.

Berry, L., D. Brookes and B. Walker. 1972. Problem of migration of European Eel (*Anguilla anguilla*). Science Progress 60: 465–485.

Bleckmann, H. and H. Münz. 1990. Physiology of lateral line mechanoreceptors in a teleost with highly branched, multiple lateral lines. Brain Behav Evolut. 35: 240–250.

Bowmaker, J.K. 2008. Evolution of vertebrate visual pigments. Vision Res. 48: 2022–2041.

Bowmaker, J.K. and D.M. Hunt. 2006. Evolution of vertebrate visual pigments. Curr. Biol. 16: R484–9.

Bowmaker, J.K. and Y.W. Kunz. 1987. Ultraviolet receptors, tetrachromatic colour vision and retinal mosaics in the brown trout (*Salmo trutta*): age-dependent changes. Vision Res. 27: 2101–2108.

Bowmaker, J.K. and E.R. Loew. 2008. Vision in Fish. *In*: R.H. Masland and T.D. Albright [eds]. The Senses: a Comprehensive Reference. 1 Vision. Elsevier, Oxford, UK. pp. 53–76.

Bowmaker, J.K., M. Semo, D.M. Hunt and G. Jeffery. 2008. Eel visual pigments revisited: The fate of retinal cones during metamorphosis. Vis Neurosci. 1–7.

Bozzano, A. 2003. Vision in the rufus snake eel, *Ophichthus rufus*: adaptive mechanisms for a burrowing life-style. Marine Biology. 143: 161–174.

Braekevelt, C.R. 1984. Retinal fine structure in the European eel *Anguilla anguilla*. Anat. Anz. Jena 157: 233–243.

Braekevelt, C.R. 1985. Retinal fine structure in the European eel *Anguilla anguilla*. IV. Photoreceptors of the yellow eel stage. Anat. Anz. 158: 23–32.

Braekevelt, C.R. 1988a. Retinal fine structure in the European eel *Anguilla anguilla*. VI. Photoreceptors of the sexually immature silver eel stage. Anat. Anz. 66: 23–31.

Braekevelt, C.R. 1988b. Retinal fine structure in the European eel *Anguilla anguilla*. VII. Pigment epithelium of the sexually mature silver eel stage. Anat. Anz. 166: 33–41.

Braekevelt, C.R. 1988c. Retinal fine structure in the European eel *Anguilla anguilla*. VIII. Photoreceptors of the sexually mature silver eel stage. Anat. Anz. 167: 1–10.

Braun, C.B., S. Coombs and R.R. Fay. 2002. What Is the nature of multisensory interaction between octavolateralis sub-systems? Brain Behav Evol. 59: 162–176.

Bretschneider, F., A.V. van den Berg and R.C. Peters. 2001. Mechanoreception: hearing and lateral line. *In*: B.G. Kapoor and T.J. Hara [eds]. Sensory Biology of Jawed Fishes: New Insights. Science Publisher, Inc., Enfield (NH), USA. pp. 215–253.

Bridges, C.D.B. 1972. The rhodopsin-porphyropsin visual system. *In*: H.J.A. Dartnall [ed]. Photochemistry of Vision: Handbook of Sensory Physiology. Springer-Verlag, Berlin, Germany. pp. 471–480.

Brown, H.R. and O.B. Ilyinsky. 1978. The ampullae of Lorenzini in the magnetic field. J. Comp. Physiol. A. 126: 333–341.

Burne, R.H. 1909. The anatomy of the olfactory organ of teleostean fishes. Proc. Zool. Soc. Lond. 2: 610–663.

Caprio, J., J.G. Brand, J.H. Teeter, T. Valentincic, D.L. Kalinoski, J. Kohbara, T. Kumazawa and S. Wegert. 1993. The taste system of the channel catfish: From biophysics to behavior. Trends Neurosci. 16: 192–197.

Carleton, K. 2009. Cichlid fish visual systems: mechanisms of spectral tuning. Integr. Zool. 4: 75–86.

Carleton, K.L., C.M. Hofmann, C. Klisz, Z. Patel, L.M. Chircus, L.H. Simenauer, N. Soodoo, R.C. Albertson and J.R. Ser. 2010. Genetic basis of differential opsin gene expression in cichlid fishes. J. Evol. Biol. 23: 840–53.

Carleton, K.L. and T.D. Kocher. 2001. Cone opsin genes of african cichlid fishes: tuning spectral sensitivity by differential gene expression. Mol. Biol. Evol. 18: 1540–50.

Carlisle, D.B. and E.J. Denton. 1959. On the metamorphosis of the visual pigments of *Anguilla anguilla*. J. Mar. Biol. Ass. UK 38: 97–102.

Carton, A.G. and J.C. Montgomery. 2003. Evidence of a rheotactic component in the odour search behavior of freshwater eels. J. Fish Biol. 62: 501–516.

Chakrabarti, P. and S. Guin. 2011. Surface architecture and histoarchitecture of the olfactory rosette of freshwater lesser spiny eel, *Macrognathus aculeatus* (Boch). Arch. Pol. Fish 19: 297–303.

Chinen, A., T. Hamaoka, Y. Yamada and S. Kawamura. 2003. Gene duplication and spectral diversification of cone visual pigments of zebrafish. Genetics 163: 663–75.

Chinen, A., Y. Matsumoto and S. Kawamura. 2005. Reconstitution of Ancestral Green Visual Pigments of Zebrafish and Molecular Mechanism of their Spectral Differentiation. Mol. Biol. Evol. 22: 1001–1010.

Chow, S., K. Kurogi, N. Mochioka, S. Kaji, O. Okazaki and K. Tsukamoto. 2009. Discovery of mature freshwater eels in the open ocean Fisheries Science 75: 257–259.

Collin, S.P. and H.B. Collin. 2001. The fish cornea: adaptations for different aquatic environments. *In*: B.G. Kapoor and T.J. Hara [eds]. Sensory Biology of Jawed Fishes—New Insights. Science Publishers Inc., Enfield, (NH), USA. pp. 57–96.

Collin, S.P. and J.C. Partridge. 1996. Retinal specialisations in the eyes of deepsea teleosts. J. Fish Biol. Suppl. A. 49: 157–174.

Coombs, S. and C.B. Braun. 2003. Information processing by the lateral line system. *In*: S.P. Collin and N.J. Marshall [eds]. Sensory Processing in the Aquatic Environment. Springer-Verlag, New York, USA. pp. 122–138.

Cottrill, P.B., W.L. Davies, M. Semo, J.K. Bowmaker, D.M. Hunt and G. Jeffery. 2009. Developmental dynamics of cone photoreceptors in the eel. BMC Dev. Biol. 9: 71.

Crnjar, R., G. Scalera, A. Bigiani, I.T. Barbarossa, P.C. Magherini and P. Pietra. 1992. Olfactory sensitivity to amino acids in the juvenile stages of the European eel *Anguilla anguilla* (L.). J. Fish Biol. 40: 567–576.

Damjanovic, I., A.L. Byzov, J.K. Bowmaker, Z. Gacic, I.A. Utina, E.M. Maximova, B. Mickovic and R.K. Andjus. 2005. Photopic Vision in Eels. Evidences of Color Discrimination. Ann. N.Y. Acad. Sci. 1048: 69–84.

Davies, W.I., S.E. Wilkie, J.A. Cowing, M.W. Hankins and D.M. Hunt. 2012. Anion sensitivity and spectral tuning of middle- and long-wavelength-sensitive (MWS/LWS) visual pigments. Cell Mol. Life Sci. 69: 2455–2464.

Davies, W.L., L.S. Carvalho, B.H. Tay, S. Brenner, D.M. Hunt and B. Venkatesh. 2009. Into the blue: gene duplication and loss underlie color vision adaptations in a deep-sea chimaera, the elephant shark *Callorhinchus milii*. Genome Res. 19: 415–426.

DeBose, J.L. and G.A. Nevitt. 2008. The use of odors at different spatial scales: coupling birds with fish. J. Chem Ecol. 34: 867–881.

Diebel, C.E., R. Proksch, C.R. Green, P. Neilson and M.M. Walker. 2000. Magnetite defines a vertebrate magnetoreceptor. Nature 406: 299–302.

Diesselhorst, G. 1938. Hörversuche an Fischen ohne Weberschen Apparat. Z. Vergl Physiol. 25: 748–783.

Dijkgraaf, S. 1963. Functioning and significance of lateral line organs. Biol. Rev. 38: 51–105.

Dittman, A. and T. Quinn. 1996. Homing in Pacific salmon: mechanisms and ecological basis. J. Exp. Biol. 199: 83–91.

Douglas, R.H. 2001. The ecology of teleost fish visual pigments: a good example of sensory adaptation to the environment? *In*: F.G. Barth and A. Schmid [eds]. Ecology of Sensing. Springer-Verlag, Berlin, Germany. pp. 215–235.

Douglas, R.H., D.M. Hunt and J.K. Bowmaker. 2003. Spectral Sensitivity Tuning in the Deep-Sea. *In*: S.P. Collin and N.J. Marshall. [eds]. Sensory Processing in Aquatic Environments. Springer-Verlag, New York, USA. pp. 323–342

Douglas, R.H. and J.C. Partridge. 1997. On the visual pigments of deep-sea fish. J. Fish Biol. 50: 68–85.

Douglas, R.H., J.C. Partridge and A.J. Hope. 1995. Visual and lenticular pigments in the eyes of demersal deep-sea fishes. J. Comp. Physiol. A. 177: 111–122.

Doving, K.B., M. Dubois-Dauphin, A. Holley and F. Jourdan. 1977. Functional anatomy of the olfactory organ of fish and the ciliary mechanism of water transport. Acta Zoologica 58: 245–255.

Duke-Elder, S. 1958. The eye in evolution. *In*: S. Duke-Elder [eds]. System of Ophthalmology. Henry Kimpton, London, UK. pp. 273–332

Fasick, J.I., T.W. Cronin, D.M. Hunt and P.R. Robinson. 1998. The visual pigments of the bottlenose dolphin (*Tursiops truncatus*). Vis. Neurosci. 15: 643–651.

Fay, R.R. and A.N. Popper. 1974. Acoustic stimulation of ear of goldfish (*Carassius auratus*). J. Exp. Biol. 61: 243–260.

Finger, T.E. 1997. Evolution of taste and solitary chemoreceptor cell systems. Brain Behav. Evol. 50: 234–43.

Finneran, J.J. and M.C. Hastings. 2000. A mathematical analysis of the peripheral auditory system mechanics in the goldfish (*Carassius auratus*). J. Acoust Soc. Am. 108: 1308–1321.

Fishelson, L. 1995. Comparative morphology and cytology of the olfactory organs in Moray eels with remarks on their foraging behavior. Anat. Rec. 243: 403–12.

Fitzgibbon, J., A. Hope, S.J. Slobodyanyuk, J. Bellingham, J.K. Bowmaker and D.M. Hunt. 1995. The rhodopsin-encoding gene of bony fish lacks introns. Gene 164: 273–277.

Flock, Å. 1964. Structure of macula utriculi with special reference to directional interplay of sensory responses as revealed by morphological polarization. J. Cell Biol. 22: 413–431.

Flock, Å. and J. Wersäll. 1962. A study of orientation of sensory hairs of receptor cells in lateral line organ of fish, with special reference to function of receptors. J. Cell Biol. 15: 19–27.

Foster, N.R. 1985. Lake trout reproductive behavior: influence of chemosensory cues from young-of-the-the-year by-products. Trans. Amer. Fish Soc. 114: 794–803.

Franz, V. 1910. Die japanischen Knochenfische der Sammlungen Haberer und Doflein. Abhandl Bayer Akad Wiss Suppl. 4: 1–132.

Garten, S. 1907. Die Veränderungen der Netzhaut durch Licht. *In*: A. Graefe and T. Saemisch [eds]. Graefe-Saemisch Handbuch der gesam. Augenheilkunde Leipzig, Germany. Vol. 2. pp. 1–30.

Gegear, R.J., L.E. Foley, A. Casselman and S.M. Reppert. 2010. Animal cryptochromes mediate magnetoreception by an unconventional photochemical mechanism. Nature 463: 804–807.

Gordon, J., R.M. Shapley and E. Kaplan. 1978. The Eel Retina. Receptor Classes and Spectral Mechanisms. J. Gen. Physiol. 71: 123–138.

Grober, M.S., A.H. Bass, G. Burd, M.A. Marchaterre, N. Segil, K. Scholz and T. Hodgson. 1987. The nervus terminalis ganglion in *Anguilla rostrata*: an immunocytochemical and HRP histochemical analysis. Brain Res. 436: 148–152.

Hain, J.H.W. 1975. The behavior of migratory eels, *Anguilla rostrata*, in response to current, salinity, and lunar period. Helgoländer wiss Meeresunters 27: 211–233.

Hama, K. 1965. Some observations on fine structure of lateral line organ of Japanese sea eel *Lyncozymba nystromi*. J. Cell Biol. 24: 193–210.

Hama, K. 1978. A study of the fine structure of the pit organ of the common Japanese sea eel *Conger myriaster*. Cell Tissue Res. 189: 375–388.

Hansen, A. and B.S. Zielinski. 2005. Diversity in the olfactory epithelium of bony fishes: development, lamellar arrangement, sensory neuron cell types and transduction components. J. Neurocytol. 34: 183–208.

Hanson, M. and M.M. Walker. 1987. Magnetic particles in European eel (*Anguilla anguilla*) and carp (*Cyprinus carpio*). Magnetic susceptibility and remanence. J. Magn. Magn. Mater. 66: 1–7.

Hanson, M. and H. Westerberg. 1986. Occurrence and properties of magnetic material in European eel (*Anguilla anguilla* L.). J. Magn. Magn. Mater. 54–57: 1467–1468.

Hanyu, I. 1959. On the falciform process vitreal vessels and other related structures of the teleost eye. I. Various types and their interrelationships. Bull Jap. Soc. Sci. Fish 25: 595–613.

Hara, T.J. 1975. Olfaction in fish. Prog. Neurobiol. 5: 271–335.

Hara, T.J. 1994. Olfaction and gustation in fish: an overview. Acta Physiol. 152: 207–217.

Harosi, F.I. 1994. An analysis of two spectral properties of vertebrate visual pigments. Vision Res. 34: 1359–67.

Hawryshyn, C.W., M.G. Arnold, D.J. Chaisson and P.C. Martin. 1989. The ontogeny of ultraviolet photosensitivity in rainbow trout (*Salmo gairdneri*). Vis Neurosci. 2: 247–54.

Hein, S.A.A. 1913. Over oogleden en fornices conjunctivae bij teleostomi. Tijds D. Nederl. Dierk Vereen Ser. 2 Dl. 12: 238–280.

Hewss, M., R.R. Melzer and U. Smola. 1998. The photoreceptors of *Muraena helena* and *Ariosoma balearicum*—a comparison of multiple bank retina of anguilliform eels (Teleostei). Zoologischer Anzeiger 237: 127–137.

Higgs, D.M. 2004. Neuroethology and sensory ecology of teleost ultrasound detection. In: G. von der Emde, J. Mogdans and B.G. Kapoor [eds]. The Senses of Fish: Adaptations for the Reception of Natural Stimuli. Kluwer Academic Publishers, Boston, USA. pp. 173–188.

Hope, A.J., J.C. Partridge, K.S. Dulai and D.M. Hunt. 1997. Mechanisms of wavelength tuning in the rod opsins of deep-sea fishes. Proc. R. Soc. Lond. B. Biol. Sci. 264: 155–63.

Hudspeth, A.J. and D.P. Corey. 1977. Sensitivity, polarity, and conductance change in response of vertebrate hair cells to controlled mechanical stimuli. Proc. Natl. Acad. Sci. USA 74: 2407–2411.

Huertas, M., L. Hagey, A.F. Hofmann, J. Cerda, A.V.M. Canário and P.C. Hubbard. 2010. Olfactory sensitivity to bile fluid and bile salts in the European eel (*Anguilla anguilla*), goldfish (*Carassius auratus*) and Mozambique tilapia (*Oreochromis mossambicus*) suggests a 'broad range' sensitivity not confined to those produced by conspecifics alone. J. Exp. Biol. 213:

Huertas, M., P.C. Hubbard, A.V.M. Canário and J. Cerda. 2007. Olfactory sensitivity to conspecific bile fluid and skin mucus in the European eel *Anguilla anguilla* (L.). J. Fish Biol. 70: 1907–1920.

Hunt, D.M., K.S. Dulai, J.C. Partridge, P. Cottrill and J.K. Bowmaker. 2001. The molecular basis for spectral tuning of rod visual pigments in deep-sea fish. J. Exp. Biol. 204: 3333–3344.

Hunt, D.M., J. Fitzgibbon, S.J. Slobodyanyuk and J.K. Bowmaker. 1996. Spectral tuning and molecular evolution of rod visual pigments in the species flock of cottoid fish in Lake Baikal. Vision Res. 36: 1217–24.

Hunt, J.J. 1992. Morphological characteristics of otoliths for selected fish in the Northwest Atlantic. J. Northw. Atl. Fish. Sci. 13: 63–75.

Inoue, J.G., M. Miya, M.J. Miller, T. Sado, R. Hanel, K. Hatooka, J. Aoyama, Y. Minegishi, M. Nishida and K. Tsukamoto. 2010. Deep-ocean origin of the freshwater eels. Biol. Lett. 6: 363–366.

Irwin, W.P. and K.J. Lohmann. 2003. Magnet-induced disorientation in hatchling loggerhead sea turtles. J. Exp. Biol. 206: 497–501.

Jannasch, H.W. 1978. Experiments in deep-sea microbiology. Oceanus 21: 50–57.

Jerkø, H., I. Turunen-Rise, P.S. Enger and O. Sand. 1989. Hearing in the eel (*Anguilla anguilla*). J. Comp. Physiol. A. 165: 455–459.

Johnsen, S. and K.J. Lohmann. 2005. The physics and neurobiology of magnetoreception. Nat. Rev. Neurosci. 6: 703–712.

Johnson, R.L., K.B. Grant, T.C. Zankel, M.F. Boehm, S.L. Merbs, J. Nathans and K. Nakanishi. 1993. Cloning and expression of goldfish opsin sequences. Biochemistry 32: 208–214.

Kalmijn, A.J. 2000. Detection and processing of electromagnetic and near-field acoustic signals in elasmobranch fishes. Phil. Trans. Roy. Soc. London. Series B: Biological Sciences 355: 1135–1141.

Kapoor, A.S. and P.P. Ojha. 1973. Functional anatomy of the nose and accessory nasal sacs in the teleost *Channa punctatus* Bloch. Acta Anat. 84: 96–105.

Karlsen, H.E. 1992. The inner ear is responsible for detection of infrasound in the perch (*Perca fluviatilis*). J. Exp. Biol. 171: 163–172.

Karlsen, H.E., R.W. Piddington, P.S. Enger and O. Sand. 2004. Infrasound initiates directional fast-start escape responses in juvenile roach *Rutilus rutilus*. J. Exp. Biol. 207: 4185–4193.

Kastelein, R.A., S. van der Heul, W.C. Verboom, N. Jennings, J. van der Veen and D. de Haan. 2008. Startle response of captive North Sea fish species to underwater tones between 0.1 and 64 kHz. Mar. Environ. Res. 65: 369–377.

Kirschvink, J.L. and J.L. Gould. 1981. Biogenic magnetite as a basis for magnetic field detection in animals. Biosystems 13: 181–201.

Kirschvink, J.L., M.M. Walker and C.E. Diebel. 2001. Magnetite-based magnetoreception. Curr. Opin. Neurobiol. 11: 462–467.

Kleckner, R.C., J.D. McCleave and G.S. Wippelhauser. 1983. Spawning of American eel, *Anguilla rostrata*, relative to thermal fronts in the Sargasso Sea. Environ. Biol. Fishes 9: 289–293.

Knights, B. 1983. Food particle-size preferences and feeding behavior in warmwater aquaculture of European eel, *Anguilla anguilla* L. Aquaculture 30: 173–190.

Kurogi, H., M. Okazaki, N. Mochioka, T. Jinbo, H. Hashimoto, M. Takahashi, A. Tawa, J. Aoyama, S.A., K. Tsukamoto, H. Tanaka, K. Gen, Y. Kazeto and S. Chow. 2011. First capture of post-spawning female of the Japanese eel *Anguilla japonica* at the southern West Mariana Ridge. Fisheries Science 77: 199–205.

Ladich, F. and A.N. Popper. 2004. Parallel evolution in fish hearing organs. *In*: G.A. Manley, A.N. Popper and R.R. Fay [eds]. Evolution of the Vertebrate Auditory System. Springer, New York, USA. pp. 95–127.

Land, M.F. 1990. Optics of the eye of marine animals. *In*: P.J. Herring, A.K. Campbell, M. Whitfield and L. Maddock [eds]. Light and life in the sea. Cambridge University Press, Cambridge, UK. pp. 149.

Leask, M.J. 1977. A physicochemical mechanism for magnetic field detection by migratory birds and homing pigeons. Nature 267: 144–145.

Lehmann, N., U. Alexiev and K. Fahmy. 2007. Linkage between the intramembrane H-bond network around aspartic acid 83 and the cytosolic environment of helix 8 in photoactivated rhodopsin. J. Mol. Biol. 366: 1129–41.

Leonard, J.B. and R.G. Summers. 1976. The ultrastructure of the integument of the American eel, *Anguilla rostrata*. Cell Tissue Res. 171: 1–30.

Lohmann, K.J., C.M.F. Lohmann and C.S. Endres. 2008. The sensory ecology of ocean navigation. J. Exp. Biol. 211: 1719–1728.

Lyall, A.H. 1957. Cone arrangement in teleost retinae. Q.J. Microsc. Sci. 98: 189–201.

Lythgoe, J.N. and G.I. Lythgoe. 1991. Fishes of the Sea. Blandford Press, London, UK.

Mathiesen, C. 1984. Structure and innervation of inner ear sensory epithelia in the European eel (*Anguilla anguilla* L.). Acta Zool. 65: 189–207.

McCleave, J.D. and R.C. Kleckner. 1982. Selective tidal steam transport in the estuarine migration of glass eels of the American ees (*Anguilla rostrata*). J. Cons. Int. Explor Mer. 40: 262–271.

McCleave, J.D. and R.C. Kleckner. 1985. Oceanic migrations of Atlantic eels (*Anguilla* spp.): adults and their offspring. Contrib. Mar. Sci. 27: 316–337.

McCleave, J.D. and J.H. Power. 1978. Influence of weak electric and magnetic fields on turning behavior in elvers of the American eel *Anguilla rostrata*. Mar. Biol. 46: 29–34.

McFarland, W.N. 1986. Light in the sea–Correlations with behaviors of fishes and invertebrates. American Zoologist 26: 389–401.

McFarland, W.N. and F.W. Munz. 1975. Part III: The evolution of photopic visual pigments in fishes. Vision Res. 15: 1071–80.

Meredith, G.E., B.L. Roberts and S. Maslam. 1987. Distribution of afferent fibers in the brainstem from end organs in the ear and lateral line in the european eel. J. Comp. Neurol. 265: 507–520.

Merrett, N.R. and P.A. Domanski. 1985. Observations of the ecology of deep-sea bottom living fishes collected off northwest Africa. II. The Moroccan slope (27 deg–34 deg N), with special reference to *Synaphobranchus kaupi*. Biol. Oceanogr 3: 349–399.

Meyer-Rochow, V.B. 1978. Skin papillae as possible electroreceptors in the deep-sea eel *Cyema atrum* (Cyemidae: Anguilloidei). Mar. Biol. 46: 277–282.

Mirzaliev, V. and E. Koloss. 1964. Characteristics of certain eye structure in river eel. Uch Zap Anat Gistol Embriol Republ Shredn Asii Kazakhstana. 1: 118–122.

Mogdans, J. and H. Bleckmann. 2001. The mechanosensory lateral line of jawed fishes. *In*: B.G. Kapoor and T.J. Hara [eds]. Sensory Biology of Jawed Fishes. Science Publishers Inc., Enfield (NH), USA. pp. 181–213.

Mogdans, J., S. Kröther and J. Engelmann. 2004. Neurobiology of the fish lateral line: adaptations for the detection of hydrodynamic stimuli in running water. *In*: G. von der Emde, J. Mogdans and B.G. Kapoor [eds]. The Senses of Fish: Adaptations for the Reception of Natural Stimuli. Kluwer Academic Publishers, Boston, USA. pp. 265–287.

Montgomery, J.C., A. Jeffs, S.D. Simpson, M. Meekan and C. Tindle. 2006. Sound as an orientation cue for the pelagic larvae of reef fishes and decapod crustaceans. *In*: J.S. Alan and W.S. David [eds]. Advances in Marine Biology. Academic Press. pp. 143–196.

Moore, A. and W.D. Riley. 2009. Magnetic particles associated with the lateral line of the European eel *Anguilla anguilla*. J. Fish Biol. 74: 1629–1634.

Mouritsen, H., U. Janssen-Bienhold, M. Liedvogel, G. Feenders, J. Stalleicken, P. Dirks and R. Weiler. 2004. Cryptochromes and neuronal-activity markers colocalize in the retina of migratory birds during magnetic orientation. Proc. Natl. Acad. Sci. USA. 101: 14294–14299.

Munz, F.W. 1958. Photosensitive pigments from the retinae of certain deep-sea fish. J. Physiol. 140: 220–225.

Munz, F.W. and W.N. McFarland. 1973. The significance of spectral position in the rhodopsins of tropical marine fishes. Vision Res. 13: 1829–74.

Myers, R.F. 1999. Micronesian Reef Fishes: A Comprehensive Guide to the Coral Reef Fishes of Micronesia. Coral Graphics, Guam, USA.

Newman, L.A. and P.R. Robinson. 2005. Cone visual pigments of aquatic mammals. Vis Neurosci. 22: 873-9.

Nishi, T., G. Kawamura and K. Matsumoto. 2004. Magnetic sense in the Japanese eel, *Anguilla japonica*, as determined by conditioning and electrocardiography. J. Exp. Biol. 207: 2965–2970.

Nishi, T., G. Kawamura and S. Sannomiya. 2005. Anosmic Japanese eel *Anguilla japonica* can no longer detect magnetic fields. Fisheries Science 71: 101–106.

Nozaki, M., I. Fujita, N. Saito, T. Tsukahara, H. Kobayashi, K. Ueda and K. Oshima. 1985. Distribution of LHRH-like immunoreactivity in the brain of the Japanese eel (*Anguilla japonica*) with special reference to the nervus terminalis. Zool. Sci. 2: 537–547.

Omura, Y., K. Tsuzuki, M. Sugiura, K. Uematsu and K. Tsukamoto. 2003. Rod cells proliferate in the eel retina throughout life. Fisheries Science 29: 924–928.

Pankhurst, N.W. 1984. Retinal development in larval and juvenile European eel, *Anguilla anguilla* (L.). Can. J. Zool. 62: 335–343.

Pankhurst, N.W. 1989. The relationship of ocular morphology to feeding modes and activity periods in shallow marine teleosts from New Zealand. Environ. Biol. Fishes 26: 201–211.

Pankhurst, N.W. and J.N. Lythgoe. 1983. Changes in vision and olfaction during sexual maturation in the European eel *Anguilla anguilla* (L.). J. Fish Biol. 23: 229–240.

Parker, G.H. 1918. A critical survey of the sense of hearing in fishes. Proc. Am. Phil. Soc. 57: 69–98.

Parry, J.W. and J.K. Bowmaker. 2000. Visual pigment reconstitution in intact goldfish retina using synthetic retinaldehyde isomers. Vision Res. 40: 2241–2247.

Parry, J.W., K.L. Carleton, T. Spady, A. Carboo, D.M. Hunt and J.K. Bowmaker. 2005. Mix and match color vision: tuning spectral sensitivity by differential opsin gene expression in Lake Malawi cichlids. Curr. Biol. 15: 1734–9.

Partridge, J.C., S.N. Archer and J.N. Lythgoe. 1988. Visual pigments in the individual rods of deep-sea fishes. J. Comp. Physiol. A. 162: 543–550.

Partridge, J.C., S.N. Archer and J. van Oostrum. 1992. Single and multiple visual pigments in deep-sea fishes. J. Mar. Biol. Assoc. UK. 72: 113–130.

Partridge, J.C. and M.E. Cummings. 1999. Adaptations of visual pigments to the aquatic environment. *In*: S.N. Archer, M.B.A. Djamgoz, E.R. Loew, J.C. Partridge and S. Valerga [eds]. Adaptive Mechanisms in the Ecology of Vision. Kluwer, Dordrecht, Netherlands. pp. 251–283.

Partridge, J.C., J. Shand, S.N. Archer, J.N. Lythgoe and W.A. van Groningen-Luyben. 1989. Interspecific variation in the visual pigments of deep-sea fishes. J. Comp. Physiol. [A] 164: 513–29.

Paxton, J.R. 2000. Fish otoliths: do sizes correlate with taxonomic group, habitat and/or luminescence? Philos. T. Roy. Soc. B. 355: 1299–1303.

Pevzner, R.A. 1978. Electron microscopic study of the taste buds of the eel, *Anguilla anguilla*. Tsitologiya 20: 1112–1118.

Phillips, J.B. and S.C. Borland. 1992. Behavioral evidence for use of a light-dependent magnetoreception mechanism by a vertebrate. Nature 359: 142–144.

Piper, H. 1906. Actionsströme vom Gehörorgan der Fische bei Schallreizung. Zentralb. f. Physiol. 20: 293–297.

Poggendorf, D. 1952. Die absoluten horschwellen des zwergwelses (*Ameiurus nebulosus*) und beitrage zur physik des Weberschen apparates der Ostariophysen. Z. Vergl. Physiol. 34: 222–257.

Pointer, M.A., L.S. Carvalho, J.A. Cowing, J.K. Bowmaker and D.M. Hunt. 2007. The visual pigments of a deep-sea teleost, the pearl eye *Scopelarchus analis*. J. Exp. Biol. 210: 2829–2835.

Popper, A.N. 1979. Ultrastructure of the sacculus and lagena in a moray eel (*Gymnothorax* sp.). J. Morphol. 161: 241–256.

Popper, A.N. and S. Coombs. 1980. Auditory mechanisms in teleost fishes: significant variations in both hearing capabilities and auditory structures are found among species of bony fishes. American Scientist 68: 429–440.

Popper, A.N. and S. Coombs. 1982. The morphology and evolution of the ear in Actinopterygian fishes. Amer. Zool. 22: 311–328.

Popper, A.N., R.R. Fay, C. Platt and O. Sand. 2003. Sound detection mechanisms and capabilities of teleost fishes. *In*: S.P. Collin and N.J. Marshall [eds]. Sensory Processing in Aquatic Environments. Springer-Verlag, New York, USA. pp. 3–38.

Popper, A.N. and D.M. Higgs. 2009. Fish: hearing, lateral lines (mechanisms, role in behavior, adaptations to life underwater). *In*: J.H. Steele, K.K. Turekian and S.A. Thorpe [eds]. Encyclopedia of Ocean Sciences. Academic Press, Oxford, UK. pp. 476–482.

Popper, A.N. and C.R. Schilt. 2008. Hearing and acoustic behavior: basic and applied considerations. *In*: J.F. Webb, R.R. Fay and A.N. Popper [eds]. Fish Bioacoustics. Springer, New York, USA. pp. 17–48.

Randall, J.E., G.R. Allen and R.C. Steene. 1990. Fishes of the Great Barrier Reef and Coral Sea. University of Hawaii Press, Hawaii, USA.

Rath, P., L.L. DeCaluwe, P.H. Bovee-Geurts, W.J. DeGrip and K.J. Rothschild. 1993. Fourier transform infrared difference spectroscopy of rhodopsin mutants: light activation of rhodopsin causes hydrogen-bonding change in residue aspartic acid-83 during meta II formation. Biochemistry 32: 10277–10282.

Retzius, G. 1881. Das Gehörorgan der Wirbelthiere: morphologisch-histologische Studien. I. Das Gehörorgan der Fische und Amphibian. Samson and Wallin, Stockholm, Sweden.

Reutter, K. and A. Hansen. 2005. Subtypes of light and dark elongated taste bud cells in fish. *In*: K. Reutter and B.G. Kapoor [eds]. Fish Chemosenses. Science Publishers Inc., Enfield (NH), USA. pp. 211–230.

Reutter, K. and M. Witt. 1993. Morphology of vertebrate taste organs and their nerve supply. *In*: S.A. Simon and S.D. Roper [eds]. Mechanisms of Taste Transduction. CRC Press, Boca Raton, Florida. USA. pp. 30–82.

Ritz, T., S. Adem and K. Schulten. 2000. A model for photoreceptor-based magnetoreception in birds. Biophys. J. 78: 707–718.

Rommel, S.A., Jr. and J.D. McCleave. 1973. Sensitivity of American eels (*Anguilla rostrata*) and Atlantic salmon (*Salmo salar*) to weak electric and magnetic fields. J. Fish. Res. Board Can. 30: 657–663.

Rousseau, K., S. Aroua, M. Schmitz, P. Elie and S. Dufour. 2009. Silvering: Metamorphosis or Puberty? *In:* G. van den Thillart, S. Dufour and J.C. Rankin [eds.]. Fish and Fisheries Series 30. Springer Science, USA. pp. 39–63.

Saitou, N. and M. Nei. 1987. The neighbor-joining method: a new method for reconstructing phylogenetic trees. Mol. Biol. Evol. 4: 406–425.

Sand, O. and P.S. Enger. 1973. Evidence for an auditory function of swimbladder in cod. J. Exp. Biol. 59: 405–414.

Sand, O., P.S. Enger, H.E. Karlsen, F. Knudsen and T. Kvernstuen. 2000. Avoidance responses to infrasound in downstream migrating European silver eels, *Anguilla anguilla*. Environ. Biol. Fish. 57: 327–336.

Sand, O., P.S. Enger, H.E. Karlsen and F.R. Knudsen. 2001. Detection of infrasound in fish and behavioral responses to intense infrasound in juvenile salmonids and European silver eels: A minireview. Am. Fish. S. S. 26: 183–193.

Sand, O. and H.E. Karlsen. 1986. Detection of Infrasound by the Atlantic Cod. J. Exp. Biol. 125: 197–204.

Sandstrom, A. 1999. Visual ecology of fish—A review with special reference to percids. Fiskeriverket Rapport 2: 45–80.

Schmidt, J. 1923. The breeding places of the eel. Philos. T. Roy. Soc. B. 211: 179–208.

Shand, J., W.L. Davies, N. Thomas, L. Balmer, J.A. Cowing, M. Pointer, L.S. Carvalho, A.E. Trezise, S.P. Collin, L.D. Beazley and D.M. Hunt. 2008. The influence of ontogeny and light environment on the expression of visual pigment opsins in the retina of the black bream, Acanthopagrus butcheri. J. Exp. Biol. 211: 1495–503.

Shapley, R. and J. Gordon. 1980. The visual sensitivity of the retina of the conger eel. Proc. R. Soc. Lond. B. Biol. Sci. 209: 317–30.

Silver, W.L. 1982. Electrophysiological responses from the peripheral olfactory system of the American eel, *Anguilla rostrata*. J. Comp. Physiol. 148: 379–388.

Sola, C. 1995. Chemoattraction of upstream migrating glass eels *Anguilla anguilla* to earthy and green odorants. Environ. Biol. Fish 43: 179–185.

Sola, C., A. Spampanato and L. Tosi. 1993. Behavioral responses of glass eels (*Anguilla anguilla*) towards amino acids. J. Fish Biol. 42: 683–691.

Sola, C. and P. Tongiorgi. 1998. Behavioral responses of glass eels of *Anguilla anguilla* to non-protein amino acids. J. Fish Biol. 53: 1253–1262.

Sola, C. and L. Tosi. 1993. Bile salts and taurine as chemical stimuli for glass eels, *Anguilla anguilla*: a behavioral study. Environ. Biol. Fishes 37: 197–204.

Sorensen, P.W. 1986. Origins of the freshwater attractant(s) of migrating elvers of the American eel, *Anguilla rostrata*. Environ. Biol. Fish 17: 185–200.

Souza, J.J., J.J. Poluhowich and R.J. Guerra. 1988. Orientation responses of american eels, *Anguilla rostrata*, to varying magnetic fields. Comp. Biochem. Physiol. A. 90: 57–61.

Sugawara, T., H. Imai, M. Nikaido, Y. Imamoto and N. Okada. 2010. Vertebrate rhodopsin adaptation to dim light via rapid meta-II intermediate formation. Mol. Biol. Evol. 27: 506–19.

Sun, H., J.P. Macke and J. Nathans. 1997. Mechanisms of spectral tuning in the mouse green cone pigment. Proc. Natl. Acad. Sci. USA 94: 8860–5.

Tesch, F.-W. 1989. Changes in swimming depth and direction of silver eels (*Anguilla anguilla* L.) from the continental shelf to the deep sea. Aquatic Living Resources 2: 9–20.

Tesch, F.W. 1977. The Eel. Chapman & Hall, London, UK.

Tesch, F.W., T. Wendt and L. Karlsson. 1992. Influence of geomagnetism on the activity and orientation of the eel, *Anguilla anguilla* (L.), as evident from laboratory experiments. Ecology of Freshwater Fish 1: 52–60.

Tricas, T.C. and S.M. Highstein. 1991. Action of the octavolateralis efferent system upon the lateral line of free-swimming toadfish, *Opsanus tau*. J. Comp. Physiol. A. 169: 25–37.

Tsukamoto, K. 1992. Discovery of the spawning area for Japanese eel. Nature 356: 789–791.

van Ginneken, V.J.T. and G.E. Maes. 2005. The European eel (*Anguilla anguilla*, Linnaeus), its lifecycle, evolution and reproduction: a literature review. Rev. Fish Biol. Fisheries 15: 367–398.

van Ginneken, V.J.T., B. Muusze, J. Klein Breteler, D. Jansma and G. van den Thillart. 2005. Microelectronic detection of activity level and magnetic orientation of yellow European eel, *Anguilla anguilla* L., in a pond. Environ. Biol. Fishes 72: 313–320.

Verrier, M.L. 1928. Recherches sur les yeux et la vision des poisons. Bull Biol. Fr. Belge Suppl. 11: 1–222.

Vilter, V. 1951. Intervention probable de la lumiere dans la naissance des structures rétinienne révélée par l'étude comparée de la rétine chez les anguilles normales et cavernicoles. C. R. Séances Soc. Biol. 145: 54–56.

Von Bartheld, C.S. 2004. The terminal nerve and its relation with extrabulbar "olfactory" projections: lessons from lampreys and lungfishes. Micro. Res. Tech. 65: 13–24.

Wagner, H.-J. 2001a. Sensory brain areas in mesopelagic fishes. Brain Behav Evol. 57: 117–133.

Wagner, H.-J. 2002. Sensory brain areas in three families of deep-sea fish (slickheads, eels and grenadiers): comparison of mesopelagic and demersal species.. Mar. Biol. 141: 807–817.

Wagner, H.-J. 2005. Role of gustation in two populations of deep-sea fish: comparison of mesopelagic and demersal species based on volumetric brain data. *In*: K. Reutter and B.G. Kapoor [eds]. Fish Chemosenses. Science Publishers Inc., Enfield (NH), USA. pp. 277–303.

Wagner, H.J. 2001b. Brain areas in abyssal demersal fishes. Brain Behav Evol. 57: 301–16.

Wagner, H.J., E. Frohlich, K. Negishi and S.P. Collin. 1998. The eyes of deep-sea fish. II. Functional morphology of the retina. Prog. Retin Eye Res. 17: 637–85.

Walker, M.M., T.E. Dennis and J.L. Kirschvink. 2002. The magnetic sense and its use in long-distance navigation by animals. Curr. Opin. Neurobiol. 12: 735–744.

Walker, M.M., C.E. Diebel, C.V. Haugh, P.M. Pankhurst, J.C. Montgomery and C.R. Green. 1997. Structure and function of the vertebrate magnetic sense. Nature 390: 371–376.

Walls, G. 1942. The Vertebrate Eye and its Adaptive Radiation. Cranbrooke, Bloomfield Hills, USA.

Wang, F.Y., W.S. Chung, H.Y. Yan and C.S. Tzeng. 2008. Adaptive evolution of cone opsin genes in two colorful cyprinids, *Opsariichthys pachycephalus* and *Candidia barbatus*. Vision Res. 48: 1695–704.

Wang, F.Y., M.Y. Tang and H.Y. Yan. 2011. A comparative study on the visual adaptations of four species of moray eel. Vision Res. 51: 1099–108.

Weadick, C.J. and B.S. Chang. 2007. Long-wavelength sensitive visual pigments of the guppy (*Poecilia reticulata*): six opsins expressed in a single individual. BMC Evol. Biol. 7 Suppl. 1: S11.

Weber, E.H. 1820. De aure et auditu hominis et animalium. Pars I. De aure animalium aquatilium. Gehard Fleischer, Leipzig, Germany.

Westin, L. 1990. Orientation mechanisms in migrating European silver eel (*Anguilla anguilla*): temperature and olfaction. Mar. Biol. 106: 175–179.

Williams, M.N. and J.M. Wild. 2001. Trigeminally innervated iron-containing structures in the beak of homing pigeons, and other birds. Brain Res. 889: 243–246.

Williamson, G.R. and P.H.J. Castle. 1975. Large-eyed specimen of eel *Anguilla bicolor* from a freshwater well. Copeia 3: 561–564.

Wiltschko, W., U. Munro, H. Ford and R. Wiltschko. 1993. Red light disrupts orientation of migratory birds. Nature 364: 525–527.

Wiltschko, W. and R. Wiltschko. 2001. Light-dependent magnetoreception in birds: the behavior of European robins, *Erithacus rubecula*, under monochromatic light of various wavelengths and intensities. J. Exp. Biol. 204: 3295–302.

Wood, P. and J.C. Partridge. 1993. Opsin substitution induced in retinal rods of the eel (*Anguilla anguilla* (L.)): a model for G-protein-linked receptors. Proc. Roy. Soc. B. 254: 227–232.

Wunder, W. 1925. Physiologische und vergleichend-anatomische Untersuchungen an der Knochenfischnetzhaut. Z. vergl Physiol. 3: 1–63.

Yamada, Y. and K. Hama. 1972. Fine structure of the lateral-line organ of the common eel, *Anguilla japonica*. Cell Tissue Res. 124: 454–464.

Yokoyama, S. 2000. Molecular evolution of vertebrate visual pigments. Prog. Retin Eye Res. 19: 385–419.

Yokoyama, S. 2008. Evolution of dim-light and color vision pigments. Annu. Rev. Genomics Hum. Genet. 9: 259–82.

Yokoyama, S., H. Zhang, F.B. Radlwimmer and N.S. Blow. 1999. Adaptive evolution of color vision of the Comoran coelacanth (*Latimeria chalumnae*). Proc. Natl. Acad. Sci. USA 96: 6279–84.

Yoshii, K., N. Kaml, K. Kurihara and Y. Kobatake. 1979. Gustatory responses of eel palatine receptors to amino acid and carboxylic acids. J. Gen. Physiol. 74: 301–317.

Yoshii, K. and K. Kurihara. 1983. Ion dependence of the eel taste response to amino acids. Brain Res. 280: 63–67.

Zacchei, A.M. and P. Tavolaro. 1988. Lateral line system during the life cycle of *Anguilla anguilla* (L.). Boll. Zool. 3: 145–153.

Zhang, H., K. Futami, N. Horie, A. Okamura, T. Utoh, N. Mikawa, Y. Yamada, S. Tanaka and N. Okamoto. 2000. Molecular cloning of fresh water and deep-sea rod opsin genes from Japanese eel *Anguilla japonica* and expressional analyses during sexual maturation. FEBS Lett. 469: 39–43.

Zimmerman, M.A. and J.D. McCleave. 1975. Orientation of elvers of American eels (*Anguilla rostrata*) in weak magnetic and electric fields. Helgoland Marine Research 27: 175–189.

Intestinal Absorption of Salts and Water

Masaaki Ando[a] and Yoshio Takei[b],*

Introduction

Since the work of Homer W. Smith (1930), it is generally accepted that teleosts in fresh water (FW) scarcely drink water; however, in seawater (SW) they drink continuously and absorb water together with monovalent ions from the intestine to compensate for osmotic water loss in marine environment. He used American eel (*Anguilla rostrata*) as an experimental animal, and described that "Eel is admirably suited to experimental work, and there is no reason to believe that results obtained upon it are not generally applicable to other marine or fresh water fish". Eels can live in both FW and SW, thus being euryhaline teleosts. Therefore, eels have been used as an experimental material in osmoregulation research. In both FW and SW environments, eels can maintain plasma osmotic pressure and electrolyte concentration at levels largely independent of the ionic concentration and composition of their environments. In order to maintain water and ion balance, eels are equipped with specialized ion- and water-transport epithelia situated in the gill, kidney, esophagus and intestine likely as in other teleosts (for

Laboratory of Physiology, Atmosphere and Ocean Research Institute, The University of Tokyo, 5-1-5 Kashiwanoha, Kashiwa, Chiba 277-8564, Japan.
[a]E-mail: ando@aori.u-tokyo.ac.jp
[b]E-mail: takei@aori.u-tokyo.ac.jp
*Corresponding author

a more extensive review of fish osmoregulation see Marshall and Grosell 2006, Evans and Claiborne 2009). In euryhaline fishes, the function of these osmoregulatory organs must change according to environmental demands, and various endogenous factors are known to play a significant role as mediators in the necessary physiological adaptation (Takei and MacCormick 2012). In this chapter, we will review initially how eels obtain water from hypertonic SW after processing through the digestive tracts, and secondly how the intestinal absorption is controlled by endogenous regulators.

Salt and Water Transport Across Digestive Tracts

Esophageal Desalination

When SW is applied directly into the intestine of SW eel *in vivo*, water is not absorbed across the intestine but exuded into the lumen by osmosis (Skadhauge 1969). Similar exudation of water is observed in the isolated SW eel intestine *in vitro* (Utida et al. 1967). Recently, we demonstrated that intestinal water exudation occurs after application of 230 mM NaCl to the luminal fluid *in vitro* (Ando et al. 2003). These results indicate that the intestine of SW eel cannot absorb water if SW (*ca.* 450 mM NaCl) enters directly into the intestine. Before arriving at the intestine, the swallowed SW must pass through the esophagus and the stomach for dilution (see Fig. 1). Hirano and Mayer-Gostain (1976) first demonstrated that the isolated SW eel esophagus desalts the luminal SW following the concentration gradient of NaCl, and such desalination is not observed in FW eel esophagus. Similar desalination is also observed in the SW flounder esophagus, although flounder desalination is observed without gradient of NaCl concentrations and inhibited by ouabain, suggesting involvement of active process (Parmelee and Renfro 1983). Ouabain-sensitive desalination is also obtained in the SW eel esophagus, corresponding to about 2/3 of total desalination (Nagashima and Ando 1994). In that article, we also demonstrated that Na^+ and Cl^- fluxes are mutually dependent. In addition, the osmotic water permeability across the SW eel esophagus was also measured (Nagashima and Ando 1994). The osmotic water permeability of the esophagus is very low (2×10^{-4} cm/s), approximately 1/10 of that in thick ascending limb of Henle in mammalian kidney which is well known as a water impermeable but ion permeable epithelium (Burg and Green 1973). Although the eel esophagus seems to be highly impermeable to water, water channel proteins, aquaporins (AQP3 and AQP1), are expressed in the European eel esophagus (Cutler and Cramb 2001), and immuno-localization of AQP3 is demonstrated in the eel esophageal epithelia (Cutler et al. 2007). Although the eel AQP3 is indeed able to transport water, urea and glycerol (MacIver et

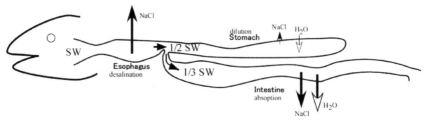

Figure 1. Salts and water transport through the digestive tracts of eels acclimated to SW. Ingested SW is desalted to 1/2 during passing through the esophagus, then to 1/3 SW in the stomach (Ando and Nagashima 1996). Under nearly isosmotic condition, water is absorbed across the intestine.

al. 2009), this water channel does not seem to contribute to water absorption in the intestine (see below). Although the role of AQP3 in the esophagus is not clear yet, NaCl concentration of the swallowed SW is indeed halved after passing through the esophagus, and is further diluted to 1/3 in the stomach of the SW eel (Ando and Nagashima 1996). The dilution in the stomach may have occurred passively by water exudation following osmotic gradient, partly due to NaCl gradient (Hirano and Mayer-Gostain 1976). All these situations are illustrated in Fig. 1.

Intestinal Ion Transport

Since fish intestine is an important osmoregulatory organ, and is relatively simple in structure among various osmoregulatory organs, many research works have been performed in the eel intestine. In 1960s, two groups used the eel intestine as an experimental material, Japanese group (Utida's laboratory) and European group (Maetz's laboratory). Using Japanese eel (*Anguilla japonica*), Utida's group demonstrated that water absorption across the intestine is enhanced after SW acclimation in both *in vitro* (Utida et al. 1967, Oide and Utida 1967) and *in vivo* (Oide and Utida 1968, Fig. 2) preparations. Figure 2 shows typical changes in body weight, and rate of drinking and absorption of water by the intestine, after transferring FW eels to SW. Similarly, using European eel (*Anguilla anguilla*), Maetz's group demonstrated in the *in vivo* system that drinking rate and Na⁺ absorption across the intestine are dependent on the external Na⁺ concentrations, and thus Na⁺ absorption is higher in fish in double-strength SW (DSW) than in SW fish (Maetz and Skadhauge 1968). Skadhauge (1969) also measured Cl⁻ flux across the intestine *in vivo*, and demonstrated that Cl⁻ absorption is higher in DSW fish than in SW fish, the order being DSW > SW > FW. From the finding that water can move against a small negative osmotic gradient across the SW eel intestine, he proposed 'solute-linked water flow' as a mechanism for water absorption (Skadhauge 1969, 1974).

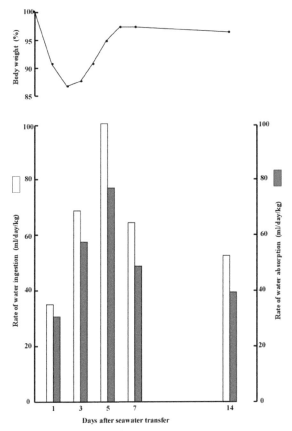

Figure 2. Time course of changes in the rate of water ingestion (white column; left ordinate) and water absorption through the intestine (black column; right ordinate) after transfer of eels from FW to SW. In FW, no phenol red is detected in the intestine (no water ingestion). Change in body weight is measured simultaneously. Cited from Oide and Utida (1968).

In 1970s, the Ussing chamber technique was applied to the eel intestine, and a serosa-negative transepithelial potential difference (PD) was observed in the SW eel intestine (Utida et al. 1972). By measuring fluxes of ^{22}Na and ^{36}Cl radioisotopes, this serosa-negative PD was shown to be caused by greater active Cl$^-$ transport than active Na$^+$ transport, where 'active' is used for ion fluxes which can not be explained by simple diffusion (Ussing's criterion) (Ando et al. 1975). Similar active Cl$^-$ transport has been reported in the intestine of the SW winter flounder (*Pseudopleuroneces americanus*) (Huang and Chen 1971). However, these results were obtained in the intact intestine, where muscle layers were not stripped off. Thus these two groups could not find interdependence of Cl$^-$ and Na$^+$ transport, and proposed independent transport system of Na$^+$ and Cl$^-$, because serosa-positive PD was observed after removing Cl⁻ from bathing media (Huang and Chen 1971, Ando et al. 1975). After stripping off the muscle layers, however, these two transports (Na⁺ and Cl⁻) were shown to be dependent mutually both in the eel (Ando and Kobayashi 1978) and in the winter flounder (Field et al. 1978). Field et al. (1978), at that time, proposed a neutral NaCl cotransport as a mechanism of NaCl absorption across the flounder intestine. Musch et al. (1982), however, demonstrated that Na⁺, Cl⁻ and K⁺ were taken altogether into the epithelium of the flounder intestine from the lumen, Na⁺-K⁺-Cl⁻ cotransport (secondary active Cl⁻ transport), and the cotransporter is now named as Na⁺-K⁺-2Cl⁻ cotransporter 2 (NKCC2). Similar Na⁺-K⁺-Cl⁻ cotransport system appears to exist in the SW eel intestine, since Cl⁻ and water absorption is coupled with K⁺ transport (Ando 1983, Ando and Utida 1986) and bumetanide-sensitive (Ando and Subramanyam 1990). Measuring Na⁺, Cl⁻ and K⁺ concentrations in the mucus layer with ion-selective microelectrodes, Simonneaux et al. (1987) also support the presence of an apical Na⁺-K⁺-Cl⁻ cotransport system in the SW eel intestine. The serosa-negative PD and the short-circuit current (Isc) observed in the SW flounder and eel intestine are explained by the apical Na⁺-K⁺-2Cl⁻ cotransport, the apical K⁺ efflux and the basolateral Cl⁻ efflux, the last two fluxes being electrogenic (Halm et al. 1985, Ando and Subramanyam 1990, Trischitta et al. 1992). Figure 3 shows a possible model for ion transport system in the SW eel intestine, and a similar model is proposed in the flounder intestine (Halm et al. 1985). When the serosa-negative PD across the intestine was first reported, nobody could believe its existence, since most transport epithelia, such as mammalian intestine and frog skin, showed serosa-positive PD. However, after finding the Na⁺-K⁺-2Cl⁻ cotransport in the SW fish intestine, similar Na⁺-K⁺-2Cl⁻ cotransport system has been reported in mammalian epithelia, such as trachea and intestine, where the cotransport is present at the basolateral membrane and contributes to a secretion of NaCl and water, thus making serosa-positive PD, and the cotransporter is now named as NKCC1. The amino acid sequence of these cotransporters (NKCC1 and NKCC2) seems to be fairly different (Soybel et al. 1995, Cutler and Cramb 2001, Isenring and Forbush 2001, Delpire and Mount 2002).

Bicarbonate ion is excreted through the SW eel intestine, and contributes to the luminal alkalinization *via* Cl⁻/HCO₃⁻ exchange in the apical membrane. The supply of HCO₃⁻ is due to endogenous generation by carbonic anhydrase within the cell and the Na⁺-HCO₃⁻ cotransporter at the basolateral membrane (Fig. 3, Ando and Subramanyam 1990). Similar HCO₃⁻ secretion in the intestine has been reported in other SW fishes, which is supposed to make a CaCO₃ precipitation (calcium cake) in the luminal fluid, that leads the luminal fluid hyposmotic to accelerate water absorption

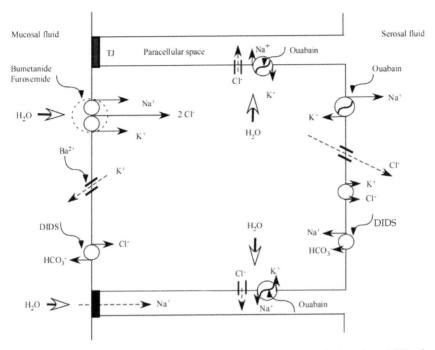

Figure 3. A hypothetical model of NaCl and water transport across the intestine of SW eel. The direction of each ion flux is indicated by solid arrow and the action of inhibitors is shown as wavy lines. Dotted arrows indicate diffusional ion fluxes (passive). Na^+, K^+, Cl^- and water fluxes are from Ando and Utida (1986), and HCO_3^- fluxes from Ando and Subramanyam (1990). TJ, tight junction. DIDS, 4,4′-diisothiocyano-2,2′-disulfonic acid disodium salt hydrate.

across the intestine. More details about the significance of HCO_3^- secretion across the SW fish intestine have been reviewed in recent years (Grosell 2006, 2011, Whittamore 2012).

Although it is well known that crypts of intestine are the site of fluid and electrolyte secretion in mammals (Welsh et al. 1982), the crypt-like structure is not generally present in fish intestine (Kapoor et al. 1975, Field et al. 1978, Frizzell et al. 1979). However, secretion of NaCl may occur in the eel intestine, since the presence of the NKCC1, a secretory cotransporter, is reported in the eel intestine (Cutler and Cramb 2001, 2002a). The presence of a distinct secretory epithelium is also suggested in the intestine of flounder (O'Grady and Wolters 1990) and killifish (Marshall et al. 2002).

In general, paracellular pathway also contributes to ion and water transport across the epithelial tissue. In the eel intestine, the paracellular conductance appears to be much higher to Na^+ than to Cl^-, since the tissue conductance mainly depends on the medium Na^+ concentration (Ando et al.

1975, Trischitta et al. 1992) and the unidirectional flux of Na^+ is much higher than that of Cl^- (Ando et al. 1975), as in mammalian intestine (Marchiando et al. 2010). Analyzing dilution potential across the intestine, which arises from the different rate of Na^+ and Cl^- diffusion through the paracellular pathway, it is shown that Cl^- permeability in the paracellular pathway is enhanced by cAMP, in goldfish (Bakker and Groot 1984, 1989, Kiliaan et al. 1989, 1993, Bakker et al. 1993), tilapia (Kiliaan et al. 1989, 1993), and flounder (Rao et al. 1984, Krasny and Frizzell 1984, Rao and Nash 1988). In the *Necturus* gallbladder, it is demonstrated that the tight junction is finely regulated by cAMP (Duffey et al. 1981) and Ca^{2+} (Palant et al. 1983). In the eel intestine, however, the paracellular Cl^- permeability seems to be mainly controlled by cGMP, since the dilution potential is decreased more greatly by 8Br-cGMP than by 8Br-cAMP (Trischitta et al. 1996). A similar decrease in the dilution potential is also observed after treatments with atrial nariuretic peptide (Trischitta et al. 1996) and nitric oxide (Trischitta et al. 2007), both of which are known to increase the intracellular cGMP.

Intestinal Nutrient Transport

In addition to NaCl absorption, glucose and amino acid absorption also accelerates water absorption across the intestine in many animals. Using the Ussing type chamber, Ferraris and Ahearn (1983, 1984) have demonstrated that D-glucose (Glc) and L-alanine (Ala) are taken up into the intestinal epithelium of the eel from the mucosal side in a concentration-dependent manner, which are composed of saturable (carrier-mediated or active) and linear (passive) components. Furthermore, using brush-border membrane vesicles (BBMVs) prepared from eel intestine, Storelli et al. (1986) have demonstrated that the saturable component of both Glc and Ala uptake depends on mucosal Na^+ concentration. The apparent affinity constants (K_{app}) for Glc and Ala uptake are similar in both studies (Ferraris and Ahearn 1984, Storelli et al. 1986). Other amino acids are also taken up into the BBMVs from SW eel in a Na^+-dependent manner (Storelli et al. 1986, 1989, Romano et al. 1989, Vilella et al. 1988, 1989, 1990). Since the eel intestinal BBMVs possess Na^+-dependent cotransporters with substrate specificities very similar to those found in mammalian intestinal brush borders (Stevens et al. 1984), these transport systems are classified, at least, into four distinct pathways (Storelli et al. 1989, Stevens 1992). These can be designated as: 1) an anionic transport pathway for glutamic and aspartic acids, 2) a cationic transport pathway for lysine and arginine, 3) a relatively specific neutral amino acid carrier for proline and α-(methylamino) isobutyric acid, and 4) a nonspecific neutral amino acid carrier for most other neutral amino acids. More detailed classification may be expected in the eel intestine after molecular identification of these transporters, just as in mammalian intestine (Bröer 2008).

Intestinal Water Transport

Water seems to follow the NaCl absorption across the intestine, since a good parallelism exists between NaCl and water absorption under various conditions in SW eel (Ando and Kondo 1993, Ando et al. 1992, 2000, Uesaka et al. 1994, 1995, 1996). This seems to support a hypothesis that water moves following osmotic gradient produced locally, named 'local osmosis' (Diamond 1964) or 'solute-linked water flow' (Skadhauge 1969, 1974). Figure 3 also shows possible routes of water flow, and implicates that water moves more efficiently when water permeability across the plasma membrane or/and tight junction is high. Indeed, osmotic water permeability across the SW eel intestine is six-time higher than that of FW eel (Ando 1975). Recently, Larsen et al. (2002, 2009) proposed 'sodium recirculation theory' by assuming the lateral intercellular space as osmotic coupling compartment in epithelial water transport. This theory assumes Na^+ back flux through the serosal plasma membrane from the absorbed fluid (serosal fluid), that contributes to dilute the absorbed fluid. Although both Diamond and Larsen imagine paracellular pathway for water absorption, water may be drawn osmotically through four pathways generally, as proposed in salmonid intestine (Madsen et al. 2011): (1) diffusion through the lipid bilayer of the enterocyte; (2) paracellular diffusion through the tight junction; (3) symport with Na^+ and glucose by the sodium-dependent glucose transporter 1 (SGLT1, see above); (4) diffusion through AQPs apically as well as laterally. To date, thirteen AQP proteins (AQP0 ~ AQP12) have been identified in mammals (Verkman 2009, 2012). AQP3, however, may not be involved in water absorption across the eel intestine, because (1) mRNA expression of AQP3 in the intestine is not different between SW and FW eels (Cutler and Cramb 2002b), and (2) no immunostaining by anti-AQP3 antibody was observed in the intestinal columnar cells of the eel (Lignot et al. 2002). Since water permeability of AQP3, an aquaglyceroporin, is known to be lowest (Verkman and Mitra 2000), other AQPs might be involved in water absorption across the eel intestine. AQP1 might be a candidate for such water channels, since Aoki et al. (2003) and Martinez et al. (2005) demonstrated that SW transfer of the eel increases AQP1 mRNA and protein levels in the intestine. Alternatively, however, water channels may be absent in the epithelial membrane, at least in the apical membrane, because water permeability in the BBMVs from eel intestine is similar to the value obtained in lipid bilayer (Alves et al. 1999). The absorbed water may move through paracellular pathway in the SW eel intestine as suggested in the mammalian (Masyuk et al. 2002, Larsen et al. 2002, 2009) and the salmonid intestine (Madsen et al. 2011). In mammalian intestine, AQP1, AQP3, AQP4 and AQP8 have been found (Koyama et al. 1999, Ramirez-Lorca et al. 1999, Masyuk et al. 2002), but AQP-mediated water transport in the small intestine is so far unknown. In the mammalian intestine, much of the increased intestinal water absorption accompanying a meal appears to be paracellular, and serves to drive the paracellular absorption of ions and nutrients by convective solvent drag, at minimal metabolic cost (Pappenheimer and Reiss 1987, Meddings and Westergaard 1989). However, such paracellular uptake of water and nutrients might be absent in fish, since Wood and Grosell (2012) have recently found independence of net water flux from paracellular permeability in the killifish (*Fundulus heteroclitus*) intestine.

Regulation of Salt and Water Transport

Long Term Regulation

Since the original finding that ovine prolactin (PRL) was necessary for the survival in FW of hypophysectomized killifish, *Fundulus heteroclitus* (Pickford and Phillips 1959), it has been well established that PRL plays an important role in maintaining hydromineral balance of euryhaline teleosts in FW (Utida et al. 1971, Loretz and Bern 1982, Foskett et al. 1983, Hirano 1986). When PRL was injected into the SW eel, the ion and water absorption across the intestine was reduced (Utida et al. 1972). In general, PRL reduces branchial salt and possibly water permeability, inhibits branchial salt extrusion, intestinal salt uptake, and urinary bladder water permeability, but stimulates urinary bladder Na^+ uptake (Hirano 1986). In many cases the response to PRL seems to be mediated by morphologic changes and is therefore probably more important during chronic, rather than acute, osmotic stress (Foskett et al. 1983).

Cortisol is the dominant adrenal (interrenal) steroid acting as both glucocorticoid and mineralocorticoid in teleost fishes, and suggested to have important roles in both FW and SW adaptation (Takei and McCormick 2012). Although the major stimulus for secretion of aldosterone (tetrapod mineralocorticoid) is the renin-angiotensin system in all tetrapod species (Kobayashi and Takei 1996), cortisol secretion is apparently not coupled to that of renin in eels (Kenyon et al. 1985, Nishimura et al. 1976). A variety of studies have demonstrated that plasma cortisol levels increase when eels are transferred from FW to SW (Hirano and Utida 1971, Forrest et al. 1973, Kenyon et al. 1985). When eels were transferred from FW to SW, plasma cortisol level increased transiently after 2 to 4 hr (Hirano and Utida 1971), and the intestinal ion and water absorption was enhanced gradually until 5 days after transfer (Oide and Utida 1968, Fig. 2). Similar enhancement in water transport was observed 24 hr after injection of adrenocorticotropic hormone (ACTH) or cortisol into the FW eel (Hirano and Utida 1968). Recently, it has been demonstrated that cortisol infusion into FW eels enhances AQP1 mRNA and protein levels in the intestine, suggesting acceleration in water absorption (Martinez et al. 2005). In general, cortisol is thought to stimulate intestinal ion and water absorption (Hirano et al. 1975), gill salt extrusion (Foskett et al. 1983), and urinary bladder salt and water reabsortion (Loretz and Bern 1983). However the effects on these tissues appear to be *via* differentiation of the respective epithelium, rather than a direct effect on the transport itself (Foskett et al. 1983). Various studies have demonstrated that cortisol secretion in teleosts is under the control of ACTH (Hirano and Utida 1971, Loretz and Bern 1983, Parwez et al. 1984, Gupta et al. 1985, Hagen et al. 2006) as it is in other vertebrates. Angiotensin II (Ang II) and Ang III are potent stimuli for cortisol secretion in eel interrenal preparation *in vitro* (Ventura 2011) as in other teleosts (Perrott and Balment 1990). Atrial natriuretic peptide (ANP) also stimulates cortisol secretion when injected into the blood of SW eels but not in FW eels (Li and Takei 2003). In the interrenal preparation from SW eels, ANP alone failed to stimulate cortisol secretion but potentiated the steroidogenic action of ACTH (Ventura et al. 2011).

Although no osmoregulatory role of growth hormone (GH) is reported in eel, there is considerable evidence for the role in SW adaptation in salmonids. Several studies have shown that mammalian GH enhances SW survival and adaptation in numerous salmonid species (Smith 1956, Komourdjian et al. 1976, Clarke et al. 1977, Miwa and Inui 1985, Bolton et al. 1987). Plasma GH levels were elevated after transferring rainbow trout from FW to 80% SW (Collie et al. 1989), while they were only transiently decreased 12 hr after transferring chum salmon from SW to FW (Ogasawara et al. 1989). By using ^{125}I-labeled salmon GH, specific binding sites for GH were demonstrated in the liver, gill, intestine and kidney of rainbow trout (Sakamoto and Hirano 1991). Since an increase in intestinal water absorption is an adaptive response critical for SW survival, the intestine represents a potential site of GH action in addition to the gill and kidney. In support of this hypothesis is the evidence that GH treatment of coho salmon causes a significant increase in intestinal Na^+-dependent proline influx (Collie and Stevens 1985). The GH effect appears to be mediated in part by insulin-like growth factor I (IGF-I) secretion (Sakamoto and Hirano 1993).

Short Term Regulation

As short-term regulators, acetylcholine (ACh), serotonin (5-HT), histamine (HA), eel atrial natriuretic peptide (eANP) and goldfish vasoactive intestinal peptide (gVIP) are known in the Japanese eel intestine (Mori and Ando 1991, Ando et al. 1992, Uesaka et al. 1995). However, all these regulators inhibit salt and water absorption, accompanied by a decrease in the short-circuit current (Isc) in a concentration-dependent manner (Fig. 4a), but

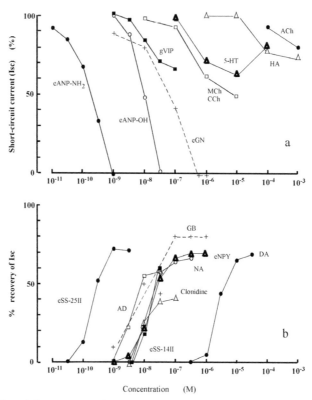

Figure 4. Effect of short-term regulators on the short-circuit current (Isc), reflecting activity on active Cl⁻ transport across the SW eel intestine (Ando et al. 1975). (a) Effect of various inhibitors. Eel atrial natriuretic peptides (C-terminally amidated (eANP-NH$_2$) and C-terminal free (eANP-OH)), goldfish vasoactive intestinal peptide (gVIP), carbachol (CCh), methacholine (MCh), serotonin (5-HT), histamine (HA), and acetylcholine (ACh) were added to the serosal fluid (solid line). In contrast, eel guanylin (eGN) was added to the mucosal fluid (dotted line). (b) Various regulators restore the inhibition induced by IBMX (10⁻⁵ M), 5-HT (10⁻⁶ M) and MCh (10⁻⁶ M). Eel somatostatin 25 II and 14 II (eSS-25 II and eSS-14 II), adrenaline (AD), noradrenaline (NA), eel neuropeptide Y (eNPY), clonidine, and dopamine (DA) were added to the serosal fluid (solid line). In contrast, guanabenz (GB) was added to the mucosal fluid (dotted line). All data are from previous studies (Mori and Ando 1991, Ando et al. 1992, 2000, 2003, Ando and Omura 1993, Uesaka et al. 1994, 1995, 1996).

no stimulatory regulators are known yet (Ando et al. 2003). Among these inhibitors, amidated eel ANP (eANP-NH$_2$) was the most potent (Fig. 4a). The role of NPs (natriuretic peptides) in intestinal water and ion absorption is described in detail in Chapter 7 of this volume. Acetylcholine and 5-HT seem to be released from nerve ends in the intestine, since electrical field stimulation (EFS) mimics the inhibitory effects of these neurotransmitters, and these EFS-induced inhibitions are inhibited by atropine (a muscarinic antagonist for ACh receptor) and ICS-205930 (a 5-HT$_3$ receptor antagonist), respectively (Mori and Ando 1991).

Although no stimulators are found yet, adrenaline (AD), noradrenaline (NA), dopamine (DA), clonidine (α_2-adrenoceptor agonist), eel somatostatin 25 II (eSS-25 II), and eel neuropeptide Y (eNPY) restore the inhibition observed after pretreatment with isobutylmethylxanthine (IBMX), 5-HT and methacholine (MCh, a muscarinic agonist for ACh receptor) (Ando and Kondo 1993, Ando and Omura 1993, Uesaka et al. 1994, 1996, Ando et al. 2003, Fig. 4b). Among these factors that restored the inhibition, eSS-25 II was the most potent (Fig. 4b).

Although all these regulators cited above act from the serosal side of the intestinal epithelia, eel guanylin (eGN) acts from the mucosal side. Guanylin has been discovered in mammals as an endogenous peptide which acts on the receptor for heat-stable enterotoxin produced by *Escherichia coli* (Currie et al. 1992). Eel GN was identified as YDECEICMFAACTGC and demonstrated to be produced in the goblet cells in the intestine (Yuge et al. 2003). Therefore, GN is considered to be secreted into the intestinal lumen from the goblet cells and to act as a luminocrine peptide (Takei and Yuge 2007). Mucosal GN inhibits NaCl and water absorption across the SW eel intestine in a concentration-dependent manner (Fig. 4a, dotted line) (unpublished data). The detailed action of GN family peptides on the intestinal absorption will be found in Chapter 7 of this volume. On the other hand, there are clonidine binding sites in the brush-border membrane of the eel intestine, which have higher affinity for guanabenz (Kim et al. 1998). Guanabenz acts from the mucosal side and restores the inhibition induced by IBMX, MCh and 5-HT in a concentration-dependent manner (Fig. 4b, dotted line) (Ando et al. 2000). Although both clonidine and guanabenz are α_2-adrenoceptor agonists, thus probably not produced in the eel, these results suggest that some endogenous factors, structurally similar to guanabenz, are present in the upper digestive tracts, such as esophagus, stomach or anterior intestine (see Fig. 1), and they act as a luminocrine substance to restore the inhibition of salt and water absorption across the middle or posterior intestine. Indeed, luminal fluid from SW eel intestine showed guanabenz-like effects and the effects were inhibited by RX821002, an inhibitor of guanabenz effects (Ando et al. 2000). Identification of guanabenz-like substance(s) from the eel digestive tracts is awaited.

Acknowledgement

The authors thank Dr. Francesca Trischitta for comments on the manuscript.

References

Alves, P., G. Soveral, R.I. Macey and T.F. Moura. 1999. Kinetics of water transport in eel intestinal vesicles. J. Membrane Biol. 171: 177–182.

Ando, M. 1975. Intestinal water transport and chloride pump in relation to sea-water adaptation of the eel, *Anguilla japonica*. Comp. Biochem. Physiol. 52A: 229 233.

Ando, M. 1983. Potassium-dependent chloride and water transport across the seawater eel intestine. J. Membrane Biol. 73: 125–130.

Ando, M., H.T. Kim, I. Takase and A. Kawahara. 2000. Imidazoline receptor contributes to ion and water transport across the intestine of the eel acclimated to sea water. Zool. Sci. 17: 307–312.

Ando, M. and M. Kobayashi. 1978. Effects of stripping of the outer layers of the eel intestine on salt and water transport. Comp. Biochem. Physiol. 61A: 497–501.

Ando, M. and K. Kondo. 1993. Noradrenalin antagonizes effects of serotonin and acetylcholine in the seawater eel intestine. J. Comp. Physiol. B. 163: 59–63.

Ando, M., K. Kondo and Y. Takei. 1992. Effects of eel atrial natriuretic peptide on NaCl and water transport across the intestine of the seawater eel. J. Comp. Physiol. B. 162: 436–439.

Ando, M., T. Mukuda and T. Kozaka. 2003. Water metabolism in the eel acclimated to sea water: from mouth to intestine. Comp. Biochem. Physiol. 136B: 621–633.

Ando, M. and K. Nagashima. 1996. Intestinal Na$^+$ and Cl$^-$ levels control drinking behavior in the seawater-adapted eel, *Anguilla japonica*. J. Exp. Biol. 199: 711–716.

Ando, M. and E. Omura. 1993. Catecholamine receptor in the seawater eel intestine. J. Comp. Physiol. B. 163: 64–69.

Ando, M. and M.V.V. Subramanyam. 1990. Bicarbonate transport systems in the intestine of the seawater eel. J. Exp. Biol. 150: 381–394.

Ando, M. and S. Utida. 1986. Effects of diuretics on sodium, potassium, chloride and water transport across the seawater eel intestine. Zool. Sci. 3: 605–612.

Ando, M., S. Utida and H. Nagahama. 1975. Active transport of chloride in eel intestine with special reference to sea water adaptation. Comp. Biochem. Physiol. 51A: 27–32.

Aoki, M., T. Kaneko, F. Katoh, S. Hasegawa, N. Tsutsui and K. Aida. 2003. Intestinal water absorption through aquaporin 1 expressed in the apical membrane of mucosal epithelial cells in seawater-adapted Japanese eel. J. Exp. Biol. 206: 3495–3505.

Bakker, R., K. Dekker, H.R. de Jonge and J.A. Groot. 1993. VIP, serotonin, and epinephrine modulate the ion selectivity of tight junctions of goldfish intestine. Am. J. Physiol. 264: R362–R368.

Bakker, R. and J.A. Groot. 1984. cAMP-mediated effects of ouabain and theophylline on paracellular ion selectivity. Am. J. Physiol. 246: G213–G217.

Bakker, R. and J.A. Groot. 1989. Further evidence for the regulation of the tight junction ion selectivity by cAMP in goldfish intestinal mucosa. J. Membrane Biol. 111: 25–35.

Bolton, J.P., N.L. Collie, H. Kawauchi and T. Hirano. 1987. Osmoregulatory actions of growth hormone in rainbow trout (*Salmo gairdneri*). J. Endocrinol. 112: 63–68.

Bröer, S. 2008. Amino acid transport across mammalian intestinal and renal epithelia. Physiol. Rev. 88: 249–286.

Burg, M.B. and N. Green. 1973. Function of the thick ascending limb of Henle's loop. Am. J. Physiol. 224: 659–668.

Clarke, W.C., S.W. Farmer and K.W. Hartwell. 1977. Effect of teleost pituitary growth hormone on growth of *Tilapia mossambica* and on growth and seawater adaptation of sockeye salmon (*Oncorhynchus nerka*). Gen. Comp. Endocrinol. 33: 174–178.

Collie, N.L., J.P. Bolton, H. Kawauchi and T. Hirano. 1989. Survival of salmonids in seawater and the time-frame of growth hormone action. Fish Physiol. Biochem. 7: 315–321.

Collie, N.L. and J.J. Stevens. 1985. Hormonal effects on L-proline transport in coho salmon (*Oncorhynchus kisutch*) intestine. Gen. Comp. Endocrinol. 59: 399–409.

Currie, M.G., K.F. Fok, J. Kato, R.J. Moore, F.K. Hamra, K.L. Duffin and C.E. Smith. 1992. Guanylin: an endogenous activator of intestinal guanylate cyclase. Proc. Natl. Acad. Sci. USA 89: 947–951.

Cutler, C.P. and G. Cramb. 2001. Molecular physiology of osmoregulation in eels and other teleosts: role of transporter isoforms and gene duplication. Comp. Biochem. Physiol. 130A: 551–564.

Cutler, C.P. and G. Cramb. 2002a. Two isoforms of the $Na^+/K^+/2Cl^-$ cotransporter are expressed in the European eel (*Anguilla anguilla*). Biochim. Biophys. Acta 1566: 92–103.

Cutler, C.P. and G. Cramb. 2002b. Branchial expression of an aquaporin 3 (AQP-3) homologue is downregulated in the European eel *Anguilla anguilla* following seawater acclimation. J. Exp. Biol. 205: 2643–2651.

Cutler, C.P., A.S. Martinez and G. Cramb. 2007. The role of aquaporin 3 in teleost fish. Comp. Biochem. Physiol. 148A: 82–91.

Delpire, E. and D.B. Mount. 2002. Human and murine phenotypes associated with defects in cation-chloride co-transport. Ann. Rev. Physiol. 64: 803–843.

Diamond, J.M. 1964. The mechanism of isotonic water transport. J. Gen. Physiol. 48: 15–42.

Duffey, M.E., B. Hanai, S. Ho and C.J. Bentzel. 1981. Regulation of epithelial tight junction permeability by cyclic AMP. Nature 294: 451–453.

Evans, D.H. and J.B. Claiborne. 2009. Osmotic and ionic regulation in fishes. *In*: D.H. Evans [ed]. Osmotic and Ionic Regulation. Cells and Animals. CRC Press, Boca Raton, USA. pp. 295–366.

Ferraris, R.P. and G.A. Ahearn. 1983. Intestinal glucose transport in carnivorous and herbivorous marine fishes. J. Comp. Physiol. B. 152: 79–90.

Ferraris, R.P. and G.A. Ahearn. 1984. Sugar and amino acid transport in fish intestine. Comp. Biochem. Physiol. 77A: 397–413.

Field, M., K.J. Karnaky Jr., P.L. Smith, J.E. Bolton and W.B. Kinter. 1978. Ion transport across the isolated intestinal mucosa of the winter flounder *Pseudopleuronectes americanus*—1. Functional and structural properties of cellular and paracellular pathways for Na and Cl. J. Membrane Biol. 41: 265–293.

Forrest, J.N. Jr., W.C. Mackay, B. Gallagher and F.H. Epstein. 1973. Plasma cortisol response to saltwater adaptation in the American eel, *Anguilla rostrata*. Am. J. Physiol. 224: 714–717.

Foskett, J.K., H.A. Bern, T.E. Machen and M. Conner. 1983. Chloride cells and the hormonal control of fish osmoregulation. J. Exp. Biol. 106: 255–281.

Frizzell, R.A., M. Field and S.G. Schultz. 1979. Sodium-coupled chloride transport by epithelial tissues. Am. J. Physiol. 236: F1–F8.

Grosell, M. 2006. Intestinal anion exchange in marine fish osmoregulation. J. Exp. Biol. 209: 2813–2827.

Grosell, M. 2011. Intestinal anion exchange in marine teleosts is involved in osmoregulation and contributes to the oceanic inorganic carbon cycle. Acta Physiol. 202: 421–434.

Gupta, O.P., B. Lahlou, J. Botella and J. Porthe-Nibelle. 1985. *In vivo* and *in vitro* studies on the release of cortisol from interrenal tissue in trout. I. Effects of ACTH and prostaglandins. Exp. Biol. 43: 201–212.

Hagen, I.J., M. Kusakabe and G. Young. 2006. Effects of ACTH and cAMP on steroidogenic acute regulatory protein and P450 11β-hydroxylase messenger RNAs in rainbow trout interrenal cells: Relationship with *in vitro* cortisol production. Gen. Comp. Endocrinol. 145: 254–262.

Halm, D.R., E.J. Krasny Jr. and R.A. Frizzell. 1985. Electrophysiology of flounder intestinal mucosa. I. Conductance properties of the cellular and paracellular pathways. J. Gen. Physiol. 85: 843–864.

Hirano, T. 1986. The spectrum of prolactin action in teleosts. *In*: C.L. Ralph [ed]. Comparative Endocrinology: Development and Directions. Liss. New York, USA pp. 53–74.

Hirano, T. and N. Mayer-Gostain. 1976. Eel esophagus as an osmoregulatory organ. Proc. Natl. Acad. Sci. USA 73: 1348–1350.

Hirano, T., M. Morisawa, M. Ando and S. Utida. 1975. Adaptive changes in ion and water transport mechanism in the eel intestine. *In*: J.W.L. Robinson [ed]. Intestinal Ion Transport. MTP, London, UK. pp. 301–317.

Hirano, T. and S. Utida. 1968. Effects of ACTH and cortisol on water movement in isolated intestine of the eel, *Anguilla japonica*. Gen. Comp. Endocrinol. 11: 373–380.

Hirano, T. and S. Utida. 1971. Plasma cortisol concentration and the rate of intestinal water absorption in the eel, *Anguilla japonica*. Endocrinol. Japon. 18: 47–52.

Huang, K.C. and T.S.T. Chen. 1971. Ion transport across the intestinal mucosa of winter flounder, *Pseudopleuronectes americanus*. Am. J. Physiol. 220: 1734–1738.

Isenring, P. and B. Forbush. 2001. Ion transport and ligand binding by the Na-K-Cl cotransporter, structure-function studies. Comp. Biochem. Physiol. 130A: 487–497.

Kapoor, B.G., H. Smit and I.A. Verighina. 1975. The alimentary canal and digestion in teleosts. *In*: F.S. Russell and C.M. Yonge [eds]. Advances in Marine Biology, vol. 13. Academic Press, London, UK. pp. 102–219.

Kenyon, C.J., A. McKeever, J.A. Oliver and I.W. Henderson. 1985. Control of renal and adrenocortical function by the renin-angiotensin system in two euryhaline teleost fishes. Gen Comp. Endocrinol. 58: 93–100.

Kiliaan, A.J., S. Holmgren, A.C. Jonsson, K. Dekker and J.A. Groot. 1993. Neuropeptides in the intestine of two teleost species (*Oreochromis mossambicus, Carassius auratus*): Localization and electrophysiological effects on the epithelium. Cell Tissue Res. 271: 123–134.

Kiliaan, A.J., H.W.J. Joosten, R. Bakker, K. Dekker and J.A. Groot. 1989. Serotonergic neurons in the intestine of two teleosts, *Carassius auratus* and *Oreochromis mossambicus*, and the effect of serotonin on transepithelial ion-selectivity and muscle tension. Neurosci. 31: 817–824.

Kim, H.T., T. Sakamoto and M. Ando. 1998. Novel [³H]clonidine binding sites in the intestine of the eel acclimated to sea water. Zool. Sci. 15: 205–212.

Kobayashi, H. and Y. Takei. 1996. The Renin-Angiotensin System. Springer-Verlag, Berlin, Heidelberg, Germany.

Komourdjian, M.P., R.L. Saunders and J.C. Fenwick. 1976. The effect of porcine somatotropin on growth, and survival in seawater of Atlantic salmon (*Salmo salar*) parr. Can. J. Zool. 54: 531–535.

Koyama, Y., T. Yamamoto, T. Tani, K. Nihei, D. Kondo, H. Funaki, E. Yoita, K. Kawasaki, N. Sato, K. Hatakeyama and I. Kihara. 1999. Expression and localization of aquaporins in rat gastrointestinal tract. Am. J. Physiol. 276: C621–C627.

Krasny, E.J. and R.A. Frizzell. 1984. Intestinal ion transport in marine teleosts. *In*: G.A. Gerencser [ed]. Chloride Transport Coupling in Biological Membrane and Epithelia. Elsevier Science, Amsterdam, Netherlands. pp. 205–218.

Larsen, E.H., J.B. Sorensen and J.N. Sorensen. 2002. Analysis of the sodium recirculation theory of solute-coupled water transport in small intestine. J. Physiol. 542: 33–50.

Larsen, E.H., N.J. Willumsen, N. Morjerg and J.N. Sorensen. 2009. The lateral intercellular space as osmotic coupling compartment in isotonic transport. Acta Physiol. 195: 171–186.

Li, Y.Y. and Y. Takei. 2003. Ambient salinity-dependent effects of homologous natriuretic peptides (ANP, VNP and CNP) on plasma cortisol levels in the eel. Gen. Comp. Endocrinol. 130: 317–323.

Lignot, J.H., C.P. Cutler, N. Hazon and G. Cramb. 2002. Immunolocalization of aquaporin 3 in the gill and the gastrointestinal tract of the European eel *Anguilla anguilla* (L.). J. Exp. Biol. 205: 2653–2663.

Loretz, C.A. and H.A. Bern. 1982. Prolactin and osmoregulation in vertebrates. An update. Neuroendocrinol. 35: 292–304.

Loretz, C.A. and H.A. Bern. 1983. Control of ion transport by *Gillichthys mirabilis* urinary bladder. Am. J. Physiol. 245: R45–R52.

MacIver, B., C.P. Cutler, J. Yin, M.G. Hill, M.L. Zeidel and W.G. Hill. 2009. Expression and functional characterization of four aquaporin water channels from the European eel (*Anguilla anguilla*). J. Exp. Biol. 212: 2856–2863.

Madsen, S.S., J.H. Olesen, K. Bedal, M.B. Engelund, Y.M. Velasco-Santamaria and C.K. Tipsmark. 2011. Functional characterization of water transport and cellular localization of three aquaporin paralogs in the salmonid intestine. Front. Physiol. 2: 1–14.

Maetz, J. and E. Skadhauge. 1968. Drinking rates and gill ionic turnover in relation to external salinities in the eel. Nature 217: 371–373.

Marchiando, A.M., W.V. Graham and J.R. Turner. 2010. Epithelial barriers in homeostasis and disease. Annu. Rev. Pathol. Mech. Dis. 5: 119–144.

Marshall, W.S. and M. Grosell. 2006. Ion transport, osmoregulation and acid-base balance. *In*: D.H. Evans and J.B. Claiborne [eds]. The Physiology of Fishes, 3rd ed. CRC Press, Boca Raton, USA. pp. 177–230.

Marshall, W.S., J.A. Howard, R.R.F. Cozzi and E.M. Lynch. 2002. NaCl and fluid secretion by the intestine of the teleost *Fundulus heteroclitus*: involvement of CFTR. J. Exp. Biol. 205: 745–758.

Martinez, A.S., C.P. Cutler, G.D. Wilson, C. Phillips, N. Hazon and G. Cramb. 2005. Regulation of expression of two aquaporin homologs in the intestine of the European eel: effects of seawater acclimation and cortisol treatment. Am. J. Physiol. 288: R1733–R1743.

Masyuk, A.I., R.A. Marinelli and N.F. La Russo. 2002. Water transport by epithelia of the digestive tract. Gastroenterol. 122: 545–562.

Meddings, J.B. and H. Westergaard. 1989. Intestinal glucose transport using perfused rat jejunum *in vivo*: model analysis and derivation of corrected kinetic constants. Clin. Sci. 76: 403–413.

Miwa, S. and Y. Inui. 1985. Effects of L-thyroxine and ovine growth hormone on smoltification of amago salmon (*Oncorhynchus rhodurus*). Gen. Comp. Endocrinol. 58: 436–442.

Mori, Y. and M. Ando. 1991. Regulation of ion and water transport across the eel intestine: effects of acetylcholine and serotonin. J. Comp. Physiol. B. 161: 387–392.

Musch, M.W., S.A. Orellana, L.S. Kimberg, M. Field, D.R. Halm, E.J. Krasny Jr. and R.A. Frizzell. 1982. Na⁺-K⁺-Cl⁻ cotransport in the intestine of a marine teleost. Nature 300: 351–353.

Nagashima, K. and M. Ando. 1994. Characterization of esophageal desalination in the seawater eel, *Anguilla japonica*. J. Comp. Physiol. 164: 47–54.

Nishimura, H., W.H. Sawyer and R.F. Nigelli. 1976. Renin, cortisol and plasma volume in marine teleost fishes adapted to dilute media. J. Endocrinol. 70: 47–59.

Ogasawara, T., T. Hirano, T. Akiyama, S. Arai and M. Tagawa. 1989. Changes in plasma prolactin and growth hormone concentrations during freshwater adaptation of juvenile chum salmon (*Oncorhynchus keta*) reared in seawater for a prolonged period. Fish Physiol. Biochem. 7: 309–313.

O'Grady, S.M. and P.J. Wolters. 1990. Evidence for chloride secretion in the intestine of the winter flounder. Am. J. Physiol. 258: C243–C247.

Oide, H. and S. Utida. 1968. Changes in intestinal absorption and renal excretion of water during adaptation to sea-water in the Japanese eel. Marine Biol. 1: 172–177.

Oide, M. and S. Utida. 1967. Changes in water and ion transport in isolated intestine of the eel during salt adaptation and migration. Marine Biol. 1: 102–106.

Palant, C.E., M.E. Duffey, B.K. Mookerje, S. Ho and C.J. Bentzel. 1983. Ca²⁺ regulation of tight-junction permeability and structure in *Necturus* gallbladder. Am. J. Physiol. 245: C203–C212.

Pappenheimer, J.R. and K.Z. Reiss. 1987. Contribution of solvent drag through intercellular junctions to absorption of nutrients by the small intestine of the rat. J. Membrane Biol. 100: 123–136.

Parmelee, J.T. and J.L. Renfro. 1983. Esophageal desalination of seawater in flounder: role of active sodium transport. Am. J. Physiol. 245: R888–R893.

Parwez, I., S.V. Goswami and B.I. Sundararaji. 1984. Effects of hypophysectomy on some osmoregulatory parameters of the catfish, *Heteropneustes fossilis* (Bloch). J. Exp. Zool. 229: 375–381.

Perrott, M.N. and R.J. Balment. 1990. The renin-angiotensin system and the regulation of plasma cortisol in the flounder, *Platichthys flesus*. Gen. Comp. Endocrinol. 78: 414–420.

Pickford, G.E. and J.G. Phillips. 1959. Prolactin, a factor promoting survival of hypophysectomized killifish in fresh water. Science 130: 454–455.

Ramirez-Lorca, R., M.L. Vizuete, J.L. Venero, M. Revuelta, J. Cano, A.A. Ilundain and M. Echevarria. 1999. Localization of aquaporin-3 mRNA and protein along the gastrointestinal tract of Wistar rats. Pflüg. Arch. 438: 94–100.

Rao, M.C. and N.T. Nash. 1988. 8-BrcAMP does not affect Na-K-Cl cotransport in winter flounder intestine. Am. J. Physiol. 255: C246–C251.

Rao, M.C., N.T. Nash and M. Field. 1984. Differing effects of cGMP and cAMP on ion transport across flounder intestine. Am. J. Physiol. 246: C167–C171.

Romano, P.M., G.A. Ahearn and C. Storelli. 1989. Na-dependent L-glutamate transport by eel intestinal BBMV: role of K[+] and Cl[-]. Am. J. Physiol. 257: R180–R188.

Sakamoto, T. and T. Hirano. 1991. Growth hormone receptors in the liver and osmoregulatory organs of rainbow trout: characterization and dynamics during adaptation to seawater. J. Endocrinol. 130: 425–433.

Sakamoto, T. and T. Hirano. 1993. Expression of insulin-like growth factor I gene in osmoregulatory organs during seawater adaptation of the salmonid fish: possible mode of osmoregulatory action of growth hormone. Proc. Natl. Acad. Sci. USA 90: 1912–1916.

Simonneaux, V., W. Humbert and R. Kirsch. 1987. Mucus and intestinal ion exchanges in the sea-water adapted eel, *Anguilla anguilla* L. J. Comp. Physiol. 157: 295–306.

Skadhauge, E. 1969. The mechanism of salt and water absorption in the intestine of the eel (*Anguilla anguilla*) adapted to waters of various salinities. J. Physiol. 204: 135–158.

Skadhauge, E. 1974. Coupling of transmural flows of NaCl and water in the intestine of the eel (*Anguilla anguilla*). J. Exp. Biol. 60: 535–546.

Smith, D.C.W. 1956. The role of endocrine organs in the salinity tolerance of trout. Mem. Soc. Endocrinol. 5: 83–101.

Smith, H.W. 1930. The absorption and excretion of water and salts by marine teleosts. Am. J. Physiol. 93: 480–505.

Soybel, D.J., S.R. Gullans, F. Maxwell and E. Delpire. 1995. Role of basolateral Na[+]-K[+]-Cl[-] cotransport in HCl secretion by amphibian gastric mucosa. Am. J. Physiol. 269: C242–C249.

Stevens, B.R. 1992. Vertebrate intestine apical membrane mechanisms of organic nutrient transport. Am. J. Physiol. 263: R458–R463.

Stevens, B.R., J.D. Kaunitz and E.M. Wright. 1984. Intestinal transport of amino acids and sugars: Advances using membrane vesicles. Annu. Rev. Physiol. 46: 417–433.

Storelli, C., S. Vilella and G. Cassano. 1986. Na-dependent D-glucose and L-alanine transport in eel intestinal brush border membrane vesicles. Am. J. Physiol. 251: R463–R469.

Storelli, C., S. Vilella, M.P. Romano, M. Maffia and G. Cassano. 1989. Brush-border amino acid transport mechanisms in carnivorous eel intestine. Am. J. Physiol. 257: R506–R510.

Takei, Y. and S.D. McCormick. 2013. Hormonal control of fish euryhalinity. *In*: S.D. McCormick, A.P. Farrell and C.J. Brauner [eds]. Fish Physiology, Vol. 32. Academic Press, San Diego, USA, pp. 69–123.

Takei, Y. and S. Yuge. 2007. The intestinal guanylin system and seawater adaptation in eels. Gen. Comp. Endocrinol. 152: 339–351.

Trischitta, F., M.G. Denaro, C. Faggio, M. Mandolfino and T. Schettino. 1996. Different effects of cGMP and cAMP in the intestine of the European eel, *Anguilla anguilla*. J. Comp. Physiol. B. 166: 30–36.

Trischitta, F., M.G. Denaro, C. Faggio and T. Schettino. 1992. Comparison of Cl⁻-absorption in the intestine of the seawater- and freshwater-adapted eel, *Anguilla anguilla*: evidence for the presence of an Na-K-Cl cotransport system on the luminal membrane of the enterocyte. J. Exp. Zool. 263: 245–253.

Trischitta, F., P. Pidala and C. Faggio. 2007. Nitric oxide modulates ionic transport in the isolated intestine of the eel, *Anguilla anguilla*. Comp. Biochem. Physiol. 148A: 368–373.

Uesaka, T., K. Yano, S. Sugimoto and M. Ando. 1996. Effects of eel neuropeptide Y on ion and water transport across the seawater eel intestine. Zool. Sci. 13: 341–346.

Uesaka, T., K. Yano, M. Yamasaki and M. Ando. 1995. Somatostatin-, vasoactive intestinal peptide-, and granulin-like peptides isolated from intestinal extracts of goldfish, *Carassius auratus*. Gen. Comp. Endocrinol. 99: 298–306.

Uesaka, T., K. Yano, M. Yamasaki, K. Nagashima and M. Ando. 1994. Somatostatin-related peptides isolated from the eel gut: effects on ion and water absorption across the intestine of the seawater eel. J. Exp. Biol. 188: 205–216.

Utida, S., S. Hatai, T. Hirano and F.I. Kamemoto. 1971. Effect of prolactin on survival and plasma sodium levels in hypophysectomized medaka *Oryzias latipes*. Gen. Comp. Endocrinol. 16: 566–573.

Utida, S., T. Hirano, H. Oide, M. Ando, D.W. Johnson and H.A. Bern. 1972. Hormonal control of the intestine and urinary bladder in teleost osmoregulation. Gen. Comp. Endocrinol. Suppl. 3: 317–327.

Utida, S., N. Isono and T. Hirano. 1967. Water movement in isolated intestine of the eel adapted to freshwater or sea water. Zool. Mag. 76: 203–204.

Ventura, A. 2011. Regulation of cortisol secretion by fast-acting hormones in eel osmoregulation. Doctoral thesis. The University of Tokyo.

Ventura, A., M. Kusakabe and Y. Takei. 2011. Distinct natriuretic peptides interact with ACTH for cortisol secretion from interrenal tissue of eels in different salinities. Gen. Comp. Endocrinol. 173: 129–138.

Verkman, A.S. 2009. Aquaporins: translating bench research to human disease. J. Exp. Biol. 212: 1707–1715.

Verkman, A.S. 2012. Aquaporins in clinical medicine. Annu. Rev. Med. 63: 303–316.

Vilella, S., G.A. Ahearn, G. Cassano, M. Maffia and C. Storelli. 1990. Lysine transport by brush-border membrane vesicles of eel intestine: interaction with neutral amino acids. Am. J. Physiol. 259: R1181–R1188.

Vilella, S., G.A. Ahearn, G. Cassano and C. Storelli. 1988. Na-dependent L-proline transport by eel intestinal brush-border membrane vesicles. Am. J. Physiol. 255: R648–R653.

Vilella, S., G. Cassano and C. Storelli. 1989. How many Na⁺-dependent carriers for L-alanine and L-proline in the eel intestine? Studies with brush-border membrane vesicles. Biochim. Biophys. Acta 984: 188–192.

Welsh, M.J., P.L. Smith, M. Fromm and R.A. Frizzell. 1982. Crypts are the site of intestinal fluid and electrolyte secretion. Science 218: 1219–1221.

Whittamore, J.M. 2012. Osmoregulation and epithelial water transport: lessons from the intestine of marine teleost fish. J. Comp. Physiol. B. 182: 1–39.

Wood, C.M. and M. Grosell. 2012. Independence of net water flux from paracellular permeability in the intestine of *Fundulus heteroclitis*, a euryhaline teleost. J. Exp. Biol. 215: 508–517.

Yuge, S., K. Inoue, S. Hyodo and Y. Takei. 2003. A novel guanylin family (guanylin, uroguanylin, and renoguanylin) in eels. J. Biol. Chem. 278: 22726–22733.

Endocrine Control of Osmoregulation

Takehiro Tsukada,[1,]* Marty Kwok-Shing Wong,[2]
Maho Ogoshi[3] and Shinya Yuge[4]

Introduction

In this chapter, we focus on recent advances in the endocrinology of the renin-angiotensin system (RAS), natriuretic peptide (NP) family, guanylin (GN) family, and adrenomedulin (AM) family in the eel, particularly in regard to their role in osmoregulation. Eels, as well as other migratory and esturarine species such as salmonids, flounders, etc., are categorized as euryhaline teleosts, which can tolerate a wide range of salinity fluctuations,

[1]Department of Anatomy, Jichi Medical University School of Medicine, 3311-1 Yakushiji, Shimotsuke, Tochigi 329-0498, Japan.
E-mail: tsukada@jichi.ac.jp
[2]Laboratory of Physiology, Department of Marine Biosciences, Atmosphere and Ocean Research Institute, the University of Tokyo, 5-1-5 Kashiwanoha, Kashiwa, Chiba 277-8564, Japan.
E-mail: martywong@aori.u-tokyo.ac.jp
[3]Ushimado Marine Institute, Faculty of Science, Okayama University, 130-17 Kashino, Ushimado, Okayama 701-4303, Japan.
E-mail: ogoshi-m@cc.okayama-u.ac.jp
[4]Department of Fisheries & Wildlife, Michigan State University, 158 Giltner Hall, East Lansing, MI 48824, USA.
E-mail: shinya.yuge@gmail.com
*Corresponding author

in contrast to stenohaline teleosts that exhibit a narrow range of salinity tolerance. The Anguillid eels have been used as experimental models to study osmoregulation since the early 20th century (Chester-Jones et al. 1966, Maetz and Skadhauge 1968, Oide and Utida 1968), and their versatile ability to acclimate to both freshwater (FW) and seawater (SW) enables researchers to examine their physiology during short-term and long-term salinity adaptation. In addition, surgical operations such as cannulation of blood vessels, intestine, urinary bladder, etc., and ablation of tissues and organs (e.g., hypophysectomy) have allowed researchers to conduct real-time monitoring of physiological parameters such as ions and hormone levels, drinking rate, blood pressure, heart rate, branchial and gastrointestinal activities (e.g., epithelial ion and water transport), and urine flow after environmental transfer or hormone treatment.

In mammals, angiotensins (ANGs), NPs, GNs, and AMs are collectively known as blood pressure and volume controllers. Briefly, ANG II is involved in elevating blood pressure in response to volume depletion and controls aldosterone secretion to promote renal Na^+ retention. ANP and BNP are secreted from the heart and lower blood pressure and volume by their potent natriuretic and diuretic effects on the kidney. GN and uroguanylin (UGN) are secreted into the intestinal lumen and induce Cl^- and water secretion in the intestine, while circulating GNs induce natriuresis and diuresis in the kidney. All AMs are vasoactive peptides secreted mainly by endothelial cells, and decrease blood pressure and volume by altering hemodynamics and renal water and Na^+ balance. ANG II, NPs, and AMs also target the brain to evoke water and salt appetite. Thus, these hormone systems are significant for cardiovascular homeostasis of terrestrial vertebrates that live in dry and gravitational environment where body fluid is depleted and accumulated in the lower extremities.

In the last few decades, many studies have been performed on the RAS, NPs, GNs, and AMs using the eel as a model species and the results have inspired new concepts in fish osmoregulation. From a comparative perspective, these hormones appear to be more significant for osmoregulation than for the control of blood pressure in teleosts. Most teleosts possess relatively low blood pressure (20–40 mmHg) as the buoyancy given by the environmental water partially neutralizes the effects of gravity (Takei 2000a). On the other hand, they face severe osmotic challenges (salt-loading in SW and water-loading in FW) and this selection pressure has driven teleosts, including eels, to develop body fluid regulatory systems that are under tight hormonal control. The four hormone systems described in this chapter respond to abrupt and chronic salinity fluctuations and act rapidly on various organs including the brain, gill, alimentary tract, and kidney to modulate the activities of ion-channels and transporters, as well as to control drinking behavior, to maintain stable body fluid osmolality (Fig. 1).

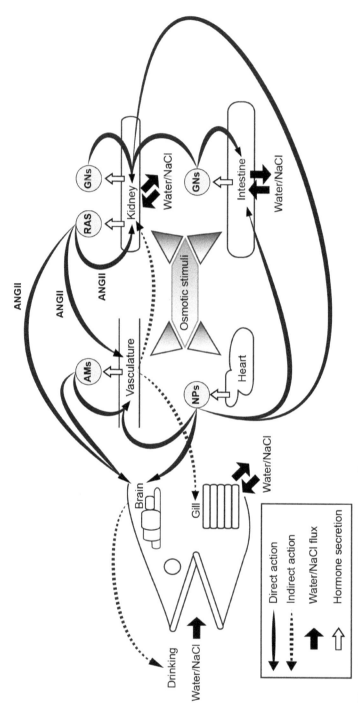

Figure 1. Secretory organs and biological actions of renin-angiotensin system (RAS), natriuretic peptides (NPs), guanylins (GNs), and adrenomedulins (AMs) in the eel.

These actions are different from, but complementary to, the spectra of well-defined osmoregulatory hormones such as prolactin (PRL), growth hormone (GH), and cortisol (Table 1). These adenohypophysial hormones and cortisol reconstruct or remodel the osmoregulatory organs by promoting *de novo* synthesis of transporters and channels when fish encounter salinity changes, and their actions are slow but long-term (Takei et al. 2006). The growth of molecular biology alongside the production of genome databases for fish have enriched our knowledge and led to ground-breaking findings in the endocrinology of fishes. Combining molecular techniques and physiological studies, we believe that research on the eel and the use of the eel as a model euryhaline fish has and will continue to make a great strides forward in the field of comparative endocrinology.

Table 1: Other osmoregulatory hormones and their functions in eel

hormone	selected biological functions	references
ACTH	intestinal water and ion permeability ↑	Hirano and Utida 1968
cortisol	intestinal water and ion permeability ↑ branchial ion permeability ↑	Hirano and Utida 1968 Mayer et al. 1967 Kamiya 1972
prolactin	esophageal water and ion permeability ↓ drinking ↓	Utida et al. 1972 Hirano et al. 1976 Kozaka et al. 2003
GH	branchial ion permeability ↑(salmonids) *GH action is a less significant in eel	Sakamoto et al. 1993 *Duan and Hirano 1991
AVT	low dose: antidiuresis (FW eel) high dose: diuresis (FW eel) vasoconstriction drinking ↓ plasma AVT level ↑(FW-SW transfer)	Babiker and Rankin 1978 Bennett and Rankin 1986 Ando et al. 2000 Henderson et al. 1985
bradykinin	blood pressure ↑ drinking ↓ plasma ANGII level ↑	Takei et al. 2001b
urotensin II/URP	blood pressure ↑ drinking ↓	Nobata et al. 2011 Nobata and Takei 2011
VIP	drinking ↓	Kozaka et al. 2003
ghrelin	drinking ↓ plasma ghrelin level ↑(FW to SW transfer)	Kozaka et al. 2003 Kaiya et al. 2006

abbreviations) ACTH: adrenocorticotropic hormone, GH: growth hormone, AVT: arginine vasotocin, URP: urotensin II-related peptide, VIP: vasoactive intestinal peptide

Renin-Angiotensin System

The renin-angiotensin system (RAS) has widespread functions in the regulation of cardiovascular tone, mineral and water balance in vertebrates (Kobayashi and Takei 1996). The RAS is a cascade peptide system that is structurally conserved among vertebrates but it is imperative to recognize that the system is also functionally diverged among different lineages (Wong and Takei 2011). In this section, studies on the RAS in the eel are summarized and comparisons are made, when appropriate, among different species of eels as well as other vertebrate models (Note: Many experiments described in this chapter were performed with cultured immature Japanese eel, *Angulla japonica*. Stage of eel life cycle (yellow/silver) is not described unless otherwise stated).

RAS Cascade

The RAS is composed of a precursor protein and several enzymes that form an intimate cascade system (Fig. 2). Angiotensinogen (AGT), which is majorly produced in liver, is the precursor protein of angiotensins. Renin, an enzyme originally identified in kidney, cleaves AGT at the *N*-terminus to produce a decapeptide called angiotensin I (ANG I). ANG I is biologically

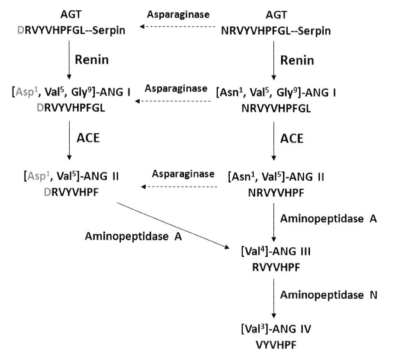

Figure 2. The Renin Angiotensin System (RAS) cascade in the eel. Angiotensinogen (AGT) is cleaved by renin to release angiotensin I (ANG I). ANG I is converted to ANG II by angiotensin converting enzyme (ACE). ANG II is further cleaved into ANG III and ANG IV by aminopeptidase A and aminopeptidase N, respectively. The asparagine residue of the AGT, ANG I, and ANG II are converted to aspartate residue by asparaginase.

inactive, and is converted to an octapeptide called angiotensin II (ANG II) by angiotensin converting enzyme (ACE) that truncates the dipeptide residue at the *C*-terminus of ANG I. ANG II is biologically active and involved in the regulation of blood volume, pressure, and drinking of vertebrates, principally *via* angiotensin type-1 (AT_1) and type-2 (AT_2) receptors. ANG II is further cleaved at the *N*-terminus to form angiotensin (2–8) (ANG III) and angiotensin (3–8) (ANG IV) by aminopeptidases.

In the eel, renin-like activity was found in the kidney and corpuscles of Stannius (Chester-Jones et al. 1966, Sokabe et al. 1966). Incubation of plasma and kidney extract produced [Asn[1], Val[5], Gly[9]]-ANG I in the Japanese and American eel (Hasegawa et al. 1983, Khosla et al. 1985) and the [Asn[1]] residue can be converted to [Asp[1]] at the *N*-terminus, resulting in the formation of significant amounts [Asp[1], Val[5], Gly[9]]-ANG I. Using HPLC separation in combination with radioimmunoassay (RIA) measurement techniques, [Asn[1], Val[5]]-ANG II, [Asp[1], Val[5]]-ANG II, [Val[4]]-ANG III, and [Val[3]]-ANG IV were all shown to be present in eel plasma (Wong and Takei 2012, Fig. 3). This is in contrast to mammals where plasma ANG III and ANG IV are maintained at almost undetectable levels (Plovsing et al. 2003). The affinities of various angiotensin subtypes towards their receptors have not

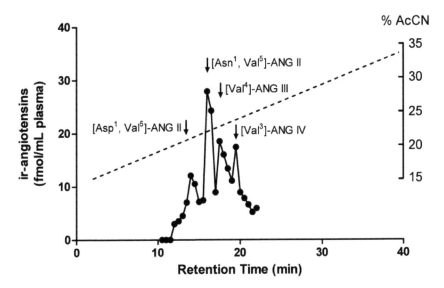

Figure 3. **A representative example of immunoreactive (ir) angiotensins in the plasma of the seawater Japanese eel.** The plasma is resolved by an HPLC system utilizing a neutral mobile phase [15–35% acetonitrile (AcCN) in 10 mM ammonium acetate, pH 7.0]. Dotted line indicates the AcCN gradient of the HPLC system. Concentrations of various angiotensin subtypes, [Asn[1], Val[5]]-ANG II, [Asp[1], Val[5]]-ANG II, [Val[4]]-ANG III, and [Val[3]]-ANG IV, are measured by radioimmunoassay and arrows indicate the elution positions of synthetic peptide standards.

been studied in fishes and it is possible that these truncated angiotensins possess specific receptors and therefore functions.

Angiotensin Receptors

Using pharmacological properties to specific antagonists, two major angiotensin receptors, AT_1 and AT_2, have been identified in mammals. Most of the well-established ANG II effects such as vasoconstriction and drinking stimulation are via AT_1 receptor signaling. However, recent studies suggest that additional receptors for various angiotensin subtypes are present. These include ANG (1–7) and ANG IV receptors that possess separate signaling pathways other than AT_1 and AT_2 receptors (for review see: Fyhrquist and Saijonmaa 2008). Most studies in fishes focused on the AT_1 receptor, and thus information regarding other angiotensin receptors is not currently available, partly due to the lack of specific blockers in fishes. Studies using mammalian AT_1 inhibitors such as saralasin and losartan in fishes produced inconsistent results and the drugs were mostly ineffective in blocking ANG II signaling (Nishimura et al. 1978, Russell et al. 2001). Losartan displaced ANG II binding in isolated trout glomeruli with an IC_{50} of 19 nM (Brown et al. 1997), but was ineffective in ANG II blocking in eel (MKS Wong and M Ando, personal communication). It was shown that amphibian AT_1 receptor was also insensitive to losartan due to the lack of amino acid sequences that are crucial for losartan to bind (Russel et al. 2001). Similarly, the putative eel AT_1 receptor also lacks the amino acid residues for losartan binding.

Although functional expression studies have not been performed on angiotensin receptors in fishes, angiotensin receptor subtypes have been identified in various organs using ligand binding assays (Marsigliante et al. 1994). A single population of AT_1-like receptors was identified in the membrane fraction of the kidney and intestine with an isoelectric point (pI value) of 6.5 and a high affinity to ANG II with Kd = 3.4 nM. In liver, two receptor subtypes were identified with pI 6.5 and 6.7; the former has binding characteristics similar to those of the kidney and intestine. Dithiothreitol reduction enhanced ANG II binding to the receptor isoform of pI 6.5 but reduced the binding to that of pI 6.7. The decrease in binding by reduction is a common characteristic of the AT_1 receptor in mammals (Fyhrquist and Saijonmaa 2008). Nevertheless, an antibody to mouse AT_1 receptor recognizes the receptor isoform of pI 6.5 but not pI 6.7, and immunoblotting showed that the receptor has a molecular weight of 75 kDa, which is comparable to those of mammals. Taken together, both isoforms possess certain characteristics of the mammalian AT_1 receptor, but further molecular characterization is required to determine their functional roles. Using the antibody against the mouse AT_1 receptor, the receptor was found to localize on the basolateral membrane of mitochondrial rich cells on the gill epithelia, basolateral membrane and intracellular sites of the absorptive epithelial cells of intestine and proximal tubular cells of the kidney (Marsigliante et al. 1994). Perfusion of gill epithelium with ANG II internalized the immunoreactive signal, which suggested a local metabolism of ANG II by the immunoreactive receptor in the chloride cells.

Biological Actions of Angiotensin II

Cardiovascular effects

In the European eel, Imbrogno et al. (2003) showed that ANG II-induced negative chronotropism and inotropism in an isolated perfused heart preparation. The ANG II-induced inhibition of cardiac contraction could be via the AT_1 receptor, as the AT_1 receptor antagonist, CV11974, abolished the effect. The effect was found to be endothelium-dependent, thus it was suggested that endothelial nitric oxide synthase (eNOS) and the cGMP signaling could be involved. However, recent studies suggested that eNOS is not present in teleosts but a soluble guanylyl cyclase (GC) and nitric oxide signaling are present in the endothelium (Donald and Broughton 2005). Genome data from various teleost species including zebrafish, medaka, stickleback, and pufferfish have also indicated that teleosts lack the *NOS3* gene that transcribes eNOS. Due to this intrinsic difference in vascular signaling between mammals and fishes, the signaling of ANG II-induced chronotropism and inotropism remains uncertain.

Intra-arterial injection of [Asp[1]]-ANG II in American yellow eel elicits a vasopressor response in a dose-dependent manner with the minimal effective dose at 1 pmol/kg (Nishimura and Sawyer 1976). Intra-arterial injection of ANG II increased dorsal aortic pressure which was not blocked by the α_1-adrenergic blocker prazosin (Bernier et al. 1999). Neither intra-arterial injection of ANG II *in vivo* nor *in vitro* perfusion of ANG II in the interrenal gland affected catecholamine secretion (Bernier et al. 1999). However, it was also shown that the non-selective α-adrenergic blockade by phentolamine lowered the ANG II-induced vasopressor response in FW eel (Oudit and Butler 1995). This indicates that the vasopressor effect of ANG II was partly integrated with catecholamine release via the α_2-adrenergic pathway. The vasopressor effect of intra-arterial injection of native [Asn[1]]-ANG II in the Japanese eel has also been shown to be higher in the dorsal aorta as opposed to the ventral aorta, indicating that ANG II preferentially constricts systemic vessels as opposed to branchial vasculature (Nobata et al. 2011).

There is a higher resting blood pressure in European eels adapted to FW than those adapted to SW (Chester-Jones et al. 1969, Tierney et al.

1995). However, a higher plasma renin activity was observed in SW eel (Henderson et al. 1976). Infusion of captopril, an ACE inhibitor, decreased blood pressure in SW eel but had no effect on the resting blood pressure of FW eel (Tierney et al. 1995). This suggests that circulating ANG II may have a role in maintaining blood pressure in a dehydrating environment. Although resting blood pressure is lower in SW eel, it does not necessarily represent low circulating renin or angiotensin levels. Papaverine, a smooth muscle relaxant that artificially induces hypotension (Bernier et al. 1999), has been shown to stimulate the endogenous RAS and elicit a 10 fold increase in plasma ANG II concentration of SW eel and a 3.5 fold increase in FW eel (Tierney et al. 1995). Therefore, the RAS is likely more responsive in SW rather than in FW, since papaverine treatment induced a greater increase in plasma ANG II levels in SW eel. In the European eel, FW to SW transfer rapidly increased plasma renin activity (Henderson et al. 1976) and plasma ANG II concentrations were higher in SW (Henderson et al. 1985, Tierney et al. 1995). However, in the Japanese eel, plasma renin activity and ANG II concentration in FW and SW are not significantly different (Sokabe et al. 1973, Tsuchida and Takei 1998), which raises the question whether the RAS status is related to SW acclimation (Table 2). The species-dependent difference implies evolutionary divergence of the functional role of the RAS in these two closely related species. Sokabe et al. (1973) demonstrated that plasma renin activities increased during the course of FW to SW transfer but were not significantly different in long-term (2–3 weeks) acclimated Japanese eel. Similarly, in American eel, a significant decrease in plasma renin activity was observed after 3 days transfer from SW to FW (Nishimura et al. 1971). In a short-term SW transfer experiment in Japanese eel, circulating ANG II levels were significantly elevated during the SW acclimation period, which peaked at 1 to 3 days but returned to pre-transfer levels after 7 days (Okawara et al. 1987, Wong and Takei 2012). Therefore, it is certain that an elevated status of the RAS is involved in SW acclimation in the Japanese eel, but this species is able to reach a new equilibrium in SW without continuous elevation of circulating ANG II. Furthermore, Japanese eel acclimated to double-strength SW was found to have a higher circulating level of ANG II, which supports the notion that the RAS is activated in a dehydrating environment (Wong and Takei 2012).

Table 2: Comparison between the plasma ANG II levels of FW and SW eels

		Anguilla anguilla	*Anguilla japonica*
Plasma	SW	32.9 ± 4.2[*][a]	102 ± 20[b]
(fmol/ml)	FW	9.7 ± 0.6[a]	81 ± 16[b]

Statistical difference between SW and FW group is indicated by asterisk (p<0.05)
References) a: Tierney et al 1995, b: Tsuchida and Takei 1998

Renal effects

In teleosts, systemic blood pressure is not a major determinant for glomerular filtration rate (GFR), since glomerular intermittency is a unique feature that controls the population of filtering, non-filtering, and non-perfused nephrons (Brown et al. 1990), which ultimately determine the actual GFR. [Asp¹]-ANG II infusion lowered the population of filtering and non-filtering nephrons in FW and SW rainbow trout, respectively (Brown et al. 1980). Incubation of FW trout kidney slices with ANG II changed the filtering surface of the glomeruli to SW type *in vitro* (Brown et al. 1990). Similar data are not currently available in eels but intra-arterial ANG II infusion significantly increased the inulin clearance, urine flow rate, sodium filtration and reabsorption rate in the American yellow eel (Nishimura and Sawyer 1976). The ANG II-induced vasopressor effect and increased GFR were not correlated, which implies a similar glomerular intermittency is present in the eel. Net sodium reabsorption was lowered by ANG II treatment, which suggests that ANG II modulates renal tubular reabsorption and secretion in addition to its role in the regulation of GFR (Nishimura and Sawyer 1976). The decrease in net sodium reabsorption is an important feature in SW fish, since it facilitates ion excretion. Taken together, ANG II could stimulate the remodeling of the nephrons in the eel from a filtering form (to excrete water) to a secretory form (to excrete salt) and its effects are important in SW acclimation.

Cortisol secretion

Exogenous renin and ANG II increased circulating levels of adrenocorticotrophic hormone (ACTH) and cortisol in FW eel (Henderson et al. 1976). Renin injection increased the circulating levels of cortisol in control eel but not in hypophysectomized eel, which implies that the stimulation of ANG II on cortisol secretion could be via the pituitary-ACTH-interrenal axis (Hirano 1969). When the European eel was transferred from FW to SW, the elevation in plasma cortisol was blocked by captopril (Kenyon et al. 1985). In the Japanese eel, ANG II enhanced cortisol production in interrenal tissue *in vitro*, which indicates that ANG II could have a direct effect on the regulation of cortisol secretion apart from its action via the pituitary-ACTH-interrenal axis (Ventura 2011).

Na⁺/K⁺-ATPase activity

Na⁺/K⁺-ATPase is often the driving force for membrane transport systems, and exists abundantly in the osmoregulatory epithelia of gill, kidney, and intestinal tissues. Branchial Na⁺/K⁺-ATPase activity was shown to be stimulated by cortisol treatment in the American yellow eel (Epstein et al. 1971). Since the RAS partly regulates cortisol secretion, it is likely that the status of the RAS serves as a 'bottle-neck' in determining Na^+/K^+-ATPase activity. Besides the indirect effect of RAS on Na^+/K^+-ATPase, *in vitro* ANG II treatment has a direct stimulatory effect on the Na^+/K^+-ATPase activity in both isolated branchial and renal tissues of the European yellow eel (Marsigliante et al. 1997, 2000). Perfusion of gills for 30 min with ANG II (0.1–100 nM) increased Na^+/K^+-ATPase activity in a dose-dependent manner in FW eel. In SW, Na^+/K^+-ATPase activities were stimulated by a low dose of ANG II but not by a high dose. In eel kidney, Na^+/K^+-ATPase activity was 2-fold higher in SW than in FW. ANG II perfusion had no effect on the kidneys of SW eel but stimulated Na^+/K^+-ATPase activity dose-dependently in the proximal and distal tubules of the kidney in FW eel to a level close to that observed in the kidneys of SW eel (Marsigliante et al. 2000). These data suggest that the RAS is involved in the reorganization of the osmoregulatory organs upon SW acclimation. On the other hand, ANG II decreased Na^+/K^+-ATPase activity in isolated enterocytes of SW eel (Marsigliante et al. 2001). The inhibition was induced by a transient increase in intracellular calcium and dependent on protein kinase C (PKC) activation. In addition to the inhibitory effect on Na^+/K^+-ATPase in enterocytes, ANG II stimulated basolateral Na^+/H^+ antiporter through AT_1-like receptors (Vilella et al. 1996). In brief, ANG II is involved in modulating transporters and ion pumps to reorganize the osmoregulatory epithelia in eel in response to rapid salinity changes.

Drinking regulation

Since drinking is the first step controlling water and electrolyte intake from the environment, it is crucial for fish osmoregulation. Indeed, in SW eels, Na^+ intake by drinking accounts for about a quarter of total Na^+ influx (Tsukada and Takei 2006). The eel is one of the model for which the measurement of drinking rate in a conscious state is well-established (Takei et al. 1998). The drinking rate of conscious eels is measured by a drop counter connected to an esophageal catheter, which is synchronized to a pulse injector that reintroduces the same amount of ingested water (80% SW is used for SW eels and FW for FW eels) into the stomach. Using this system, many dipsogens (a substance that causes thirst) and antidipsogens (a substance that suppresses thirst) were identified in the eel and the regulatory mechanisms were studied (for details see chapter "Regulation of Drinking").

ANG II is a well-known dipsogen in vertebrates (Kobayashi and Takei 1996). Injection of ANG II or stimulation of the endogenous RAS by papaverine increased drinking in eels (Takei et al. 1979, Takei et al. 1988, Tierney et al. 1995, Tsuchida and Takei 1999, Ando et al. 2000, Kozaka et al.

2003). In the Japanese eel, hypovolemia caused by hemorrhage stimulated the endogenous RAS and enhanced drinking (Hirano 1974, Tsuchida and Takei 1998). The dipsogenic signal by ANG II was shown to be via the hindbrain, since removal of the forebrain did not impair the ANG II-induced drinking (Takei et al. 1979). The systemic injection of ANG II produced a biphasic effect on drinking in the eel (Ando et al. 2000): a stimulatory phase during the initial 10 min followed by an inhibitory phase that lasted for more than 30 min via the area postrema in the hindbrain, but intracranial injection only enhanced drinking rate without the inhibitory phase (Ogoshi et al. 2008). Using Evans blue as a tracer, blood-contacting neurons were identified in the area postrema (Mukuda et al. 2005). These blood-contacting neurons were not immunoreactive to an ANG II antiserum, but studies on a receptor level have not yet been performed. On the other hand, there is conflicting data on the role of ANG II on the regulation of drinking in SW eel. Intra-arterial infusion of captopril significantly reduced the blood pressure, plasma ANG II, and drinking rate of SW eel (Takei and Tsuchida 2000). However, intra-arterial infusion of an anti-ANG II antiserum, which significantly lowered the plasma ANG II, had no effect on either drinking rate or blood pressure.

Natriuretic Peptide Family

Since the discovery of atrial natriuretic peptide (ANP) from the mammalian heart and its action as a potent natriuretic factor on renal function (de Bold et al. 1981), there has been significant effort in characterizing and identifying a functional role for the NP family in teleost fish. The NP family was first identified in the Japanese eel and a number of investigations using eels combined with homologous assay systems revealed that the eel NP family acts on a variety of tissues to maintain body fluid homeostasis. However, the physiological function of NP's has evolved differently from mammalian NP's.

Molecular Structure of NPs

Natriuretic peptides (NPs) are peptide hormones with 22–36 amino acid residues and play important roles in cardiovascular and body fluid homeostasis. The molecular structure is characterized by a 17-amino acid intramolecular ring, which is formed by a disulfide bond between two cysteine residues (Fig. 4). NP was initially purified and sequenced in the human atria (Kangawa and Matsuo 1984) and named atrial natriuretic peptide (ANP: current approved symbol from HUGO Gene Nomenclature Committee database is 'NPPA'). Five years after the discovery of human

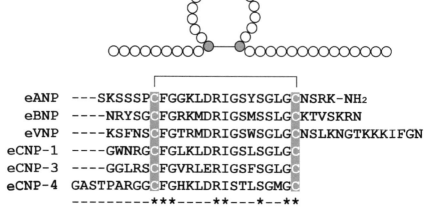

Figure 4. **Molecular structure of natriuretic peptides (NPs) in the Japanese eel.** Two cysteine residues (gray filling) are used for a disulfide bond. Asterisks indicate conserved amino acids among NPs. All NPs have an intramolecular ring structure within the molecule. C-terminal end of eANP is amidated (-NH$_2$). All eCNPs lack a tail sequence after the intramolecular ring. CNP2 is not expressed in Japanese eel.

ANP, eel ANP (eANP) was isolated from the atrium (Takei et al. 1989) (Fig. 4). Circulating eANP-(1-27) has sequence homology of ca. 60% to that of human ANP, and the vasodepressor effect of eANP in eels was 110 times more potent than that of non-homologous rat ANP (Takei et al. 1989). Interestingly, unlike tetrapod ANPs, eel and some other teleost ANPs are amidated at the C-terminal residue (Takei et al. 1997), and the amidated eANP (eANP-NH$_2$) can stimulate cGMP production more effectively than non-amidated eANP through eel natriuretic peptide receptor-A (eNPRA: approved symbol 'NPR1') (Kashiwagi et al. 1999).

One year after eANP was sequenced, C-type natriuretic peptide (CNP: approved symbol 'NPPC') was identified from eel brain (Takei et al. 1990). CNP is the most conserved member in the NP family (ca. 80% homology throughout vertebrates) and does not possess the C-terminal 'tail' sequence after the intramolecular ring (Fig. 4). Since CNP is expressed mainly in the brain, it was suggested that it acts as a paracrine factor in the central nervous system (CNS) (Fowkes and McArdle 2000).

In mammals, B-type natriuretic peptide (BNP: approved symbol 'NPPB') is predominantly expressed in the heart and acts as a circulating hormone similar to ANP. Although BNP had not been identified in teleost fish in 1990's, a distinct NP with a long C-terminal 'tail' sequence extending from the intramolecular ring, named ventricular natriuretic peptide (VNP: approved symbol 'NPPV'), was purified and sequenced from eel ventricle (Takei et al. 1991). *VNP* is present in eel, salmon, bichir, and sturgeon (early-diverged species) but absent in medaka, zebrafish, and pufferfish (recently-evolved species), suggesting that the *VNP* was lost before the divergence of some more recent species. VNP does not possess the amino acid residues that characterize BNP but its localization and functions are analogous to those of BNP. Therefore, *VNP* was initially thought to be an ortholog of *BNP*. In 2004, however, Kawakoshi et al. cloned *BNP* from the heart of the talapia and pufferfish and eel *BNP* (*eBNP*) was later sequenced by Inoue et al. (2005). Furthermore, Inoue et al. (2003) also showed that the ancestral *CNP* had diverged into 4 subtypes (*CNP1* to *CNP4*) in teleost fish and all four types of *CNPs* are expressed in the medaka and pufferfish tissues. Since *CNP1* (originally *CNP*), *CNP3* and *CNP4*, were found to be expressed in eels (Nobata et al. 2010) there are at least 6 types of NPs in eels (Fig. 4).

Tissue Distribution of NPs and Regulation of Release

Distribution of mRNA of all 6 eel NPs in different tissues was examined by reverse transcription-PCR (Nobata et al. 2010). Briefly, *eANP* and *eVNP* are exclusively expressed in the heart, while *eCNP1* and *eCNP4* are specifically expressed in the brain. *eBNP* is strongly expressed in the heart, and to a lesser extent, in the pituitary. A unique distribution pattern was observed for *eCNP3* that was strong expression in pituitary followed by intestine and brain, and weak expression in heart and pancreas.

A homologous RIA was developed to measure eANP and it was revealed that the concentration of eANP in the atrium is ca. 200 times higher than that of ventricle (Table 3) (Takei et al. 1992). eVNP is produced in both atrium and ventricle, and its concentration is also higher in the atrium (Table 3). However, total eVNP content is highest in the ventricle because of its large size (Takei and Balment 1993, Takei et al. 1994a). Electron microscopy showed electron-dense secretory granules in the cardiomyocytes of eel

Table 3: Natriuretic peptide tissue concentration of SW and FW eels

Tissue (concentration)		eANP	eVNP	eCNP-1
Atrium	SW	56101 ± 10283^a	3444 ± 502^a	~3[c]
(fmol/mg)	FW	10627 ± 1647^a	2544 ± 310^a	~5[c]
Ventricle	SW	41 ± 13^a	1265 ± 94^a	~2[c]
(fmol/mg)	FW	56 ± 8^a	1397 ± 83^a	~5[c]
Plasma	SW	60.7 ± 3.6^b	55.0 ± 3.7^b	~20[c]
(fmol/ml)	FW	68.1 ± 2.9^b	51.6 ± 3.0^b	~120[c]

References) a: Takei and Balment 1993, b: Kaiya and Takei 1996a, c: Takei et al. 2001

atrium and ventricle (Takei 2000b), which suggested that eANP and eVNP are stored in the granules similar to mammalian cardiac NPs. Precursors of eANP and eVNP are pro-ANP and pro-VNP (both 14 kDa) (Takei et al. 1994b, 1997). Since the circulating form of eANP is eANP-NH$_2$, the eel pro-ANP in the granules is cleaved and amidated when it is secreted. However, eel pro-VNP is processed in a different manner since mature eVNP (1–36), truncated form (1–25), and pro-VNP are all present in the circulation. The mature eVNP (1–36) is the major circulating form, and the amount of pro-VNP is approximately one quarter of the total circulating eVNP.

Circulating eANP and eVNP concentrations were measured in chronically-cannulated eels in FW and SW. The basal eANP and eVNP were not different between FW and SW adapted eels (Table 3) (Kaiya and Takei 1996a). However, the plasma eANP and eVNP levels transiently increased during the first few hours of FW to SW transfer, but their levels remained unchanged following a reverse transfer from SW to FW (Kaiya and Takei 1996b). Hypertonic saline injection into the circulation, and surprisingly mannitol injection that increased plasma osmolality but decreased plasma Na$^+$ concentration, were profound stimuli for eANP and eVNP secretion in the eel (Kaiya and Takei 1996c). Volume expansion by bolus injection of isotonic saline also increased plasma eANP and eVNP levels, but the increment was relatively small when compared to that of osmotic stimuli (Kaiya and Takei 1996c). These data indicate that eANP and eVNP secretion respond primarily to an acute increase in plasma osmolality (as opposed to plasma Na$^+$ concentration alone).

This is very intriguing in terms of comparative endocrinology, since volume loading is a major stimulus for ANP secretion in mammals. Although the mRNA expression is considerably lower, *eANP* and *eVNP* mRNAs were detected in extracardiac tissues including epithelium of entire intestine, brain, gill, red body (rete mirabile), kidney and interrenal (Loretz et al. 1997, Takei et al. 1997). *eCNP1* is mainly expressed in the brain, but also locally produced by the heart, liver, gills, intestine, and kidney to a lesser extent (Takei et al. 2001a). Despite the low gene expression levels in the heart, plasma eCNP1 concentration is high in FW eels (ca. 100 fmol/ml), and the plasma eCNP1 level diminishes to one-fifth in SW eels, which is consistent with the decrease in eCNP1 content in the atrium and ventricle (Takei et al. 2001). Circulating CNP (ortholog of teleost *CNP4*) is usually not found in mammals, but in elasmobranchs that only possess only CNP (ortholog of teleost *CNP3*), it was found that the CNP acts as both a circulating and a paracrine hormone (Kawakoshi et al. 2001). It appears that eCNP1 still retains some ancestral characteristics of CNP in the eel, although CNP3, from which ANP, BNP and VNP were generated by tandem duplication, still holds the feature of a circulating hormone (Miyanishi et al. 2011).

Plasma and tissue concentrations of eBNP, eCNP3 and eCNP4, as well as their secretory mechanism have not been determined in eels.

Natriuretic Peptide Receptors

Three natriuretic peptide receptors (NPRs) have been identified in vertebrates, and they are categorized into two groups according to whether they possess an intracellular guanylyl cyclase (GC) domain or not. Membrane-bound GC-coupled NPRs are involved in the cGMP mediated intracellular signaling cascade and are known to be the 'biological receptors'. The receptors are composed of 1) an extracellular ligand-binding domain, 2) a membrane-spanning domain, 3) an intracellular kinase-like domain, and 4) a GC domain, and they usually form homo-tetramers with disulfide bonds (Fig. 5). The GC-coupled receptors are further divided into two subgroups based on their ligand specificities: NPRA and NPRB (approved symbol: NPR1 and NPR2). ANP and BNP bind NPRA, while CNP is specific for NPRB. The third receptor, named NPRC (approved symbol: NPR3), does not possess the kinase-like and GC domains and thus lacks intrinsic GC activity (Fig. 5). Unlike GC-coupled receptors, NPRC forms homo-dimers and has broad ligand selectivity. The NPRC ligand-receptor complex is internalized into the cytoplasm and the ligand is degraded enzymatically, while NPRC appears to be recycled. Therefore, NPRC essentially behaves as a 'clearance receptor' to regulate the local concentration of NPs that are available to bind the biological receptors (Maack 1992).

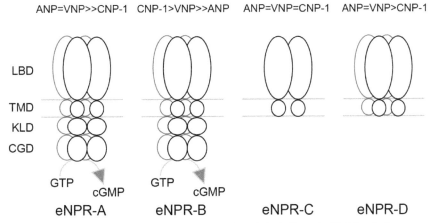

Figure 5. Natriuretic peptide receptors (NPRs) and their ligand affinities in Japanese eels. eNPRA and eNPRB are composed of an extracellular ligand-binding domain (LBD), a membrane-spanning domain (MSD), an intracellular kinase-like domain (KLD), and a guanylyl cyclase domain (GCD), and form homo-tetramers with disulfide bonds. NPRC and NPRD lack KLD and GCD. Relative affinity to each natriuretic peptide is shown above each receptor.

Eel *NPRA* (*eNPRA*) and *NPRB* (*eNPRB*) have been cloned from the cDNA libraries of the kidney and gill, respectively (Katafuchi et al. 1994, Kashiwagi et al. 1999). Although the homology of the extracellular ligand-binding domain is less conserved when compared to mammalian NPRA and NPRB (41% and 44% identity in amino acid sequence), the intracellular kinase-like and GC domains share 78% and 82% homology between eel and human NPRA and NPRB, respectively. eNPRA expressed in COS-7 cells has affinities to eANP, eANP-NH$_2$, eVNP (K$_d$ = 0.1 nM) and eCNP1 (K$_d$ = 1 nM) (Kashiwagi et al. 1999). Interestingly, eNPRA does not possess the third cysteine residue that is well-conserved among mammalian NPRAs, and it is essential to form the second intramolecular disulfide loop. Therefore, it was thought that the lack of a second loop results in high receptor affinities to eANP and eVNP, although it was shown that the cysteine residue is important for ANP binding to bovine NPRC (Iwashina et al. 1994). In terms of GC activity, eANP-NH$_2$ and eVNP can generate cGMP efficiently at physiological concentration, while a relatively high concentration of non-amidated eANP and eCNP1 is required for cGMP production (Kashiwagi et al. 1999). Therefore, it was suggested that the biologically active form is eANP-NH$_2$, and non-amidated eANP can be an endogenous antagonist.

RNA protection analysis revealed that *eNPRA* is widely expressed in various tissues, and relative strong expression levels were found in osmoregulatory organs including gill, intestine, kidney, and urinary bladder. An eNPRA variant, which lacks 9 amino acid residues (VFTKTGYYK) within the kinase-like domain, has no GC activity. The variant is also expressed in various tissues but its expression level is significantly lower than *eNPRA* (Kashiwagi et al. 1999). High expression levels of *eNPRB* were observed in gill, liver, and atrium, followed by moderate levels in brain, ventricle, intestine, interrenal, and kidney (Katafuchi et al. 1994, Ventura et al. 2011). The expression of *eNPRB* is salinity dependent and transfer of the Japanese eel from FW to SW markedly decreases its expression in gill, liver, and atrium (Katafuchi et al. 1994). Binding assay on eNPRB shows that eCNP1 is the most potent ligand to increase GC activity, and the potency is at least 500 times stronger than that of eANP. Since eVNP can stimulate cGMP production moderately through eNPRB, eVNP is thought to be a ligand for both eNPRA and eNPRB.

NPRC, a GC-deficient receptor, has been cloned from a cDNA library of eel gills (Takashima et al. 1995). Eel NPRC (eNPRC) shares approximately 60% identity with mammalian NPRC and all five cysteine residues in the extracellular ligand-binding domain are conserved. These structures are responsible for the formation of the first and second loop structures and homo-dimer. However, interestingly, site-directed mutagenesis analysis revealed that eNPRC uses the second cysteine residue within the first loop for dimerization, while mammalian NPRC uses the fifth cysteine residue within base of the extracellular domain. Although the disulfide linkage between the dimeric structures is different, the affinities of eNPRC to eANP, eCNP1, and eVNP are almost identical to those of mammals. *eNPRC is strongly expressed in gill and heart* and to lesser extents in other tissues, and the expression levels are generally down-regulated in SW except in the anterior intestine where the expression is up-regulated (Takashima et al. 1995, Mishina and Takei 1997). Another GC-deficient receptor, *NPRD*, was cloned from a cDNA library of the brain in the Japanese eel (Kashiwagi et al. 1995). Eel NPRD (eNPRD) and eNPRC share 70% identity at the amino acid level and most structural domains including the short cytoplasmic tail. Ligand-binding properties of eNPRD expressed in COS-7 cells are similar to that of eNPRC, except that the eCNP1 affinity is slightly lower than those of eANP, eANP-NH$_2$, and eVNP. However, biochemical and pharmacological analysis revealed that eNPRD exists as a homo-tetramer like eNPRA and eNPRB. In addition, *eNPRD is only expressed in brain and gills* in comparison to the widely distributed *eNPRC* (Kashiwagi et al. 1995). These results suggest that eNPRD is not merely a clearance receptor, although its physiological roles have not been investigated.

Osmoregulatory Actions of Natriuretic Peptides

Physiological functions of NPs have been studied in a number of fish models including eels. However, since the piscine NPs were not been fully characterized until the 2000's, early functional studies used heterologous rat NPs. Since the amino acid sequences of NPs and the structures of NPRs are not well-conserved among the different classes of vertebrates, the physiological responses and the potencies examined by a heterologous assay system are conflicting and the results should be interpreted with caution, therefore, in the following section only those studies using homologous NP's are discussed.

Drinking

Like mammals, NPs are antidipsogenic in the eel (Tsuchida and Takei 1998). In SW eels, intra-arterial injection of eNPs strongly inhibits drinking, and the potency varies among various NPs: eANP $\geq$ eVNP > eBNP = eCNP3 > eCNP1 $\geq$ eCNP4 (Miyanishi et al. 2011). The antidipsogenic effect of NPs was also observed in FW eels, although FW eels scarcely drink environmental water (Miyanishi et al. 2011). The antidipsogenic action of NPs is short-lived, and the drinking rate usually returns to control rates within 1 hr post-injection (Takei and Balment 1993). The short-term action is consistent with the rapid metabolic clearance rate of NPs in the circulation (Nobata et al. 2010). eANP is the most potent antidipsogen known in the eel and the potency is 2–3 orders higher than that of other hormones such as bradykinin, arginine vasotocin, and eel intestinal pentapeptide isolated from the eel intestine (Ando 2000, Takei et al. 2001). Site of administration also makes a difference to the level of potency and is a further 2–3 orders of magnitude higher when administered intracranially (Kozaka et al. 2003). The target site for the NP-induced antidipsogenic effect was strongly suggested to be the area postrema (AP) located on the dorsal surface of the medulla oblongata (hindbrain) of the eel (Ando 2000, Tsukada et al. 2007, Nobata and Takei 2011), since 1) systemic injection of eANP inhibits neuronal activity of the medulla oblongata, 2) the AP is a circumventricular organ, which lacks a blood-brain-barrier and acts as an interface between the circulation and CNS, 3) immunohistochemistry of eNPRA showed a dense signal localized at the AP region, and 4) heat-coagulative and chemical lesion of the AP region diminishes the ANP-induced antidipsogenic effect in SW eels.

Intestine

The intestine is an important organ for fish osmoregulation, in particular, for water and electrolyte absorption. In eel, the intestine is divided into 3 segments (anterior, middle, and posterior), and monovalent ions (Na^+ and Cl^-) are actively absorbed in the anterior and middle intestine with water (Tsukada et al. 2005). Electrophysiological analysis showed that the isolated anterior and middle intestine bathed in normal Ringer solutions express a spontaneous serosa-negative transepithelial potential difference (TEP), which accounts for active Cl^- transport from mucosa to serosa (Ando and Hara 1994). Serosal administration of eNPs reduces the short-circuit current (I_{sc}), which indicates an inhibition of coupled Na^+-Cl^- uptake from mucosa, in a dose-dependent manner and the potencies were: eANP-NH_2 > eVNP > eANP >> eCNP1 (Loretz and Takei 1997). Although the TEP and I_{sc} in the SW eel intestine were higher than those of the FW eel, the inhibitory action of eNPs and their sensitivities are the same in both salinities (Loretz and Takei 1997). The inhibitory action of eANP was also examined *in situ* by monitoring the changes in Na^+ and Cl^- concentrations in the anterior/middle intestine of SW eels after systemic eANP infusion (Tsukada et al. 2005). The data showed that hourly eANP infusion strongly inhibited Na^+, Cl^-, and water absorption, whereas the ionic concentration and fluid volume in the esophageal segment were not influenced by the eANP infusion. Similar to the antidipsogenic action of eANP, its action on the intestine was also short-lived, and the absorption rate recovered to the normal level by subsequent hourly infusion of saline. The action of eANP was not blocked by tetrodotoxin, indicating that enteric nerves in the submucosa were not involved in the eANP-mediated effects (Ando et al. 1992). ANP stimulated cGMP accumulation in the intestine of the winter flounder (O'Grady 1989), and cGMP analogue, 8-bromoguanosine 3', 5'-cyclic monophosphate, induced an inhibitory effect on intestinal NaCl absorption that mimicked the actions of ANP (Ando et al. 1992). Therefore, eANP and eVNP are thought to act directly on the intestinal epithelial cells and affect the ion transport (presumably Na^+-Cl^- and/or Na^+-K^+-$2Cl^-$ cotransporters on the apical membranes) through the cGMP-dependent signaling cascade. It was also suggested that NP-ergic cells located in the eel intestine may produce eNPs locally to regulate the intestinal ion transport (Loretz and Takei 1997, Loretz et al. 1997).

Gills and kidney

The gill is a major osmoregulatory organ that actively transports Na^+ and Cl^- through mitochondrial-rich cells on the branchial epithelia. However, since the clearance receptor, eNPRC, is predominant in eel gills (Mishina and Takei 1997), the gill does not show noticeable biological responses to eNPs. Rather, the gill is considered to be an organ that regulates circulating eNP levels *via* eNPRC. The branchial eNP clearance can be explained by the fact that the plasma eANP and eVNP levels in the ventral aorta (before branchial perfusion) are higher than those in the dorsal aorta (after branchial perfusion) in cannulated conscious FW eels (Kaiya and Takei 1996a). There is a small population of eNPRB in the FW eel gill and eCNP1 can produce cGMP in the branchial membrane fraction. However, the physiological function of eCNP1 in the gills has not been elucidated.

In mammals, ANP targets the kidney to induce natriuresis and diuresis, which results in a decrease in blood volume. Interestingly, eANP is natriuretic, but antidiuretic in SW eels (Takei and Kaiya 1998), therefore, net Na^+ clearance (urine volume x Na^+ concentration) remains unchanged before and after eANP treatment. The antidiuretic action of eANP may increase blood volume in eel. However, since eANP, as well as eBNP and eVNP treatments, increases hematocrit values in SW eels (indicating a decrease in blood volume) (Miyanishi et al. 2011), the role of eANP-induced antidiuresis on the regulation of blood volume is still unclear. The natriuretic and antidiuretic effects of eANP were not observed in FW eels (Takei and Kaiya 1998).

Although direct effects of eNPs on the gills and kidney have not been demonstrated in eels, eNPs may indirectly modulate the activity of these organs. Intra-arterial injection of eNPs transiently decreased blood pressure (eANP > eVNP > eBNP > eCNP4 > eCNP1 = eCNP3 in SW eel, eANP = eVNP = eBNP > eCNP1 = eCNP3 = eCNP4 in FW eel) (Nobata et al. 2010, Miyanshi et al. 2011). The hypotensive action is partially mediated by the beta-adrenergic system, since pretreatment of propranolol (non-selective beta adrenergic antagonist), but not phentolamine (non-selective alpha adrenergic antagonist) and atropine (cholinergic antagonist), attenuated the eNP-induced hypotension (Nobata et al. 2010). The hypotensive effects of NPs at the ventral and dorsal aorta are similar, which suggests that eNPs influence both branchial and systemic circulation equally. Therefore, the vasodilatory effects of eNPs may affect the branchial and renal hemodynamics, which in turn may alter the branchial perfusion and glomerular filtration rate.

Natriuretic Peptides and Physiological Relevance

According to the effects of eNPs on the osmoregulatory organs mentioned above, the actions of eNPs are fast but short-lived. eANP and eVNP reduce Na^+ influx by inhibition of drinking and intestinal Na^+ absorption in the eel. Although the inhibitory effects were observed in both FW and SW eels (Loretz and Takei 1997, Tsukada and Takei 2001, Miyanishi et al. 2011), the reduction of Na^+ loading is physiologically significant in terms of SW acclimation since SW fish are always exposed to a risk of hypernatremia (increase in plasma Na^+ concentration). Indeed, systemic eANP or eVNP infusion induced hyponatremia (decrease in plasma Na^+ concentration) in SW eels but not in FW eels (Tsukada and Takei 2001, Miyanishi et al. 2011), and the reduction of plasma Na^+ concentration can be explained by the reduction of oral Na^+ intake and intestinal Na^+ uptake (Tsukada et al. 2005). Other studies have shown that the hyponatremic effect of eANP disappears when the eel was forced to drink SW at a normal rate during eANP infusion (Tsukada et al. 2005). Interestingly, the elevated drinking after SW transfer (known as 'chloride response') is diminished soon after the transient increase in plasma eANP and eVNP (Hirano 1974, Kaiya and Takei 1996b). Since the drinking pattern is just opposite to that of plasma eANP and eVNP level, eANP and eVNP may act to counter the abrupt increase in plasma Na^+ concentration by reducing oral and intestinal Na^+ uptake in the initial phase of SW acclimation. Furthermore, eANP induces cortisol secretion from the interrenal of the eel (Li and Takei 2003). Cortisol in fish is important for long-term SW acclimation (Balment et al. 1987) as it is known to increase Na^+/K^+-ATPase in branchial mitochondrial-rich cells and angiogenesis in the intestinal tract to increase active transport of water and ions. Therefore, eANP released just after SW transfer may also induce cortisol secretion and indirectly promote long-term SW acclimation of eels. In addition to the osmoregulatory actions of eANP and eVNP during SW transfer, eANP and eVNP may also be important in regulating drinking in long term acclimated SW eels, since the drinking rate of SW eels increased when plasma eANP and eVNP were neutralized with anti-eANP and eVNP antiserum.

Guanylin Family

In marine teleost fish, the intestine plays a pivotal role as an osmoregulatory organ in regard to water and NaCl absorption from imbibed SW. Thus, it is hypothesized that intestinal salt and water absorption is controlled by intestinal fluid regulatory hormones in SW fish.

Guanylin hormones were expected to be the most probable candidates based on the historical finding in mammals that oral intake of hypertonic saline resulted in greater natriuresis than intravenous administration in the rabbit and human (Lennane et al. 1975a, b, Carey 1978). In this experiment, presumably, the increase in luminal Na^+ levels was detected by an intestinal sodium sensor, and then, an intestinal natriuretic factor was released according to the sensor's signal and conveyed to the kidney to induce natriuresis. To date, the primary candidate for this natriuretic factor is the guanylin family (Forte et al. 2000, Forte 2004, Fonteles and do Nascimento 2011). Indeed, intestinal production and/or expression of the guanylin peptides in rats is up- and down-regulated by oral intake of high and low salt, respectively (Li et al. 1996, Kita et al. 1999, Carrithers et al. 2002). It is unlikely that ANP is this factor, as plasma ANP level remained unchanged after oral sodium loading (Saville et al. 1988) and ANP is produced scantly in the intestine relative to the heart (González Bosc et al. 2000, Takei and Hirose 2002).

The mechanism of intestinal Na^+ sensing and subsequent release of a natriuretic factor from the intestine (Lennane et al. 1975a, b, Carey 1978) could be more critical for body salt regulation in SW fish than in FW fish and mammals, because the alimentary tracts of SW fish are routinely exposed to hypertonic fluids coming from the continuously ingested SW. Especially in eels, an intestinal NaCl sensing system has been suggested to exist (Hirano 1974, Ando and Nagashima 1996). The evidence drove researchers to study fish guanylin family on eels, particularly Japanese and European eels.

Guanylin Peptides and their Receptors

Guanylin peptides (GNs) are hormones that consist of 15–17 amino acid residues with two disulfide bonds (Fig. 6) and bind to the guanylyl cyclase C (GC-C) receptor to activate cGMP production (Fig. 7) (Forte et al. 2000, Forte 2004, Takei and Yuge 2007, Fonteles and do Nascimento 2011). In mammals, two hormone genes, *guanylin* (*GN*: approved symbol '*GUCA2A*') (Currie et al. 1992) and *uroguanylin* (*UGN*: approved symbol '*GUCA2B*') (Hamra et al. 1993), and one receptor gene, *GC-C* (approved symbol '*GUCY2C*') (Schulz et al. 1990), have been identified, and both hormones bind to the same GC-C. The well-known functions of the guanylin system are to stimulate Cl⁻ secretion via the cystic fibrosis transmembrane conductance regulator

(A)

Eel guanylin (GN)

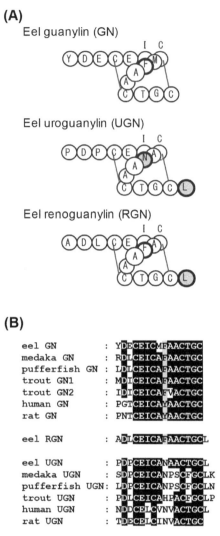

Eel uroguanylin (UGN)

Eel renoguanylin (RGN)

(B)

```
eel GN          : YDECEICMFAACTGC
medaka GN       : RDLCEICAFAACTGC
pufferfish GN   : LDLCEICAFAACTGC
trout GN1       : MDICEICAFAACTGC
trout GN2       : IDICEICAFVACTGC
human GN        : PGTCEICAMAACTGC
rat GN          : PNTCEICAMAACTGC

eel RGN         : ADLCEICAFAACTGCL

eel UGN         : PDPCEICANAACTGCL
medaka UGN      : SDPCEICANPSCFGCLK
pufferfish UGN  : LDPCEICANPSCFGCLN
trout UGN       : PDLCEICAHPACFGCLP
human UGN       : NDDCELCVNVACTGCL
rat UGN         : TDECELCINVACTGC
```

Figure 6. Structure of eel guanylin (GN), uroguanylin (UGN) and renoguanylin (RGN) (A), and alignment of mature peptide sequences of guanylin genes in fish and mammals (B). Two guanylin-like genes are found in rainbow trout, and their amino acid sequences exhibit 81% identity. However, it is not known whether this is derived from the salmonid genome tetraploidy or one of them is another type of guanylin. GeneBank Accession Numbers: Japanese eel GN, AB080640 (European eel GN, FM173260); medaka GN, BJ877267; pufferfish, deduced from scaffold_178 in the *Takifugu* genome database 2012; rainbow trout GN1, BX866654; rainbow trout GN2, deduced from the contig 13015 in the salmon EST database 2012; human GN, M95174; rat GN M93005; Japanese eel RGN, AB080641 (European eel RGN, FM173262); Japanese eel UGN, AB080642 (European eel UGN, FM173261); medaka UGN, BJ512940; pufferfish UGN, deduced from scaffold_178 in the *Takifugu* genome database 2012; rainbow trout UGN, BX309850; human UGN, U34279; rat UGN, U41322.

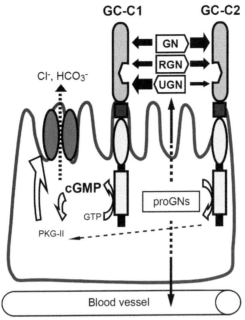

Figure 7. A predicted model of action of guanylin (GN), uroguanylin (UGN) and renoguanylin (RGN) on guanylyl cyclase C1 and 2 (GC-C1 and GC-C2) receptors in an eel enterocyte. The width of arrows indicates potency of each guanylin for cGMP production on each GC-C. A cystic fibrosis transmembrane conductance regulator (CFTR) Cl⁻ channel at the mucosal side is presumably activated by the guanylin-GC-C cGMP signaling.

(CFTR) Cl⁻ channel in association with water secretion in the intestine (Chao et al. 1994, Cuthbert et al. 1994, Volant et al. 1997, Joo et al. 1998, Forte et al. 2000) and to induce natriuresis, kaliuresis and diuresis in the kidney (Greenberg et al. 1997, Fonteles et al. 1998, Forte et al. 2000, Fonteles and do Nascimento 2011). In both Japanese and European eels, however, three different *guanylin* genes (Comrie et al. 2001a, Yuge et al. 2003, Cramb et al. 2005, Kalujnaia et al. 2009) and two isoforms of *GC-C* genes (*GC-C1* and *GC-C2*) (Comrie et al. 2001b, Yuge et al. 2006, Kalujnaia et al. 2009) have been identified. The third *guanylin* was named '*reno*'*guanylin* (reno = renal) because of its abundant mRNA expression in the kidney (Yuge et al. 2003). In the following, we do not distinguish between Japanese and European eels, when similar findings are reported.

Since four cysteine residues are conserved among all vertebrate species (Fig. 6B) (Takei and Yuge 2007), the putative mature peptides of all three eel guanylins could form a tertiary structure with two disulfide bonds like the mammalian guanylin and uroguanylin (Fig. 6A). Eel guanylin and uroguanylin were regarded as the mammalian counterparts based on the characteristic ninth amino acid residue (Fig. 6). It is believed that the aromatic amino acid residue (Tyr/Phe) of guanylin at the ninth position is a cleavage site where a chymotrypsin-like endopeptidase present in brush border membranes of renal proximal tubules acts to inactivate guanylin (Arao et al. 1994, Forte et al. 2000), while the non-aromatic amino acid residue (Asn/His) enables uroguanylin to act on the renal tubules without being degraded by this enzyme (Forte et al. 2000). In addition, the presence of additional one or two amino acid residues at the C-terminus is observed in uroguanylins of various vertebrate species including eels (Fig. 6) (Takei and Yuge 2007). Leu at the 16th position is considered to stabilize the peptide conformation (Schulz et al. 1998), which might also contribute to the stability of uroguanylin in the renal tubules. Interestingly, the putative mature sequence in renoguanylin shares features of both guanylin and uroguanylin such as an aromatic amino acid (Phe) residue at the ninth position and an additional Leu residue at the C-terminus (Fig. 6) (Yuge et al. 2003, Cramb et al. 2005, Kalujnaia et al. 2009). However, the renoguanylin sequence is more similar to the guanylin sequence than the uroguanylin sequence (Fig. 6B). Clustering renoguanylin into the guanylin group is also shown in phylogenetic analyses based on prepropeptide sequences in vertebrates (Takei and Yuge 2007). To date, this unique *renoguanylin* has not been identified in other animals, even in other teleost fish.

Both eel GC-C1 and GC-C2 receptors possess a considerable number of motifs characteristic to the mammalian GC-C, consisting of four major domains in their deduced amino acid sequences (Yuge et al. 2006, Kalujnaia et al. 2009), and exhibit ability of cGMP production in response to synthetic eel guanylins (Fig. 7) (Yuge et al. 2006). However, phylogenetic analyses have not determined which eel *GC-C* is an ortholog of the mammalian *GC-C* (Yuge et al. 2006). The two GC-Cs of Japanese eel share 67% identity at the whole amino acid sequences, but the identity is higher (78%) in the intracellular domain (kinase-like domain + cyclase catalytic domain) and lower (53%) in the extracellular ligand-binding domain (Yuge et al. 2006). Almost the same characterization is reported in the *GC-Cs* of the European eel (Kalujnaia et al. 2009). In COS cell lines expressing each gene, both eel GC-Cs generated cGMP after administration of synthetic eel guanylins, but exhibited differences in ligand-dependent cGMP production; the order of potency was uroguanylin > guanylin ≥ renoguanylin for GC-C1, guanylin ≥ renoguanylin > uroguanylin for GC-C2 (Fig. 7) (Yuge et al. 2006). Eel GC-C1 therefore appears to be similar to the mammalian GC-C, as mammalian GC-C also prefers uroguanylin to guanylin in terms of cGMP production (Hamra et al. 1993, Greenberg et al. 1997, Hamra et al. 1997, Forte et al. 2000). However, both eel GC-Cs are less responsive to a heat-stable enterotoxin produced by an enterotoxigenic *E. coli*, which is known to be a super-agonist of guanylins, than the mammalian GC-C (Yuge et al. 2006) (see Forte et al. 2000, Forte 2004, Lin et al. 2010). So far, no renoguanylin-selective *GC-C* has been found, but based on the available information, renoguanylin appears to be closer to guanylin than to uroguanylin.

mRNA Expression and Cellular Levels

The alimentary tract, especially the intestine, is the major expression site of all three *guanylins* and two *GC-Cs* in eel (Comrie et al. 2001a, b, Yuge et al. 2003, Cramb et al. 2005, Yuge et al. 2006, Kalujnaia et al. 2009). mRNA expressions of eel *uroguanylin, renoguanylin, GC-C1* and *GC-C2* are also detected abundantly in the kidney; in particular *renoguanylin* (Comrie et al. 2001a, b, Yuge et al. 2003, Yuge et al. 2006, Kalujnaia et al. 2009). *Uroguanylin* and *GC-C2* transcripts are also detectable in the liver and esophagus and in esophagus and stomach, respectively, in the Japanese eel (Yuge et al. 2003, 2006). This tissue distribution of mRNA expression is also observed in the European eel for *uroguanylin*, but not *GC-C2* (Comrie et al. 2001a, b). The tissue distribution patterns of mRNA expression may also place *renoguanylin* at the intermediate position between *guanylin* and *uroguanylin*. Because both hormone and receptor mRNAs are co-expressed in the intestine and kidney, the guanylins are predicted to act on these sites in a paracrine and/or autocrine fashion (also called "luminocrine" in this case) in eel. This assumption is partly supported by the immunohistochemical finding that immunoreactive guanylin is found in some goblet cells in eel intestinal mucosa (Yuge et al. 2003). As the goblet cells are known to be typical exocrine cells (Specian and Oliver 1991), guanylin is presumably secreted from these intestinal cells into the lumen and act locally on the cell surface GC-C receptors in the mucosal epithelia in eel. Although cells producing eel uroguanylin and renoguanylin and cellular localization of eel GC-C1 and GC-C2 in intestine remain to be identified, the luminocrine actions of all three guanylins on the intestine were proposed by functional analyses in the Ussing chamber experiments (see below for details). On the other hand, no study has been performed on circulating guanylins in eel. In mammals, guanylin and uroguanylin are detected in the circulation (Date et al. 1996, Fan et al. 1996, Forte 2004, Fonteles and do Nascimento 2011). Especially the circulating uroguanylin, which should be stable in renal tubules due to its resistance to the cleavage enzyme, is postulated to be the intestinal natriuretic factor.

Salinity-dependent Changes in Tissue Transcripts

Gene expression of *guanylin, uroguanylin, GC-C1* and *GC-C2* was up-regulated in the intestine of SW eels compared to FW eels (Comrie et al. 2001a, b, Yuge et al. 2003, 2006, Cramb et al. 2005, Takei and Yuge 2007,

Kalujnaia et al. 2009), suggesting that the intestinal guanylin system plays an important role in SW acclimation of eels. The up-regulation was greater in the anterior as opposed to the posterior segment of the intestine in SW Japanese eels (Yuge et al. 2003, 2006), whereas there was no marked difference among segments in SW European eels (Kalujnaia et al. 2009). The increase in *uroguanylin* mRNA expression was much greater in the intestine of SW European eels than that of *guanylin* mRNA expression (Cramb et al. 2005, Kalujnaia et al. 2009), while no such difference was observed in SW Japanese eels (Yuge et al. 2003). *Uroguanylin* transcripts were also increased in the kidney of SW Japanese eels (Yuge et al. 2003). In contrast, *renoguanylin* mRNA expression was not markedly increased in the SW eel intestine as found in the other *guanylins*, and was unchanged in SW and FW eel kidney (Yuge et al. 2003, Cramb et al. 2005, Takei and Yuge 2007, Kalujnaia et al. 2009). In the liver, esophagus and stomach, none of the hormone and the receptor genes exhibited significant expression changes between SW and FW Japanese eels (Yuge et al. 2006). In the Japanese eel, the intestinal mRNA expression of *guanylin* and *uroguanylin* was enhanced 24 hr, but not 3 hr, after SW transfer, with the exception of an increase in *uroguanylin* mRNA expression at 3 h post transfer in the posterior intestine (Yuge et al. 2003). In the same transfer experiment there was no change in the *GC-C1* and *GC-C2* transcripts within 24 hr (Yuge et al. 2006). Namely, in the eel guanylin system, the hormone transcripts were more rapidly responsive to the environmental salinity increase than the receptor transcripts (Takei and Yuge 2007), although some different cases appeared to occur in the European eel (Kalujnaia et al. 2009). Before measurement of luminal and plasma concentrations of guanylins is reported, it is too early to apply a short and long acting role for the guanylins (Takei and Hirose 2002, Takei and Loretz 2006). However, the marked up-regulation of hormone genes after SW acclimation (Comrie et al. 2001a, b, Yuge et al. 2003, 2006, Cramb et al. 2005, Takei and Yuge 2007, Kalujnaia et al. 2009) would suggest that the guanylin system possesses at least the long-acting hormone feature in SW eel. Moreover, in European eel, the *guanylins-GC-C* mRNA expression was analyzed in both yellow (non-migratory) and silver (migratory and sexually matured) eels acclimated to FW and SW in order to examine if the expression is increased in eels at the silver stage when they are naturally ready to migrate into sea from rivers. In fact, especially *uroguanylin* and *GC-C2* mRNAs expression in intestine and/or kidney appears to be higher in silver eels than in yellow eels in the same FW or SW condition, while these results seem to be vary depending on the batch of experiment (Comrie et al. 2001a, b, Cramb et al. 2005, Kalujnaia et al. 2009). This research is interesting but remains to be further investigated.

Taken together, guanylin and uroguanylin peptides may exert their osmoregulatory roles via GC-C1 and GC-C2 receptors in the intestine for SW acclimation. It is unclear how uroguanylin contributes to SW adaptation in the kidney. Moreover, renoguanylin may possess different function(s) from those of guanylin and uroguanylin, based on mRNA expression studies.

Osmoregulatory Functions of Guanylins in the Seawater Eel Intestine

Osmoregulatory actions of guanylins were investigated in the SW eel intestine, in which both the hormone and receptor genes were expressed and up-regulated in SW, by electrophysiological experiments using Ussing chambers and by use of synthetic eel guanylins (Yuge and Takei 2007). All three guanylins decreased short circuit current (I_{sc}) in a dose-dependent manner when administrated to the apical side of eel intestinal segments mounted in Ussing chambers, but not to the basolateral side (Note: Transepithelial potential is expressed as the potential of serosa relative to zero potential in mucosa, and the serosa-to-mucosa I_{sc} is expressed as a positive value). Thus, it was considered that in the eel intestine, all three guanylins acted on the mucosal epithelia in a luminocrine fashion and both GC-C1 and GC-C2 receptors are located on the mucosal surface of the epithelia. This is consistent with the postulation of luminal secretion of guanylin from the goblet cells in the eel intestine (Yuge et al. 2003) as described above.

Efficacies of guanylins appeared to be higher in the SW eels than in FW eels, and higher in the middle and posterior parts of SW eel intestine (Yuge and Takei 2007). As the lower part of the intestine is known to play a major role in water absorption in SW eels (Aoki et al. 2003, Tsukada and Takei 2006), a relationship of the guanylin action to water absorption may be worth considering in these segments. No marked differences in the efficacies were observed among three guanylins in the middle segment, while guanylin showed a slightly stronger effect on the posterior segment. The latter result may be partly related to the abundant mRNA expression of the guanylin-preferred receptor, GC-C2, in this segment (Yuge et al. 2006).

It is assumed that the potential target and action of the guanylin is a CFTR-like channel and Cl^- and HCO_3^- secretion via this channel in the intestine of SW eels (Yuge and Takei 2007). The guanylin-induced reduction of I_{sc} was inhibited in the presence of a Cl^- channel blocker, 5-nitro-2-(3-phenylpropylamino)-benzoic acid (NPPB), on the mucosal side (Yuge and Takei 2007). In mammals, similar results have been observed, and measurements of transepithalial Cl^- and HCO_3^- movements (Cl^-, by use of its radioactive ion; HCO_3^-, by pH-Stat) have concluded that guanylin and uroguanylin induce Cl^- and HCO_3^- secretion via CFTR in the intestine (Cuthbert et al. 1994, Guba et al. 1996, Joo et al. 1998). In general, HCO_3^- is known to be transported through CFTR (Seidler et al. 1997). In the SW eel intestine, although more detailed experiments are necessary, similar effects on intestinal ions transport might occur after mucosal guanylin administration. However, the regulatory mechanism of the guanylin action in SW eels may be different from that in mammals (Takei and Yuge 2007). In the mammalian intestine, mucosal CFTR plays a pivotal role in Cl⁻ secretion from cells to lumen, which is associated with fluid secretion, sometimes diarrhea (Sheppard and Welsh 1999). In this situation, intracellular Cl⁻ ions are supplied from the blood via a secretory type Na⁺-K⁺-2Cl⁻-cotransporter (NKCC1) located on the serosal membrane (Gamba 2005). In contrast, in the intestine of SW fish, water absorption follows NaCl absorption and is mediated by an absorptive type of NKCC (NKCC2) located on the mucosal side (Marshall and Grosell 2006), but the Cl⁻ and water secretion system is unclear, or maybe undesirable (Marshall and Singer 2002, Takei and Yuge 2007). These facts may be related to an electrophysiological difference in the Ussing chamber experiment between the SW fish intestine (serosa-negative) and the mammalian intestine (serosa-positive).

If the guanylin-induced Cl⁻ and HCO_3^- secretion occurs in the SW eel intestine, there would be three possible osmoregulatory significances of the actions of guanylin especially in the middle and posterior segments of the intestine (Takei and Yuge 2007, Yuge and Takei 2007) (Fig. 8). Firstly, like ANP, guanylins might fine-tune NaCl absorption by excreting an excessive amount of Cl⁻ into the intestine. Secondly, guanylin may play a role in supplying Cl⁻ into the lumen to cope with a shortage of luminal Cl⁻ that can be absorbed by a double ratio to Na⁺ via the mucosal NKCC2.

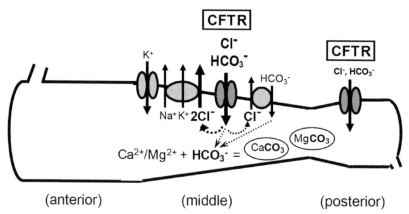

(anterior)   (middle)   (posterior)

Figure 8. Hypothetical osmoregulation of guanylin in the seawater eel intestine. The guanylin-GC-C cGMP signaling cascade may activate cystic fibrosis transmembrane conductance regulator (CFTR) Cl⁻ channel at the mucosa, and induce Cl⁻ and HCO_3^- secretion. Cl⁻ secretion may result in inhibition of NaCl absorption. Secreted Cl⁻ might be re-used for the apical absorption type of Na⁺-K⁺-2Cl⁻-co-transporter (NKCC2) and the apical Cl⁻/HCO_3^- exchanger, leading to facilitating water absorption together with Na⁺ and/or Cl⁻. Secreted HCO_3^- can precipitate luminal Ca^{2+} and Mg^{2+} as carbonate complexes, $CaCO_3$ and $MgCO_3$, respectively, and reduce the luminal fluid osmolality for water absorption.

As water is absorbed mainly in concert with the NKCC2-mediated NaCl absorption in SW fish intestine (Marshall and Grosell 2006), a higher luminal Cl⁻ level may be necessary in comparison to Na^+. Thirdly, HCO_3^- secretion via the mucosal CFTR could contribute to precipitate the luminal divalent cations, Ca^{2+} and Mg^{2+}, as carbonate complexes, $CaCO_3$ and $MgCO_3$. These divalent ions are scarcely absorbed in fish intestine, and the consequence is an increase in the luminal fluid osmolality, which is a disadvantage for effective water absorption (Wilson et al. 2002, Tsukada and Takei 2006). The removal of free Ca^{2+} and Mg^{2+} reduces the luminal osmolality, and then, makes the intestinal water absorption more facile (Wilson et al. 2002). In fact, these carbonate complexes are commonly observed as white precipitates in the SW eel intestine (Takei and Yuge 2007).

Other Significances

The augmentation of intestinal *guanylin* and *GC-C* mRNAs expression in a SW environment supports the notion of the guanylin family as a whole body NaCl regulator from a comparative standpoint, and would further endorse the postulation of an intestinal Na^+ sensor and an intestinal natriuretic factor that were previously proposed (Lennane et al. 1975a, b, Carey 1978). However, in the eel, the roles for the guanylins in the intestine-kidney endocrine axis remain to be investigated. In the kidney, local actions of uroguanylin and renoguanylin may also be conceivable. In the eel kidney, due to the possible peptide degradation at the ninth Phe residue by the chymotrypsin-like endopeptidase (see above), renoguanylin may not be present in the urine and not act on the luminal surface of renal tubules. Nonetheless, it was reported that in a heterologous experiment, in rat renal tubules, perfused eel renoguanylin is capable of regulating some ion transport (Lessa et al. 2009). The esophagus is also known as an osmoregulatory organ in SW eels (Hirano and Mayer-Gostan 1976, Nagashima and Ando 1994), and is an interesting site of *uroguanylin* and *GC-C2* mRNA expression. However, this aspect has not yet been investigated. Furthermore, recently, three novel aspects have been reported in the mammalian guanylin system, which we have yet to be considered in the eel. These include 1) GC-C-independent and CFTR-independent HCO_3^- secretion in intestine (Sellers et al. 2008); 2) regulation of epithelial cell proliferation in intestine (Forte 2004); and 3) the possible presence of a novel receptor, maybe not related to the GC-C type receptor, for uroguanylin and guanylin (Forte 2004, Giblin et al. 2006).

As all guanylin actions are to enhance salt secretion, the guanylin family is suggested to decrease blood pressure in mammals (Forte et al. 2004). For example, *uroguanylin*-knockout mice show elevated blood pressure (Lorenz et al. 2003). Therefore, the guanylin family could also be hormones for integrative regulation of salt, blood volume and pressure like the RAS,

NP family, and AM family. Interactions between GN and these families are intriguing. In fact, synergetic effects of guanylins and ANP on natriuresis were suggested in rat kidney (Santos-Neto et al. 1999, 2006, Fonteles and do Nascimento 2011). Moreover, as both of the hormone actions share GC-cGMP signaling from different sides of the intestinal epithelia, it is important to know how these hormone families co-work for the intestinal NaCl transport in eel.

Adrenomedullin Family

The studies on the RAS, NP family and guanylin family have revealed novel functions in fish osmoregulation in comparison to mammals. In addition to the new function, studies in fish sometimes result in the discovery of new hormones and this is exemplified by VNP/CNP1-3 and renoguanylin. Moreover, the discovery of a new family member in fish has led to the discovery of a new hormone (its ortholog) in mammals. In this section, the adrenomedullin (AM) family is introduced as the representative case. Though a single member of AM family was initially identified in humans, other subtypes and the presence of an AM 'family' were identified for the first time in teleost species whose genome database was available such as pufferfish and medaka (Ogoshi et al. 2003). Later, the findings were brought to eels and humans research using "reverse phylogenic" methods which is an approach often used to seek new hormones and functions from an evolutionary standpoint (Takei et al. 2007). Studies of AM family hormones in eels have revealed that novel AM members have potent effects on cardiovascular and body fluid regulation (Nobata et al. 2008, Ogoshi et al. 2008).

Molecular Structure and Evolution of the Adrenomedullin Family

AM is a hormone that is involved in cardiovascular and body fluid regulation. It was initially isolated from human pheochromocytoma cells of an adrenal medulla origin (Kitamura et al. 1993). AM is processed from a preproadrenomedullin precursor after the cleavage of signal peptide (Fig. 9). The *N*-terminal 20 amino acid residues of the prohormone in mammals was reported to have specific physiological function and named as proadrenomedullin *N*-terminal 20 peptide (PAMP) (Kitamura et al. 1994). The mature form of human AM with a single disulfide bridge between two cysteine residues is cleaved off at the RXXR site within the *N*-terminal region and the five consecutive arginine residues in the middle of proAM. A glycine residue present proximal to the arginine residues in the *C*-terminus is used for the amidation of mature AM. Cleavage of arginine residues by

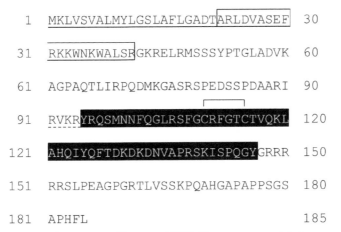

```
  1   MKLVSVALMYLGSLAFLGADTARLDVASEF   30

 31   RKKWNKWALSRGKRELRMSSSYPTGLADVK   60

 61   AGPAQTLIRPQDMKGASRSPEDSSPDAARI   90

 91   RVKRYRQSMNNFQGLRSFGCRFGTCTVQKL  120

121   AHQIYQFTDKDKDNVAPRSKISPQGYGRRR  150

151   RRSLPEAGPGRTLVSSKPQAHGAPAPPSGS  180

181   APHFL                          185
```

Figure 9. Amino acid sequence of human AM(1). The mature peptide is reversed and proadrenomedullin N-terminal 20 peptide is boxed. Signal peptide is underlined and RXXR cleavage site is underlined with a broken line. The disulfide bond is bracketed.

carboxypeptidase and subsequent amidation of the glycine residue gives rise to the physiologically active AM.

In teleost fishes, five members (*AM1* to *AM5*: approved symbol '*ADM1* to *ADM5*') were identified from the genome database of pufferfish (*Takifugu rubripes*) (Ogoshi et al. 2003) and Japanese medaka (*Oryzias latipes*) (Ogoshi et al. 2006) with reference to the human *AM* gene, the ortholog of teleost *AM1*. These findings led to the identification of four members of *AM* genes in the Japanese eel, eel *AM1* (*eAM1*), *eAM2*, *eAM3*, and *eAM5*, but the existence of the *AM4* gene is unknown (Nobata et al. 2008) (Fig. 10). The structural characteristics, such as a putative intramolecular ring and C-terminal amidation signal, were conserved in all *eAMs*. The PAMP-like sequence and consecutive arginine residues in the C-terminus, which are major characteristics of *AM1*, were conserved in *eAM1*. Eel *AM2* and *eAM3* were determined according to the putative mature sequences and the C-terminal glycine residue followed by a stop codon, which are some highly conserved characteristics. A characteristic of *AM5* is that the C-terminal glycine residue is followed by a single arginine residue prior to a stop codon, and is conserved in *eAM5*. Multiple members of the *AM* family were also identified in mammals. *AM2* was identified by two research groups at the same year and therefore was also known as intermedin (Roh et al. 2004, Takei et al. 2004). The *AM5* gene was discovered as an ortholog of the teleost *AM5* gene in several tetrapod species, but mutation was found in the human ortholog with deletion of two nucleotides and the *AM5* gene is undetectable in the mouse and rat (Ogoshi et al. 2006).

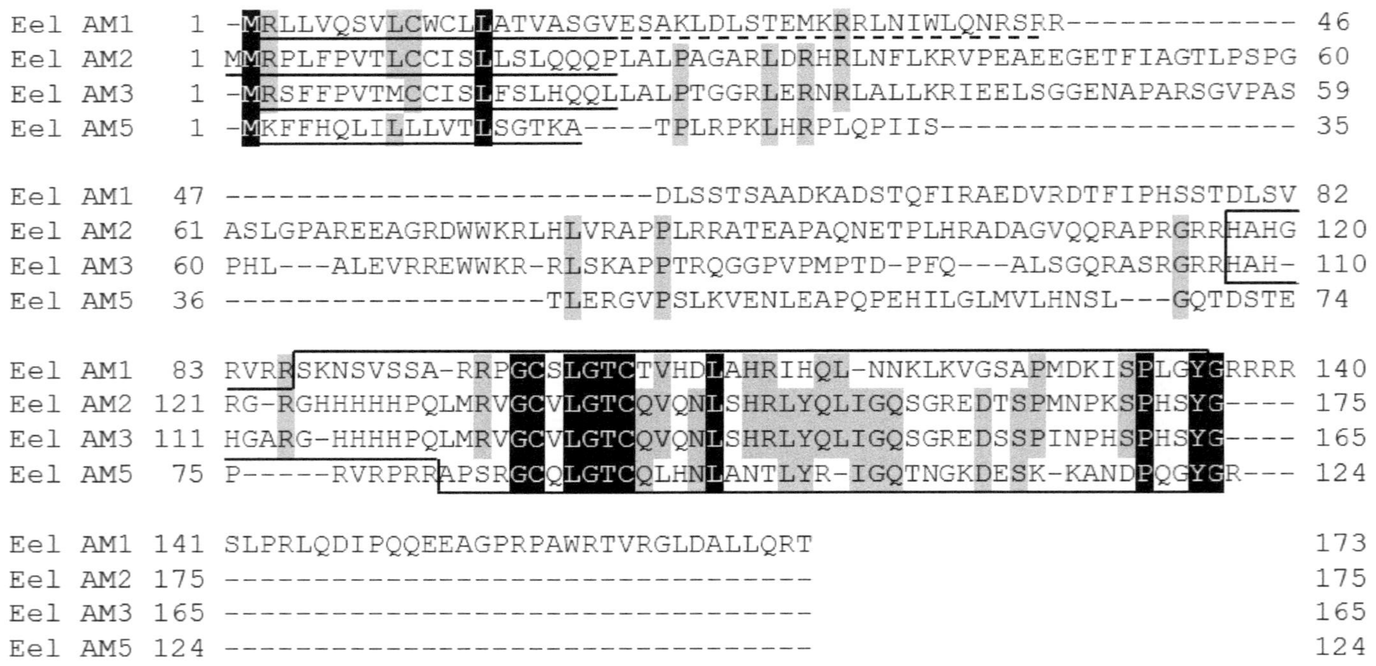

```
Eel AM1    1  -MRLLVQSVLCWCLLATVASGVESAKLDLSTEMKRRLNIWLQNRSRR-----------   46
Eel AM2    1  MMRPLFPVTLCCISLSLQQQPLALPAGARLDRHRLNFLKRVPEAEGETFIAGTLPSPG   60
Eel AM3    1  -MRSFFPVTMCCISLFSLHQQLLALPTGGRLERNRLALLKRIEELSGGENAPARSGVPAS   59
Eel AM5    1  -MKFFHQLILLLVTLSGTKA----TPLRPKLHRPLQPIIS-------   35

Eel AM1   47  -------------DLSSTSAADKADSTQFIRAEDVRDTFIPHSSTDLSV   82
Eel AM2   61  ASLGPAREEAGRDWKRLHIVRAPPLRRATEAPAQNETPLHRADAGVQQRAPRGRRHAHG  120
Eel AM3   60  PHL---ALEVRREWWKR-RLSKAPPTRQGGPVPMPTD-PFQ---ALSGQRASRGRRHAH-  110
Eel AM5   36  -------TLERGVPSLKVENLEAPQPEHILGLMVLHNSL---GQTDSTE   74

Eel AM1   83  RVRRSKNSVSSA-RRPGCSLGTCTVHDIAHRIHQL-NNKLKVGSAPMDKISPLGYGRRRR  140
Eel AM2  121  RG-RGHHHHHPQLMRVGCVLGTCQVQNLSHRLYQLIGQSGREDTSPMNPKSPHSYG----  175
Eel AM3  111  HGARG-HHHHPQLMRVGCVLGTCQVQNLSHRLYQLIGQSGREDSSPINPHSPHSYG----  165
Eel AM5   75  P----RVRPRRAPSRGCQLGTCQLHNLANTLYR-IGQTNGKDESK-KANDPQGYGR---  124

Eel AM1  141  SLPRLQDIPQQEEAGPRPAWRTVRGLDALLQRT  173
Eel AM2  175  --------  175
Eel AM3  165  --------  165
Eel AM5  124  --------  124
```

Figure 10. Precursor sequences of eel AMs. Putative signal peptides are underlined and a PAMP-like sequence is underlined with a broken line. Putative mature sequences are boxed. The sequences conserved among four AMs are reversed and among three AMs are shaded.

Comparative genomic analyses have revealed the molecular evolution of *AM* family in vertebrates (Fig. 11). *AM1/AM4* and *AM2/AM3* were generated by the whole genome duplication (3rd round duplication, 3R; Vandepoele et al. 2004) that occurred in the teleost lineage, and the duplicated counterpart of *AM5* seems to have disappeared during the evolution of euteleosts (Ogoshi et al. 2006). Therefore, three members of *AMs* (*AM1*, *AM2* and *AM5*), which have existed prior to divergence of the teleost and tetrapod lineages, are preserved in mammals. An *AM* gene has also been identified in lampreys which did not experience the 3R genome duplication (Wong and Takei 2009). Molecular phylogenetic analyses have revealed that the precursor sequence of lamprey *AM* was closely related to those of *AM1* of other species, but the putative mature sequence was grouped along with *AM2*. The ancestral gene of the *AM* family appears to have both *AM1* and *AM2* characteristics that are conserved in the lamprey *AM* gene. An *AM5*-like gene was not detected in lampreys, suggesting that it was generated after the divergence of the cyclostome lineage.

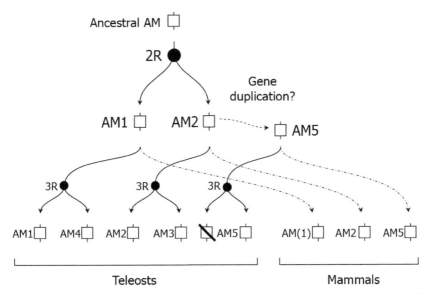

Figure 11. A scheme suggesting possible evolutionary history of vertebrate AM family. 2R and 3R: second- and third-round whole genome duplication.

Tissue Distribution of Adrenomedulins

Eel *AM* family genes are expressed in the whole body as they are in humans (Nobata et al. 2008), suggesting their importance and wide-spread functions in various organs and tissues. Strong expression of *eAM1* was observed in the heart, kidney, liver, spleen and red body. *eAM2* was particularly abundant in the pituitary and spleen. *eAM3* was expressed in the kidney, esophagus, liver and spleen. *eAM5* was highly expressed in the spleen and red body. Since the red body contains the rete mirabile, a vasculature-rich tissue in the swimbladder, *eAM1* and *eAM5* may be expressed in blood vessels similar to that of mammalian *AM*. Little difference has been reported in the expression levels of each member of the AM family between FW and SW adapted eels (Nobata et al. 2008).

Immunohistochemical studies of fish AM have only been reported in sea lampreys (Pombal et al. 2008). AM-like immunoreactive cells are located in the hypothalamus, diencephalon and mesencephalon in the brain of lampreys, as is the case in mammals. However, it is not clear whether the antibody against human AM is specific to lamprey AM, since the sequence identity between human AM and lamprey AM is only 34% (Wong and Takei 2009).

Biological Actions of Adrenomedullins

In mammals, the functions of AM have been examined since its discovery and it was revealed that AM is involved in vasodilation (Kitamura et al. 2002), hypotension (Shindo et al. 2000), ACTH and oxytocin secretion (Mimoto et al. 2001), body fluid regulation (Israel and Diaz 2000, Murphy and Samson 1995, Samson and Murphy 1997), and antimicrobial action (Allaker et al. 1999). Synthetic eAM1, eAM2 or eAM5 injected into the dorsal aorta of the eel induced potent hypotension (Nobata et al. 2008). The effects were more potent on the dorsal aorta than the ventral aorta, which indicates that AMs may be involved in decreasing the resistance of systemic vessels more profoundly than that of gill circulation. The hypotensive effects of eAMs were not different between FW and SW eel. Similar *in vivo* injection experiments have been performed in mice, which showed that the potency of AMs was AM1>AM2>AM5 (Takei et al. 2008). In contrast, eAM2 and eAM5 were equally more potent than eAM1 in the eel. The effects of eAM3, the duplicated counterpart of eAM2, have not been investigated. Moreover, intra-arterial injection of AM2 and AM5, but not AM1 enhanced drinking (Ogoshi et al. 2008). AM2 and AM5 infusion induced mild antidiuresis, while AM1 caused antinatriuresis. Therefore, the AM family is involved in cardiovascular and body fluid regulation in the eel. In other teleost species, intracerebroventricular (icv) injection of AM2 inhibited food intake in goldfish (*Carassius auratus*) (Martinez-Alvarez et al. 2009). Similarly, icv injection of mammalian AM2 inhibited food and water intake in rats (Taylor et al. 2005).

Receptors for the Adrenomedullin Family

The functional mechanism of the AM family is largely unclear, since most of their effects cannot be concluded simply by their ligand-binding affinities to the known receptors. The receptor components of the AM family have not been reported in the eel. In mammals, AM acts through the combination of the calcitonin receptor-like receptor (CLR: approved symbol '*CALCRL*') and the receptor activity-modifying protein 2 (RAMP2) or RAMP3 and activates the cAMP pathway (McLatchie et al. 1998) (Fig. 12). Teleost *CLR* and *RAMPs* were identified only in mefugu pufferfish (*Takifugu obscurus*) (Nag et al. 2006) with 3 types of *CLR* and 5 types of *RAMPs* thus far reported. The potency of AM1, AM2 and AM5 were examined with all combination patterns, and it was found that AM1 promotes cAMP production in combinations of CLR1-RAMP2/3/5 and AM2 and AM5 in a combination of CLR1-RAMP3. However, the cAMP accumulation on CLR1-RAMP3 by AM2 or AM5 was less potent than that by AM1. Therefore, the order of potency for cAMP accumulation on CLR1-RAMP3 observed in mefugu is

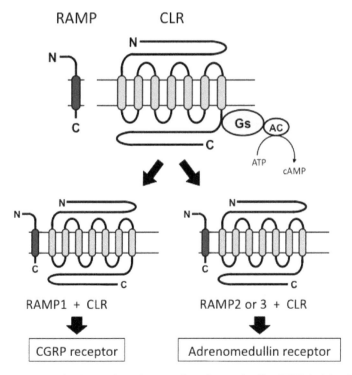

Figure 12. Receptor for AM and another member of superfamily, CGRP (calcitonin gene-related peptide), in mammals. CLR: calcitonin receptor-like receptor, RAMP: receptor activity-modifying protein, Gs: Gs alpha subunit, AC: adenylate cyclase.

not consistent with the order for hypotensive actions observed in the eel. In addition, icv injection of AM2 induces oxytocin release in rats, but this effect was not completely blocked by co-administration of antagonists to the AM receptor and calcitonin gene-related peptide (CGRP) receptor that were both reported to have binding affinities for the AM2 ligand (Hashimoto et al. 2007). These results suggest the existence of a novel AM2-specific receptor. Therefore, further receptor analysis is required for a full understanding of AM functions and their regulatory mechanisms in eels.

Perspectives

This chapter introduced four hormonal systems involved in eel osmoregulation. The osmoregulatory hormones are secreted in a various tissues and modulate ion and water fluxes via target organs in response to internal and/or external osmotic stimuli (Fig. 1). Other than these peptide hormones, arginine vasotocin (AVT), bradykinin, urotensin II, urotensin II-related peptide (URP), vasoactive intestinal peptide (VIP), and ghrelin have also been shown to have osmoregulatory actions in the eel (Table 1). Although their physiological roles are not fully understood, the reported actions in eels are partially overlapped with those of the RAS, NPs, GNs and AMs (see Table 1). Briefly, AVT is the neurohypophysial hormone of the vasopressin lineage in teleosts, and it inhibits drinking (Ando et al. 2000) and controls urine flow in eels (Babiker and Rankin 1978). Bradykinin, an active component of the kalikrein-kinin system, is antidipsogenic and hypertensive, and could be involved in the control of plasma ANG II levels in SW eels (Takei et al. 2001). The urophysial urotensin II and its paralog URP increase blood pressure in SW eels through both central and peripheral actions (Nobata et al. 2011). Brain-gut peptides, VIP and ghrelin, inhibit drinking in the eel when administered intracranially, and the potency of the antidipsogenic effect of ghrelin is higher than that of ANP (Kozaka et al. 2003).

The established osmoregulatory hormones including adrenocorticotropic hormone (ACTH), cortisol, and prolactin also play pivotal roles in eel osmoregulation (Table 1). ACTH and cortisol are well-known as SW-adaptation hormones and they increase ion and water transport in the eel intestine by promoting vascularization and reconstruction of the epithelial cell layer in the gastro-intestinal tract (Hirano and Utida 1968). Cortisol also potentiates Na^+/K^+-ATPase activity in branchial mitochondrion-rich cells (Mayer et al. 1967, Kamiya 1972). Conversely, prolactin antagonizes cortisol actions by lowering water and ion permeabilities in the gastro-intestinal tract, which results in promotion of FW adaptation (Utida et al. 1972, Hirano et al. 1976). Interestingly, growth hormone (GH) appears to be less significant in eels when compared with salmonids as plasma GH levels of SW and FW eels are unchanged and the hypophysectomized FW eel can survive a direct exposure to SW (Duan and Hirano 1991).

In summary, due to migration eels experience drastic osmotic changes in their life cycle, and as such have evolved many endocrine systems in order to overcome the challenges of rapid changes in environmental salinities. As osmotic balance is a result of the integration of various endocrine factors acting on target tissues, it is important to consider that some of them, if not all, regulate body fluid and ion balance cooperatively and/or antagonistically. In this post-genome era, integration between bioinformatics and the conventional physiological and integrative biological approaches in the eel will be key to addressing these questions. We expect that the eel will be a spotlight experimental model for deciphering the complex hormonal crosstalk in osmoregulation.

Acknowledgments

We are deeply indebted to Prof. Yoshio Takei (University of Tokyo, Japan), who is our mentor and guided us to the field of comparative endocrinology, for reviewing this manuscript. We also thank Dr. Gary W. Anderson (University of Manitoba, Canada) for helpful comments and revising the language of the manuscript.

References

Allaker, R.P., C. Zihni and S. Kapas. 1999. An investigation into the antimicrobial effects of adrenomedullin on members of the skin, oral, respiratory tract and gut microflora. FEMS Immunol. Med. Microbiol. 23: 289–293.

Ando, M. 2000. Some factors affecting drinking behavior and their interactions in seawater-acclimated eels, *Anguilla japonica*. Zool. Sci. 17: 171–178.

Ando, M. and I. Hara. 1994. Alteration of sensitivity to various regulators in the intestine of the eel following seawater acclimation. Comp. Biochem. and Physiol. 109A: 447–453.

Ando, M. and K. Nagashima. 1996. Intestinal Na$^+$ and Cl$^-$ levels control drinking behavior in the seawater-adapted eel, *Anguilla japonica*. J. Exp. Biol. 199: 711–716.

Ando, M., K.A. Kondo and Y. Takei. 1992. Effects of eel atrial natriuretic peptide on NaCl and water transport across the intestine of the seawater eel. J. Comp. Physiol. B. 162: 436–439.

Ando, M., Y. Fujii, T. Kadota, T. Kozaka, T. Mukuda, I. Takase and A. Kawahara. 2000. Some factors affecting drinking behavior and their interactions in seawater-acclimated eels, *Anguilla japonica*. Zool. Sci. 17: 171–178.

Aoki, M., T. Kaneko, F. Katoh, S. Hasegawa, N. Tsutsui and K. Aida. 2003. Intestinal water absorption through aquaporin 1 expressed in the apical membrane of mucosal epithelial cells in seawater-adapted Japanese eel. J. Exp. Biol. 206: 3495–3505.

Arao, M., T. Yamaguchi, T. Sugimoto, M. Fukase and K. Chihara. 1994. Characterization of a chymotrypsin-like hydrolytic activity in the opossum kidney cell. Biochem. Cell Biol. 72: 157–162.

Babiker, M.M. and J.C. Rankin. 1978. Neurohypophysial hormonal control of kidney function in the European eel (*Anguilla anguilla* L.) adapted to sea-water or fresh water. J. Endocrinol. 76: 347–358.

Balment, R.J., N. Hazon and M.N. Perrott. 1987. Control of corticosteroid secretion and its relation to osmoregulation in lower vertebrates. *In*: R. Kirsch and B. Lahlou [eds]. Comparative Physiology of Environmental Adaptation, Vol. 1. Karger, Basel, Switzerland. pp. 92–102.

Bennett, M.B. and J.C. Rankin. 1986. The effects of neurohypophysial hormones on the vascular resistance of the isolated perfused gill of the European eel, *Anguilla anguilla* L. Gen. Comp. Endocrinol. 64: 60–66.

Bernier, N.J., J.E. Mckendry and S.F. Perry. 1999. Blood pressure regulation during hypotension in two teleost species: differential involvement of the renin-angiotensin and adrenergic systems. J. Exp. Biol. 202: 1677–1690.

Brown, J.A., J.A. Oliver, I.W. Henderson and B.A. Jackson. 1980. Angiotensin and single nephron glomerular function in the trout *Salmo gairdneri*. Am. J. Physiol. 239: R509–514.

Brown, J.A., C.J. Gray and S.M. Taylor. 1990. The renin-angiotensin system and glomerular function of teleost fish. Prog. Clin. Biol. Res. 342: 528–533.

Brown, J.A., S.K. Pope, S. Amer, C.S. Cobb, R. Williamson, G. Parkyn and S.J. Aves. 1997. Angiotensin receptors in teleost fish glomeruli. *In*: S. Kawashima and S. Kikuyama [eds]. Proceeding of the 13th International Congress on Comparative Endocrinology, Monduzzi International proceedings division, Bologna, Italy. pp. 1313–1319.

Carey, R.M. 1978. Evidence for a splanchnic sodium input monitor regulating renal sodium excretion in man: Lack of dependence upon aldosterone. Circ. Res. 43: 19–23.

Carrithers, S.L., B.A. Jackson, W.Y. Cai, R.N. Greenberg and C.E. Ott. 2002. Site-specific effects of dietary salt intake on guanylin and uroguanylin mRNA expression in rat intestine. Regul. Pept. 107: 87–95.

Chao, A.C., F.J. de Sauvage, Y.J. Dong, J.A. Wagner, D.V. Goeddel and P. Gardner. 1994. Activation of intestinal CFTR Cl⁻ channel by heat-stable enterotoxin and guanylin *via* cAMP-dependent protein kinase. EMBO J. 13: 1065–1072.

Chester-Jones, I., I.W. Henderson, D.K.O. Chan, J.C. Rankin, W. Mosley, J.J. Brown, A.F. Lever, J.I. Robertson and M. Tree. 1966. Pressor activity in extracts of the corpuscles of Stannius from the European eel (*Anguilla anguilla* L.). J. Endocrinol. 34: 393–408.

Chester-Jones, I., D.K.O. Chan and J.C. Rankin. 1969. Renal function in the European eel (*Anguilla anguilla* L.): changes in blood pressure and renal function of the freshwater eel transferred to sea-water. J. Endocrinol. 43: 9–19.

Comrie, M.M., C.P. Cutler and G. Cramb. 2001a. Cloning and expression of guanylin from the European eel (*Anguilla anguilla*). Biochem. Biophys. Res. Commun. 281: 1078–1085.

Comrie, M.M., C.P. Cutler and G. Cramb. 2001b. Cloning and expression of two isoforms of guanylate cyclase C (GC-C) from the European eel (*Anguilla anguilla*). Comp. Biochem. Physiol. B. 129: 575–586.

Cramb, G., A.S. Martinez, I.S. McWilliam and G.D. Wilson. 2005. Cloning and expression of guanylin-like peptides in teleost fish. Ann. NY Acad. Sci. 1040: 277–280.

Currie, M.G., K.F. Fok, J. Kato, R.J. Moore, F.K. Hamra, K.L. Duffin and C.E. Smith. 1992. Guanylin: an endogenous activator of intestinal guanylate cyclase. Proc. Natl. Acad. Sci. USA 89: 947–951.

Cuthbert, A.W., M.E. Hickman, L.J. MacVinish, M.J. Evans, W.H. Colledge, R. Ratcliff, P.W. Seale and P.P.A. Humphrey. 1994. Chloride secretion in response to guanylin in colonic epithelia from normal and transgenic cystic fibrosis mice. Br. J. Pharmacol. 112: 31–36.

Date, Y., M. Nakazato, H. Yamaguchi, M. Miyazato and S. Matsukura. 1996. Tissue distribution and plasma concentration of human guanylin. Intern. Med. 35: 171–175.

de Bold A.J., H.B. Borenstein, A.T. Veress and H. Sonnenberg. 1981. A rapid and potent natriuretic response to intravenous injection of atrial myocardial extracts in rats. Life Sci. 28: 89–94.

Donald, J.A. and B.R.S. Broughton. 2005. Nitric oxide control of lower vertebrate blood vessels by vasomotor nerves. Comp. Biochem. Physiol. A. 142: 188–197.

Duan, C. and T. Hirano. 1991. Plasma kinetics of growth hormone in the Japanese eel, *Anguilla japonica*. Aquaculture 95: 179–188.

Epstein, F.H., M. Cynamon and W. McKay. 1971. Endocrine control of Na-K-ATP and seawater adaptation in *Anguilla rostrata*. Gen. Comp. Endocrinol. 16: 232–238.

Fan, X., F.K. Hamra, R.H. Freeman, S.L. Eber, W.J. Krause, R.W. Lim, V.M. Pace, M.G. Currie and L.R. Forte. 1996. Uroguanylin: cloning of preprouroguanylin cDNA, mRNA expression in the intestine and heart and isolation of uroguanylin and prouroguanylin from plasma. Biochem. Biophys. Res. Commun. 219: 457–462.

Fonteles, M.C. and N.R. do Nascimento. 2011. Guanylin peptide family: history, interactions with ANP, and new pharmacological perspectives. Can. J. Physiol. Pharmacol. 89: 575–585.

Fonteles, M.C., R.N. Greenberg, H.S.A. Monteiro, M.G. Currie and L.R. Forte. 1998. Natriuretic and kaliuretic activities of guanylin and uroguanylin in the isolated perfused rat kidney. Am. J. Physiol. 275: F191–F197.

Forte, L.R. 2004. Uroguanylin and guanylin peptides: pharmacology and experimental therapeutics. Pharmacol. Therapeut. 104: 137–162.

Forte, L.R., R.M. London, W.J. Krause and R.H. Freeman. 2000. Mechanisms of guanylin action via cyclic GMP in the kidney. Annu. Rev. Physiol. 62: 673–695.

Fowkes, R.C. and C.A. McArdle. 2000. C-type natriuretic peptide: an important neuroendocrine regulator? Trends Endocrinol. Metab. TEM 11: 333–338.

Fyhrquist, F. and O. Saijonmaa. 2008. Renin-angiotensin system revisited. J. Intern. Med. 264: 224–236.

Gamba, G. 2005. Molecular physiology and pathophysiology of electroneutral cation-chloride cotransporters. Physiol. Rev. 85: 423–493.

Giblin, M.F., H. Gali, G.L. Sieckman, N.K. Owen, T.J. Hoffman, W.A. Volkert and L.R. Forte. 2006. *In vitro* and *in vivo* evaluation of ^{111}In-labeled *E. coli* heat-stable enterotoxin analogs for specific targeting of human breast cancers. Breast Cancer Res. Treat. 98: 7–15.

González Bosc, L.V., M.P. Majowicz and N.A. Vidal. 2000. Effects of atrial natriuretic peptide in the gut. Peptides 21: 875–887.

Greenberg, R.N., M. Hill, J. Crytzer, W.J. Krause, S.L. Eber, F.K. Hamra and L.R. Forte. 1997. Comparison of effects of uroguanylin, guanylin, and Escherichia coli heat-stable enterotoxin STa in mouse intestine and kidney: evidence that uroguanylin is an intestinal natriuretic hormone. J. Invest. Med. 45: 276–283.

Guba, M., M. Kuhn, W.G. Forssmann, M. Classen, M. Gregor and U. Seidler. 1996. Guanylin strongly stimulates rat duodenal HCO_3^- secretion: proposed mechanism and comparison with other secretagogues. Gastroenterology 111: 1558–1568.

Hamra, F.K., L.R. Forte, S.L. Eber, N.V. Pidhorodeckyj, W.J. Krause, R.H. Freeman, D.T. Chin, J.A. Tompkins, K.F. Fok, C.E. Smith, K.L. Duffin, N.R. Siegel and M.G. Currie. 1993. Uroguanylin: structure and activity of a second endogenous peptide that stimulates intestinal guanylate cyclase. Proc. Natl. Acad. Sci. USA 90: 10464–10468.

Hamra, F.K., S.L. Eber, D.T. Chin, M.G. Currie and L.R. Forte. 1997. Regulation of intestinal uroguanylin/guanylin receptor-mediated responses by mucosal acidity. Proc. Natl. Acad. Sci. USA 94: 2705–2710.

Hasegawa, Y., T. Nakajima and H. Sokabe. 1983. Chemical structure of angiotensin formed with kidney renin in the Japanese eel, *Anguilla japonica*. Biomed. Res. 4: 417–420.

Hashimoto, H., S. Hyodo, M. Kawasaki, M. Shibata, T. Saito, H. Suzuki, H. Otsubo, T. Yokoyama, H. Fujihara, T. Higuchi, Y. Takei and Y. Ueta. 2007. Adrenomedullin 2 (AM2)/ intermedin is a more potent activator of hypothalamic oxytocin-secreting neurons than AM possibly through an unidentified receptor in rats. Peptides 28: 1104–1112.

Henderson, I.W., V. Jotisankasa, W. Mosley and M. Oguri. 1976. Endocrine and environmental influences upon plasma cortisol concentrations and plasma renin activity of the eel, *Anguilla anguilla* L. J. Endocrinol. 70: 81–95.

Henderson, I.W., N. Hazon and K. Hughes. 1985. Hormones, ionic regulation and kidney function in fishes. Symp. Soc. Exp. Biol. 39: 245–265.

Hirano, T. 1969. Effects of hypophysectomy and salinity change on plasma cortisol concentration in the Japanese eel, *Anguilla japonica*. Endocrinol. Jpn. 16: 557–560.

Hirano, T. 1974. Some factors regulating water intake by the eel, *Anguilla japonica*. J. Exp. Biol. 61: 737–747.

Hirano, T. and S. Utida. 1968. Effects of ACTH and cortisol on water movement in isolated intestine of the eel, *Anguilla japonica*. Gen. Comp. Endocrinol. 11: 373–380.

Hirano, T. and N. Mayer-Gostan. 1976. Eel esophagus as an osmoregulatory organ. Proc. Natl. Acad. Sci. USA 73: 1348–1350.

Hirano, T., M. Morisawa, M. Ando and S. Utida. 1976. Adaptive changes in ion and water transport mechanism in the eel intestine. *In*: J.W.L. Robinson [ed]. Intestinal Ion Transport. MTP Press, Lancaster, UK. pp. 301–317.

Imbrogno, S., M.C. Cerra and B. Tota. 2003. Angiotensin II-induced inotropism requires an endocardial endothelium-nitric oxide mechanism in the *in-vitro* heart of *Anguilla anguilla*. J. Exp. Biol. 206: 2675–2684.

Inoue, K., K. Naruse, S. Yamagami, H. Mitani, N. Suzuki and Y. Takei. 2003. Four functionally distinct C-type natriuretic peptides found in fish reveal evolutionary history of the natriuretic peptide system. Proc. Natl. Acad. Sci. USA 100: 10079–10084.

Inoue, K., T. Sakamoto, S. Yuge, H. Iwatani, S. Yamagami, M. Tsutsumi, H. Hori, M.C. Cerra, B. Tota, N. Suzuki, N. Okamoto and Y. Takei. 2005. Structural and functional evolution of three cardiac natriuretic peptides. Mol. Biol. Evol. 22: 2428–2434.

Israel, A. and E. Diaz. 2000. Diuretic and natriuretic action of adrenomedullin administered intracerebroventricularly in conscious rats. Regul. Pept. 89: 13–18.

Iwashina, M., T. Mizuno, S. Hirose, T. Ito and H. Hagiwara. 1994. His145-Trp146 residues and the disulfide-linked loops in atrial natriuretic peptide receptor are critical for the ligand-binding activity. J. Biochem. 115: 563–567.

Joo, N.S., R.M. London, H.D. Kim, L.R. Forte and L.L. Clarke. 1998. Regulation of intestinal Cl⁻ and HCO₃⁻ secretion by uroguanylin. Am. J. Physiol. 274: G633–G644.

Kaiya, H. and Y. Takei. 1996a. Atrial and ventricular natriuretic peptide concentrations in plasma of freshwater- and seawater-adapted eels. Gen. Comp. Endocrinol. 102: 183–190.

Kaiya, H. and Y. Takei. 1996b. Changes in plasma atrial and ventricular natriuretic peptide concentrations after transfer of eels from freshwater to seawater or vice versa. Gen. Comp. Endocrinol. 104: 337–345.

Kaiya, H. and Y. Takei. 1996c. Osmotic and volaemic regulation of atrial and ventricular natriuretic peptide secretion in conscious eels. J. Endocrinol. 149: 441–447.

Kaiya, H., T. Tsukada, S. Yuge, H. Mondo, K. Kangawa and Y. Takei. 2006. Identification of eel ghrelin in plasma and stomach by radioimmunoassay and histochemistry. Gen. Comp. Endocrinol. 148: 375–382.

Kalujnaia, S., G.D. Wilson, A.L. Feilen and G. Cramb. 2009. Guanylin-like peptides, guanylate cyclase and osmoregulation in the European eel (*Anguilla anguilla*). Gen. Comp. Endocrinol. 161: 103–114.

Kamiya, M. 1972. Hormonal effect on Na-K-ATPase activity in the gill of Japanese eel, *Anguilla japonica*, with special reference to seawater adaptation. Endocrinol. Jpn. 19: 489–493.

Kangawa, K. and H. Matsuo. 1984. Purification and complete amino acid sequence of alpha-human atrial natriuretic polypeptide (alpha-hANP). Biochem. Biophys. Res. Commun. 118: 131–139.

Kashiwagi, M., T. Katafuchi, A. Kato, H. Inuyama, T. Ito, H. Hagiwara, Y. Takei and S. Hirose. 1995. Cloning and properties of a novel natriuretic peptide receptor, NPR-D. E. J. Biochem. 233: 102–109.

Kashiwagi, M., K. Miyamoto, Y. Takei and S. Hirose. 1999. Cloning, properties and tissue distribution of natriuretic peptide receptor-A of euryhaline eel, *Anguilla japonica*. Eur. J. Biochem. 259: 204–211.

Katafuchi, T., A. Takashima, M. Kashiwagi, H. Hagiwara, Y. Takei and S. Hirose. 1994. Cloning and expression of eel natriuretic-peptide receptor B and comparison with its mammalian counterparts. Eur. J. Biochem. 222: 835–842.

Kawakoshi, A., S. Hyodo and Y. Takei. 2001. CNP is the only Natriuretic Peptide in an Elasmobranch Fish, Triakis scyllia. Zool. Sci. 18: 861–868.

Kawakoshi, A., S. Hyodo, K. Inoue, Y. Kobayashi and Y. Takei. 2004. Four natriuretic peptides (ANP, BNP, VNP and CNP) coexist in the sturgeon: identification of BNP in fish lineage. J. Mol. Endocrinol. 32: 547–555.

Kenyon, C.J., A. McKeever, J.A. Oliver and I.W. Henderson. 1985. Control of renal and adrenocortical function by the renin-angiotensin system in two euryhaline teleost fishes. Gen. Comp. Endocrinol. 58: 93–100.

Khosla, M.C., H. Nishimura, Y. Hasegawa and F.M. Bumpus. 1985. Identification and synthesis of [1-asparagine, 5-valine, 9-glycine] angiotensin I produced from plasma of American eel *Anguilla rostrata*. Gen. Comp. Endocrinol. 57: 223–233.

Kita, T., K. Kitamura, J. Sakata and T. Eto. 1999. Marked increase of guanylin secretion in response to salt loading in the rat small intestine. Am. J. Physiol. 277: G960–G966.

Kitamura, K., K. Kangawa, M. Kawamoto, Y. Ichiki, S. Nakamura, H. Matsuo and T. Eto. 1993. Adrenomedullin: a novel hypotensive peptide isolated from human pheochromocytoma. Biochem. Biophys. Res. Commun. 192: 553–560.

Kitamura, K., K. Kangawa, Y. Ishiyama, H. Washimine, Y. Ichiki, M. Kawamoto, N. Minamino, H. Matsuo and T. Eto. 1994. Identification and hypotensive activity of proadrenomedullin N-terminal 20 peptide (PAMP). FEBS Lett. 351: 35–37.

Kitamura, K., K. Kangawa and T. Eto. 2002. Adrenomedullin and PAMP: discovery, structures, and cardiovascular functions. Microsc. Res. Tech. 57: 3–13.

Kobayashi, H. and Y. Takei. 1996. The Renin-Angiotensin System: Comparative Aspects. Springer, Berlin, Germany.

Kozaka, T., Y. Fujii and M. Ando. 2003. Central effects of various ligands on drinking behavior in eels acclimated to seawater. J. Exp. Biol. 206: 687–692.

Lennane, R.J., W.S. Peart, R.M. Carey and J. Shaw. 1975a. A comparison of natriuresis after oral and intravenous sodium loading in sodium-depleted rabbits: evidence for a gastrointestinal or portal monitor of sodium intake. Clin. Sci. Mol. Med. 49: 433–436.

Lennane, R.J., R.M. Carey, T.J. Goodwin and W.S. Peart. 1975b. A comparison of natriuresis after oral and intravenous sodium loading in sodium-depleted man: evidence for a gastrointestinal or portal monitor of sodium intake. Clin. Sci. Mol. Med. 49: 437–440.

Lessa, L.M., J.B. Amorim, M.C. Fonteles and G. Malnic. 2009. Effect of renoguanylin on hydrogen/bicarbonate ion transport in rat renal tubules. Regul. Pept. 157: 37–43.

Li, Z., J.W. Knowles, D. Goyeau, S. Prabhakar, D.B. Short, A.G. Perkins and M.F. Goy. 1996. Low salt intake down-regulate the guanylin signaling pathway in rat distal colon. Gastroenterology 111: 1714–1721.

Li, Y.Y. and Y. Takei. 2003. Ambient salinity-dependent effects of homologous natriuretic peptides (ANP, VNP, and CNP) on plasma cortisol level in the eel. Gen. Comp. Endocrinol. 130: 317–323.

Lin, J.E., M. Valentino, G. Marszalowicz, M.S. Magee, P. Li, A.E. Snook, B.A. Stoecker, C. Chang and S.A. Waldman. 2010. Bacterial heat-stable enterotoxins: translation of pathogenic peptides into novel targeted diagnostics and therapeutics. Toxins (Basel). 2: 2028–2054.

Lorenz, J.N., M. Nieman, J. Sabo, L.P. Sanford, J.A. Hawkins, N. Elitsur, L.R. Gawenis, L.L. Clarke and M.B. Cohen. 2003. Uroguanylin knockout mice have increased blood pressure and impaired natriuretic response to enteral NaCl load. J. Clin. Invest. 112: 1244–1254.

Loretz, C.A. and Y. Takei. 1997. Natriuretic peptide inhibition of intestinal salt absorption in the Japanese eel: physiological significance. Fish Physiol. Biochem. 17: 319–324.

Loretz, C.A., C. Pollina, H. Kaiya, H. Sakaguchi and Y. Takei. 1997. Local synthesis of natriuretic peptides in the eel intestine. Biochem. Biophys. Res. Commun. 238: 817–822.

Maack, T. 1992. Receptors of atrial natriuretic factor. Annu. Rev. Physiol. 54: 11–27.

Maetz, J. and E. Skadhauge. 1968. Drinking rates and gill ionic turnover in relation to external salinities in the eel. Nature. 217: 371–373.

Marshall, W.S. and T.D. Singer. 2002. Cystic fibrosis transmembrane conductance regulator in teleost fish. Biochim. Biophys. Acta. 1566: 16–27.

Marshall, W.S. and M. Grosell. 2006. Ion transport, osmoregulation and acid-base balance in homeostasis and reproduction. *In*: D.H. Evans and J.B. Claiborne [eds]. The Physiology of Fishes. CRC Press, Boca Raton, USA. pp. 177–230.

Marsigliante, S., T. Verri, S. Barker, E. Jimenez, G.P. Vinson and C. Storelli. 1994. Angiotensin II receptor subtypes in eel (*Anguilla anguilla*). J. Mol. Endocrinol. 12: 61–69.

Marsigliante, S., A. Muscella, G.P. Vinson and C. Storelli. 1997. Angiotensin II receptors in the gill of sea water- and freshwater-adapted eel. J. Mol. Endocrinol. 18: 67–76.

Marsigliante, S., A. Muscella, S. Barker and C. Storelli. 2000. Angiotensin II modulates the activity of the Na$^+$/K$^+$ATPase in eel kidney. J. Endocrinol. 165: 147–156.

Marsigliante, S., A. Muscella, S. Greco, M.G. Elia, S. Vilella and C. Storelli. 2001. Na$^+$/K$^+$-ATPase activity inhibition and isoform-specific translocation of protein kinase C following angiotensin II administration in isolated eel enterocytes. J. Endocrinol. 168: 339–346.

Martinez-Alvarez, R.M., H. Volkoff, J.A. Munoz-Cueto and M.J. Delgado. 2009. Effect of calcitonin gene-related peptide (CGRP), adrenomedullin and adrenomedullin-2/intermedin on food intake in goldfish (*Carassius auratus*). Peptides 30: 803–807.

Mayer, N., J. Meatz. D.K.O, Chan, M. Foster and I. Chester Jones. 1967. Cortisol, a sodium excreting factor in the eel (*Anguilla anguilla* L.) adapted to sea water. Nature London 214: 1118.

McCormick, S.D. 2001. Endocrine Control of Osmoregulation in Teleost Fish. American Zoologist 41: 781–794.

McLatchie, L.M., N.J. Fraser, M.J. Main, A. Wise, J. Brown, N. Thompson, R. Solari, M.G. Lee and S.M. Foord. 1998. RAMPs regulate the transport and ligand specificity of the calcitonin-receptor-like receptor. Nature 393: 333–339.

Mimoto, T., T. Nishioka, K. Asaba, T. Takao and K. Hashimoto. 2001. Effects of adrenomedullin on adrenocorticotropic hormone (ACTH) release in pituitary cell cultures and on ACTH and oxytocin responses to shaker stress in conscious rat. Brain Res. 922: 261–266.

Mishina, S. and Y. Takei. 1997. Characterisation of natriuretic peptide receptors in eel gill. J. Endocrinol. 154: 415–422.

Miyanishi, H., S. Nobata and Y. Takei. 2011. Relative antidipsogenic potencies of six homologous natriuretic peptides in eels. Zool. Sci. 28: 719–726.

Mommsen, T.P., M.M. Vijayan and T.W. Moon. 1999. Cortisol in teleosts: dynamics, mechanisms of action, and metabolic regulation. Reviews in Fish Biology and Fisheries 9: 211–268.

Mukuda, T., Y. Matsunaga, K. Kawamoto, K. Yamaguchi and M. Ando. 2005. "Blood-contacting neurons" in the brain of the Japanese eel *Anguilla japonica*. J. Exp. Zool. A Comp. Exp. Biol. 303: 366–376.

Murphy, T.C. and W.K. Samson. 1995. The novel vasoactive hormone, adrenomedullin, inhibits water drinking in the rat. Endocrinology 136: 2459–2463.

Nag, K., A. Kato, T. Nakada, K. Hoshijima, A.C. Mistry, Y. Takei and S. Hirose. 2006. Molecular and functional characterization of adrenomedullin receptors in pufferfish. Am. J. Physiol. Regul. Integr. Comp. Physiol. 290: R467–478.

Nagashima, K. and M. Ando. 1994. Characterization of esophageal desalination in the seawater eel, *Anguilla japonica*. J. Comp. Physiol. 164: 47–54.

Nicoll, C.S. 1981. Role of prolacin in water and ion balance in vertebrates. *In*: R.B. Jaffe [ed]. Prolactin. Elsevier, New York, USA pp. 127–166.

Nishimura, H. and W.H. Sawyer. 1976. Vasopressor, diuretic, and natriuretic responses to angiotensins by the American eel, *Anguilla rostrata*. Gen. Comp. Endocrinol. 29: 337–348.

Nishimura, H., W.H. Sawyer and R.F. Nigrelli. 1971. Effects of changes in extrarenal salinity on renin activity in plasma and kidneys of two euryhaline teleost fishes, *Anguilla rostrata* and *Opsanus tau*. Physiologist 14: 202.

Nishimura, H., V.M. Norton and F.M. Bumpus. 1978. Lack of specific inhibition of angiotensin II in eels by angiotensin antagonists. Am. J. Physiol. 235: H95–103.

Nobata, S and Y. Takei. 2011. The area postrema in hindbrain is a central player for regulation of drinking behavior in Japanese eels. Am. J. Physiol. Regul. Integr. Comp. Physiol. 300: R1569–1577.

Nobata, S., M. Ogoshi and Y. Takei. 2008. Potent cardiovascular actions of homologous adrenomedullins in eels. Am. J. Physiol. Regul. Integr. Comp. Physiol. 294: R1544–1553.

Nobata, S., A. Ventura, H. Kaiya and Y. Takei. 2010. Diversified cardiovascular actions of six homologous natriuretic peptides (ANP, BNP, VNP, CNP1, CNP3, and CNP4) in conscious eels. Am. J. Physiol. Regul. Integr. Comp. Physiol. 298: R1549–1559.

Nobata, S., J.A. Donald, R.J. Balment and Y. Takei. 2011. Potent cardiovascular effects of homologous urotensin II (UII)-related peptide and UII in unanesthetized eels after peripheral and central injections. Am. J. Physiol. Regul. Integr. Comp. Physiol. 300: R437–446.

Oide, H. and S. Utida. 1968. Changes in intestinal absorption and renal excretion of water during adaptation to sea-water in the Japanese eel. Marine Biology 1: 172–177.

Okawara, Y., T. Karakida, M. Aihara, K. Yamaguchi and H. Kobayashi. 1987. Involvement of angiotensin II in water intake in the Japanese eel, *Anguilla japonica*. Zool. Sci. 4: 523–528.

Ogoshi, M., K. Inoue and Y. Takei. 2003. Identification of a novel adrenomedullin gene family in teleost fish. Biochem. Biophys. Res. Commun. 311: 1072–1077.

Ogoshi, M., K. Inoue, K. Naruse and Y. Takei. 2006. Evolutionary history of the calcitonin gene-related peptide family in vertebrates revealed by comparative genomic analyses. Peptides 27: 3154–3164.

Ogoshi, M., S. Nobata and Y. Takei. 2008. Potent osmoregulatory actions of homologous adrenomedullins administered peripherally and centrally in eels. Am. J. Physiol. Regul. Integr. Comp. Physiol. 295: R2075–2083.

O'Grady, S.M. 1989. Cyclic nucleotide-mediated effects of ANF and VIP on flounder intestinal ion transport. Am. J. Physiol. 256: C142–146.

Oudit, G.Y. and D.G. Butler. 1995. Angiotensin II and cardiovascular regulation in a freshwater teleost, *Anguilla rostrata* LeSueur. Am. J. Physiol. Regul. Integr. Comp. Physiol. 269: R726–735.

Plovsing, R.R., C. Wamberg, N.C. Sandgaard, J.A. Simonsen, N.H. Holstein-Rathlou, P.F. Hoilund-Carlsen and P. Bie. 2003. Effects of truncated angiotensins in humans after double blockade of the renin system. Am. J. Physiol. Regul. Integr. Comp. Physiol. 285: R981–991.

Pombal, M.A., J.M. Lopez, C. de Arriba Mdel, A. Gonzalez and M. Megias. 2008. Distribution of adrenomedullin-like immunoreactivity in the brain of the adult sea lamprey. Brain Res. Bull. 75: 261–265.

Roh, J., C.L. Chang, A. Bhalla, C. Klein and S.Y. Hsu. 2004. Intermedin is a calcitonin/calcitonin gene-related peptide family peptide acting through the calcitonin receptor-like receptor/receptor activity-modifying protein receptor complexes. J. Biol. Chem. 279: 7264–7274.

Russell, M.J., A.M. Klemmer and K.R. Olson. 2001. Angiotensin signaling and receptor types in teleost fish. Comp. Biochem. Physiol. A Mol. Integr. Physiol. 128: 41–51.

Sakamoto, T., S.D. McCormick and T. Hirano. 1993. Osmoregulatory actions of growth hormone and its mode of action in salmonids: a review. Fish Physiol. Biochem. 11: 155–164.

Samson, W.K. and T.C. Murphy. 1997. Adrenomedullin inhibits salt appetite. Endocrinology 138: 613–616.

Santos-Neto, M.S., A.F. Carvalho, L.R. Forte and M.C. Fonteles. 1999. Relationship between the actions of atrial natriuretic peptide (ANP), guanylin and uroguanylin on the isolated kidney. Braz. J. Med. Biol. Res. 32: 1015–1019.

Santos-Neto, M.S., A.F. Carvalho, H.S. Monteiro, L.R. Forte and M.C. Fonteles. 2006. Interaction of atrial natriuretic peptide, urodilatin, guanylin and uroguanylin in the isolated perfused rat kidney. Regul. Pept. 136: 14–22.

Saville, M.A., P.G. Geer, B.C. Wang, R.J. Jr. Leadley and K.L. Goetz. 1988. A high-salt meal produces natriuresis in humans without elevating plasma atriopeptin. Proc. Soc. Exp. Biol. Med. 188: 387–393.

Schulz, A., S. Escher, U.C. Marx, M. Meyer, P. Rösch, W.G. Forssmann and K. Adermann. 1998. Carboxy-terminal extension stabilizes the topological stereoisomers of guanylin. J. Pept. Res. 52: 518–525.

Schulz, S., C.K. Green, P.S.T. Yuen and D.L. Garbers. 1990. Guanylyl cyclase is a heat-stable enterotoxin receptor. Cell 63: 941–948.

Seidler, U., I. Blumenstein, A. Kretz, D. Viellard-Baron, H. Rossmann, W.H. Colledge, M. Evans, R. Ratcliff and M. Gregor. 1997. A functional CFTR protein is required for mouse intestinal cAMP-, cGMP- and Ca^{2+}-dependent HCO_3^- secretion. J. Physiol. London 505: 411–423.

Sellers, Z.M., E. Mann, A. Smith, K.H. Ko, R. Giannella, M.B. Cohen, K.E. Barrett and H. Dong. 2008. Heat-stable enterotoxin of *Escherichia coli* (STa) can stimulate duodenal HCO_3^- secretion *via* a novel GC-C- and CFTR-independent pathway. FASEB J. 22: 1306–1316.

Sheppard, D.N. and M.J. Welsh. 1999. Structure and function of the CFTR chloride channel. Physiol. Rev. 79: S23–S45.

Shindo, T., H. Kurihara, K. Maemura, Y. Kurihara, T. Kuwaki, T. Izumida, N. Minamino, K.H. Ju, H. Morita, Y. Oh-hashi, M. Kumada, K. Kangawa, R. Nagai and Y. Yazaki. 2000. Hypotension and resistance to lipopolysaccharide-induced shock in transgenic mice overexpressing adrenomedullin in their vasculature. Circulation 101: 2309–2316.

Sokabe, H., S. Mizogami, T. Murase and F. Sakai. 1966. Renin and euryhalinity in the Japanese eel, *Anguilla japonica*. Nature 212: 952–953.

Sokabe, H., H. Oide, M. Ogawa and S. Utida. 1973. Plasma renin activity in Japanese eels (*Anguilla japonica*) adapted to seawater or in dehydration. Gen. Comp. Endocrinol. 21: 160–167.

Specian, R.D. and M.G. Oliver. 1991. Functional biology of intestinal goblet cells. Am. J. Physiol. 260: C183–C193.

Takashima, A., T. Katafuchi, M. Shibasaki, M. Kashiwagi, H. Hagiwara, Y. Takei and S. Hirose. 1995. Cloning, properties, site-directed mutagenesis analysis of the subunit structure, tissue distribution and regulation of expression of the type-C eel natriuretic peptide receptor. Eur. J. Biochem. 227: 673–680.

Takei, Y. 2000a. Comparative physiology of body fluid regulation in vertebrates with special reference to thirst regulation. The Japanese journal of physiology 50: 171–186.

Takei, Y. 2000b. Structural and functional evolution of the natriuretic peptide system in vertebrates. Int. Rev. Cytol. 194: 1–66.

Takei, Y. and R.J. Balment. 1993. Biochemistry and physiology of a family of eel natriuretic peptides. Fish Physiol. Biochem. 11: 183–188.

Takei, Y. and H. Kaiya. 1998. Antidiuretic effect of eel ANP infused at physiological doses in conscious, seawater-adapted eels, *Anguilla japonica*. Zool. Sci. 15: 399–404.

Takei, Y. and T. Tsuchida. 2000. Role of the renin-angiotensin system in drinking of seawater-adapted eels *Anguilla japonica*: a reevaluation. Am. J. Physiol. Regul. Integr. Comp. Physiol. 279: R1105–1111.

Takei, Y. and S. Hirose. 2002. The natriuretic peptide system in eels: a key endocrine system for euryhalinity? Am. J. Physiol. 282: R940–R951.

Takei, Y. and S. Yuge. 2007. The intestinal guanylin system and seawater adaptation in eels. Gen. Comp. Endocrinol. 152: 339–351.

Takei, Y., T. Hirano and H. Kobayashi. 1979. Angiotensin and water intake in the Japanese eel, *Anguilla japonica*. Gen. Comp. Endocrinol. 38: 466–475.

Takei, Y., J. Okubo and K. Yamaguchi. 1988. Effect of cellular dehydration on drinking and plasma angiotensin II level in the eel, *Anguilla japonica*. Zool. Sci. 5: 43–51.

Takei, Y., A. Takahashi, T.X. Watanabe, K. Nakajima and S. Sakakibara. 1989. Amino acid sequence and relative biological activity of eel atrial natriuretic peptide. Biochem. Biophy. Res. Commun. 164: 537–543.

Takei, Y., A. Takahashi, T.X. Watanabe, K. Nakajima, S. Sakakibara, T. Takao and Y. Shimonishi. 1990. Amino acid sequence and relative biological activity of a natriuretic peptide isolated from eel brain. Biochem. Biophys. Res. Commun. 170: 883–891.

Takei, Y., A. Takahashi, T.X. Watanabe, K. Nakajima and S. Sakakibara. 1991. A novel natriuretic peptide isolated from eel cardiac ventricles. FEBS Lett. 282: 317–320.

Takei, Y., K. Ando and M. Kawakami. 1992. Atrial natriuretic peptide in eel plasma, heart and brain characterized by homologous radioimmunoassay. J. Endocrinol. 135: 325–331.

Takei, Y., M. Ueki and T. Nishizawa. 1994a. Eel ventricular natriuretic peptide: cDNA cloning and mRNA expression. J. Mol. Endocrinol. 13: 339–345.

Takei, Y., A. Takahashi, T.X. Watanabe, K. Nakajima and K. Ando. 1994b. Eel ventricular natriuretic peptide: isolation of a low molecular size form and characterization of plasma form by homologous radioimmunoassay. J. Endocrinol. 141: 81–89.

Takei, Y., M. Ueki, A. Takahashi and T. Nishizawa. 1997. Cloning, sequence analysis, tissue-specific expression, and prohormone isolation of eel atrial natriuretic peptide. Zool. Sci. 14: 993–999.

Takei, Y., T. Tsuchida and A. Tanakadate. 1998. Evaluation of water intake in seawater adaptation in eels using a synchronized drop counter and pulse injector system. Zool. Sci. 15: 677–682.

Takei, Y., K. Inoue, K. Ando, T. Ihara, T. Katafuchi, M. Kashiwagi and S. Hirose. 2001a. Enhanced expression and release of C-type natriuretic peptide in freshwater eels. Am. J. Physiol. Regul. Integr. Comp. Physiol. 280: R1727–1735.

Takei, Y., T. Tsuchida, Z. Li and J.M. Conlon. 2001b. Antidipsogenic effects of eel bradykinins in the eel *Anguilla japonica*. Am. J. Physiol. Regul. Integr. Comp. Physiol. 281: R1090–1096.

Takei, Y., K. Inoue, M. Ogoshi, T. Kawahara, H. Bannai and S. Miyano. 2004. Identification of novel adrenomedullin in mammals: a potent cardiovascular and renal regulator. FEBS Lett. 556: 53–58.

Takei, Y., A. Kawakoshi, T. Tsukada, S. Yuge, M. Ogoshi, K. Inoue, S. Hyodo, H. Bannai and S. Miyano. 2006. Contribution of comparative fish studies to general endocrinology: structure and function of some osmoregulatory hormones. J. Exp. Zool. A. Comp. Exp. Biol. 305: 787–98.

Takei, Y., M. Ogoshi and K. Inoue. 2007. A 'reverse' phylogenetic approach for identification of novel osmoregulatory and cardiovascular hormones in vertebrates. Front Neuroendocrinol. 28: 143–160.

Takei, Y., H. Hashimoto, K. Inoue, T. Osaki, K. Yoshizawa-Kumagaye, M. Tsunemi, T.X. Watanabe, M. Ogoshi, N. Minamino and Y. Ueta. 2008. Central and peripheral cardiovascular actions of adrenomedullin 5, a novel member of the calcitonin gene-related peptide family, in mammals. J. Endocrinol. 197: 391–400.

Takei, Y. and C.A. Loretz. 2006. Endocrinology. *In*: D.H. Evans and J.B. Claiborne [eds]. The Physiology of Fishes, 3rd edition. CRC Press, Boca Raton, USA. pp. 271–318.

Taylor, M.M., S.L. Bagley and W.K. Samson. 2005. Intermedin/adrenomedullin-2 acts within central nervous system to elevate blood pressure and inhibit food and water intake. Am. J. Physiol. Regul. Integr. Comp. Physiol. 288: R919–927.

Tierney, M.L., G. Luke, G. Cramb and N. Hazon. 1995. The role of the renin-angiotensin system in the control of blood pressure and drinking in the European eel, *Anguilla anguilla*. Gen. Comp. Endocrinol. 100: 39–48.

Tsuchida, T. and Y. Takei. 1998. Effects of homologous atrial natriuretic peptide on drinking and plasma ANG II level in eels. Am. J. Physiol. 275: R1605–1610.

Tsuchida, T. and Y. Takei. 1999. A potent dipsogenic action of homologous angiotensin II infused at physiological doses in eels. Zool. Sci. 16: 479–483.

Tsukada, T. and Y. Takei. 2001. Relative potency of three homologous natriuretic peptides (ANP, CNP and VNP) in eel osmoregulation. Zool. Sci. 18: 1253–1258.

Tsukada, T. and Y. Takei. 2006. Integrative approach to osmoregulatory action of atrial natriuretic peptide in seawater eels. Gen. Comp. Endocrinol. 147: 31–38.

Tsukada, T., J.C. Rankin and Y. Takei. 2005. Involvement of drinking and intestinal sodium absorption in hyponatremic effect of atrial natriuretic peptide in seawater eels. Zool. Sci. 22: 77–85.

Tsukada, T., S. Nobata, S. Hyodo and Y. Takei. 2007. Area postrema, a brain circumventricular organ, is the site of antidipsogenic action of circulating atrial natriuretic peptide in eels. J. Exp. Biol. 210: 3970–3978.

Utida, S., T. Hirano, M. Ando, D.W. Johnson and H.A. Bern. 1972. Hormonal control of the intestine and urinary bladder in teleost osmoregulation. Gen. Comp. Endocrinol. (Suppl. 3) 317.

Vandepoele, K., W. De Vos, J.S. Taylor, A. Meyer and Y. Van de Peer. 2004. Major events in the genome evolution of vertebrates: paranome age and size differ considerably between ray-finned fishes and land vertebrates. Proc. Natl. Acad. Sci. USA 101: 1638–1643.

Ventura, A. 2011. Regulation of Cortisol Secretion by Fast-Acting Hormones in Eel Osmoregulation. Ph.D. Thesis, the University of Tokyo, The University of Tokyo Press, Tokyo, Japan.

Ventura, A., M. Kusakabe and Y. Takei. 2011. Salinity-dependent in vitro effects of homologous natriuretic peptides on the pituitary-interrenal axis in eels. Gen. Comp. Endocrinol. 173: 129–138.

Vilella, S., V. Zonno, S. Marsigliante, L. Ingrosso, A. Muscella, M.M. Ho, G.P. Vinson and C. Storelli. 1996. Angiotensin II stimulation of the basolateral located Na^+/H^+ antiporter in eel (*Anguilla anguilla*) enterocytes. J. Mol. Endocrinol. 16: 57–62.

Volant, K., O. Grishina, M. Descroix-Vagne and D. Pansu. 1997. Guanylin-, heat-stable enterotoxin of Escherichia coli- and vasoactive intestinal peptide-induced water and ion secretion in the rat intestine *in vivo*. Eur. J. Pharmacol. 11: 217–227.

Wendelaar Bonga, S.E. 1997. The stress response in fish. Physiol. Rev. 77: 591–625.

Wilson, R.W., J.M. Wilson and M. Grosell. 2002. Intestinal bicarbonate secretion by marine teleost fish—why and how? Biochim. Biophys. Acta. 1566: 182–193.

Wong, M.K.S. and Y. Takei. 2009. Cyclostome and chondrichthyan adrenomedullins reveal ancestral features of the adrenomedullin family. Comp. Biochem. Physiol. B. Biochem. Mol. Biol. 154: 317–325.

Wong, M.K.S. and Y. Takei. 2011. Characterisation of native angiotensin from an anciently diverged serine-protease inhibitor (SERPIN) in lamprey. J. Endocrinol. 209: 127–137.

Wong, M.K.S. and Y. Takei. 2012. Changes in plasma angiotensin subtypes in Japanese eel acclimated to various salinities from deionized water to double-strength seawater. Gen. Comp. Endocrinol. 178: 250–258.

Yuge, S., K. Inoue, S. Hyodo and Y. Takei. 2003. A novel guanylin family (guanylin, uroguanylin, and renoguanylin) in eels: possible osmoregulatory hormones in intestine and kidney. J. Biol. Chem. 278: 22726–22733.

Yuge, S., S. Yamagami, K. Inoue, N. Suzuki and Y. Takei. 2006. Identification of two functional guanylin receptors in eel: multiple hormone-receptor system for osmoregulation in fish intestine and kidney. Gen. Comp. Endocrinol. 149: 10–20.

Yuge, S. and Y. Takei. 2007. Regulation of ion transport in eel intestine by the homologous guanylin family of peptides. Zoolog. Sci. 24: 1222–1230.

Regulation of Drinking

Shigenori Nobata[a],* and Masaaki Ando[b]

Introduction

In teleost fishes, including the eel, the composition of body fluids is influenced by the environmental water, since, either in freshwater (FW) or in seawater (SW), the fish are in direct contact with the environmental medium *via* the body surface (Marshall and Grosell 2006). Nevertheless, body fluid osmolality is maintained near a setpoint value that is typical of each species regardless of environmental salinity. The setpoint values are achieved by balancing the gain and loss of water and ions. In regard to water, in FW water excretion is more important than ingestion because these fishes continuously face the osmotic gain of water. On the other hand, SW fishes must acquire water constantly to compensate for the osmotic water loss across body surfaces. Although the relative importance of water ingestion is different between SW and FW fishes, it is drinking that is a primary means of water intake both in SW and FW fishes.

In marine teleosts, regulation of drinking is especially important in body fluid homeostasis (Takei and Balment 2009). As migratory fishes, both catadromous and anadromous species move between SW and FW twice in their lifetimes, and the osmoregulatory machineries are strongly regulated

Atmosphere and Ocean Research Institute, University of Tokyo, 5-1-5 Kashiwanoha, Kashiwa, Chiba 277-8564, Japan.
[a]E-mail: nobata@aori.u-tokyo.ac.jp
[b]E-mail: ando@aori.u-tokyo.ac.jp
*Corresponding author

in response to changes in the external and internal environments. When FW-acclimated eels (FW eels) are directly transferred to SW, they usually survive and acclimate to SW in 1–2 weeks (Kamiya and Utida 1968, Utida et al. 1971). Eels have an excellent osmotic adaptability that is sustained by robust and flexible machineries, among which regulation of drinking is one of the most important factors underlying this adaptability.

The drinking behavior of eels has long been a target of intensive research in order to understand the significance of drinking in fishes, and now much knowledge has been accumulated. Progress in this research field can be attributed to the establishment of a unique method for real-time measurement of drinking in conscious eels (Hirano 1974), together with the hardiness to tolerate surgical intervention and the distinguished adaptability of eels. As a fish oral cavity is always filled with environmental water for breathing, drinking is achieved without the thirst-motivated search for water that is seen in terrestrial animals. Therefore, eels can ingest water by reflex swallowing just after the perception of relative changes in external and internal water and ion concentrations. Knowledge of drinking regulation in eels may help to understand the complex drinking mechanisms of terrestrial animals.

Accumulated evidence shows that physiological parameters, such as plasma osmolality, blood pressure and volume influence drinking behavior, and that drinking rate is controlled by integrated neural and hormonal signals. In this chapter, drinking behavior of eels is reviewed comprehensively, from methodology for experiments to physiological significance, central regulation, regulators and the future prospects of research in this field.

Measurement of Drinking Rate in Eels

To measure drinking rate in fishes, radioactive substances or dyes that are not absorbed by the gut have been used as tracers (Smith 1930, Evans 1968). Using these methods yields only average drinking rates over a period of time, and cannot be used to measure real-time drinking rate. A new method was established to measure drinking in the eel by Hirano (1974), in which ingested water is externalized through an esophageal fistula and drinking rate can be measured continuously with time. The eel is well-suited for esophageal cannulation as it has an exceptionally long esophagus among teleosts. Thanks to this method, the so-called "chloride response" and the induction of transient drinking by peripheral angiotensin II (see below) were discovered. However, eels with the esophageal fistula become gradually dehydrated in SW because the water supply to the intestine is interrupted.

This problem was solved by reintroducing ingested water into the stomach using a pulse injector (Takei et al. 1998). Figure 1 is a schematic representation of this successful experimental system. The experimental eels have catheters in the esophagus and stomach, the external ends of which are connected with a drop counter and pulse injector, respectively. When the eel drinks, each drop from the esophageal catheter is detected by the drop counter and the drinking rate is calculated and recorded by the monitoring system. The pulse injector is driven by a signal from the drop counter, and introduces an equal volume of fluid into the stomach through the stomach catheter. In SW-acclimated eels (SW eels), 50% SW is reintroduced into the stomach because ingested SW is desalted and diluted to half-initial concentration in the esophagus (Hirano and Mayer-Gostan 1976, Ando and Nagashima 1996). If the pulse injector is turned off, SW eels become gradually dehydrated (which is defined as "artificial dehydration" in this chapter) and finally die due to the dehydration in both cellular and extracellular compartments (Takei et al. 1998, Nobata and Takei 2011).

In order to examine the effects of various stimuli on drinking, catheters can be inserted into the various blood vessels and brain sites. Blood pressure is measured *via* arterial catheters. And blood and urine are collected through

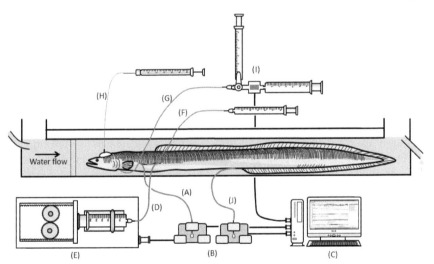

Figure 1. Schematic representation of system for measurement of drinking volume in eels. Ingested water is collected from the esophageal cannula (A), and the water drops are detected by the drop counter (B) and recorded in the computer (C). Synchronously, partially-desalted seawater (corresponding to water naturally desalted in the esophagus) is reintroduced into the stomach through the stomach cannula (D) by the pulse injector (E). Various stimuli can be injected or infused through the cannulae inserted into the dorsal aorta (F), ventral aorta (G) and brain (H). In addition, blood pressure and urine volume are measured through the cannulae inserted into the blood vessels (I) and bladder (J), and are recorded in the computer (C).

catheters inserted into the blood vessels and urinary bladder, respectively. Although such operation seems highly stressful, the eels maintain normal plasma cortisol levels after recovery from anesthesia, usually more than 18 hr after operation (Li and Takei 2003). These cortisol levels are similar to the cortisol levels in other fishes in non-stressed states (Olson and Farrell 2006). The data from our laboratory that is introduced in this chapter are obtained from cultured Japanese eels (*Anguilla japonica*) of ca. 200 g body mass that are sexually immature.

Characteristics of Drinking Behavior in Eels

The relative importance of drinking differs among teleosts in different habitats such as the sea, river and estuarine regions. Therefore, it is likely that the physiological significance of drinking changes dramatically during migration in eels.

Drinking in SW and FW Eels

SW eels actively drink surrounding water and dilute it gradually in the esophagus and stomach, and the ingested SW is finally diluted to isotonicity and absorbed together with monovalent ions (Na^+ and Cl^-) in the intestine (Takei and Loretz 2011). When ingested SW is drained from the esophagus (artificial dehydration), plasma osmolality increases and blood volume decreases, and the eel dies after ca. 5 days in SW, indicating the importance of drinking and subsequent absorption of water in the intestine for SW adaptation (Takei et al. 1998). In SW, insufficient drinking leads to dehydration because osmotic water loss exceeds the gain by drinking, whereas excess drinking induces hypernatremia because of limited capacity to excrete excess ions by the gills. Therefore, drinking is strictly fine-tuned in response to change in the body fluid balance in SW eels.

Conversely, FW eels can obtain sufficient water across the gills without drinking, according to the osmotic gradient between environmental water and plasma. Therefore, it has been long accepted that FW eels hardly drink water. However, drinking rate in FW eels is quite different among individuals, and FW eels can maintain proper body fluid osmolality even if they drink much water as SW eels do. Excess drinking may endanger FW eels because of hypervolemia and hyponatremia. However, hypervolemia can be overcome by an increase in hypotonic urine production, and hyponatremia can be compensated by the active uptake of ions by the gills from environmental FW. As a result, body fluid balance seems to be maintained irrespective of drinking in FW eels.

Acute Response After Transfer of FW Eels to SW

When the surrounding water is changed from FW to SW, Japanese eels immediately start a burst of strong drinking that is slowly reduced over time. Then, drinking rate gradually increases again (Fig. 2A) and is stabilized at higher levels than that in FW (Fig. 2B). As the drinking rate is higher in the eels with artificial dehydration (Takei et al. 1998), it is likely that the inhibition following the initial burst of drinking is caused partly by distension of the stomach and by the detection of ions in the intestine (Hirano 1974, Ando and Nagashima 1996). Similar immediate drinking

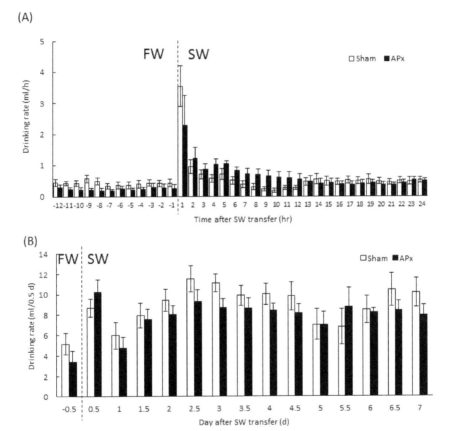

Figure 2. Time course of drinking rate for 1d (A) and 1wk (B) after seawater (SW) transfer of freshwater (FW)-acclimated sham controls (open bar) and area postrema (AP)-lesioned eels (APx eels) (closed bar). Dotted line represents the time when the external environment is changed from FW to SW. Water intake is strongly enhanced in both groups for 1hr after the transfer. Thereafter, drinking rate undergoes reduction and then stabilization 1d or later after the transfer. Copious drinking in SW eels is not influenced by AP-lesioning (Nobata and Takei 2011).

response to SW transfer have also been observed in European eels (Kirsch and Mayer-Gostan 1973).

The burst of drinking induced immediately after SW transfer is named the "chloride response", because it is induced by chloride ions in SW (Hirano 1974); eels responded to NaCl, KCl, $CaCl_2$, choline-Cl and tetraethylammonium-Cl, but not to Na_2SO_4, $NaHCO_3$, or to hyperosmotic stimulus such as mannitol. In addition, other halide-containing salts such as KBr and NaI also induce the burst of drinking, indicating that a chemoreceptor sensitive to halide ions is involved (Hirano 1974).

The chemoreceptor seems to exist in the oral cavity because the chloride response is evoked by stimulating the oral cavity locally (Fig. 3) (unpublished data). Although the physiological significance of this response has not yet been defined, eels may be able to anticipate the long-term consequences of seawater exposure, namely, the necessity to drink. As the chloride response can be induced by 30% SW in the oral cavity (Fig. 3), eels may recognize estuarine waters at the river mouth by sensing small changes in Cl⁻ concentration. Thus, silver eels that migrate into the sea for spawning may be able to forestall the future dehydration by way of the chloride response.

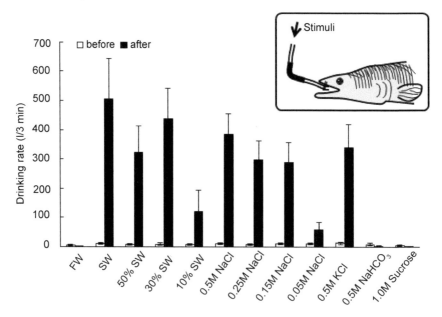

Figure 3. Induction of burst drinking by infusion of various solutions into the oral cavity in FW eels. The infusion is performed through a vinyl tube that is positioned on the maxilla of eels (boxed drawing). Two mL of test solutions were infused at rate of 0.1 mL/sec. The drinking rate during the 3 min before infusion (open bar) was compared with that after infusion (closed bar) (unpublished data).

Central Regulation

Central regulation of drinking in eels has been investigated for more than 40 years. Early work on surgical ablation of brain regions and cranial nerve showed that the forebrain is not necessary, but that the vagus nerve is required for the SW drinking response in eels (Satou and Hirano 1969). More recently, central mechanisms that regulate drinking have been elucidated by histological methods such as immunohistochemistry and tracing of the neural network in conjunction with ablation studies. Accumulating evidence consistently shows that the hindbrain plays an important role in the regulation of drinking in eels. This is in contrast to tetrapods in which the forebrain is necessary for inducing thirst. In eels, the following steps are assumed to be involved in regulation of drinking: (1) circulating factors or afferent neural signals are received in the brain; (2) the signals are transmitted to the motor nucleus; and lastly, (3) pharyngeal and esophageal muscles responsible for drinking are stimulated by motor neurons. The area postrema (AP) is the receptive site for circulating factors, and the spino-occipital motor nucleus (NSO) and glossopharyngeal-vagal motor complex (GVC) are the origin of the motor nuclei.

Involvement of the Medulla Oblongata in Drinking

Ablation studies showed that the forebrain and midbrain are not necessary for water intake, because the eel with only the cerebellum and medulla oblongata intact can respond to SW transfer by drinking (Hirano et al. 1972). Transection of the IXth (glossopharyngeal) and Xth (vagus) cranial nerves, projecting from the medulla oblongata, shorten the survival time in SW after hydromineral disturbance but not in FW (Mayer-Gostan and Hirano 1976). In particular, vagotomy induces high mortality with low drinking rate in SW (Hirano et al. 1972). The vagus nerve innervates all four gill arches and the viscera, including the esophagus and stomach. Therefore, the vagotomized eel may not be able to sense the salinity change and regulate the drinking-related muscles, resulting in failure of drinking. Peripheral injection of angiotensin II (Ang II), a well-known dipsogen in vertebrates, promotes water intake in eels without forebrain and midbrain (Takei et al. 1979). Altogether, these findings suggest that the medulla oblongata plays an important role in regulation of drinking in eels.

Detection of Circulating Factors at the Area Postrema (AP)

The AP is one of the sensory circumventricular organs such as the subfornical organ (SFO) and organum vasculosum of the lamina terminalis (OVLT) in mammals (McKinley and Johnson 2004). In eels, the AP is located along the dorsal midline of the medulla oblongata (Fig. 4), but the SFO and OVLT

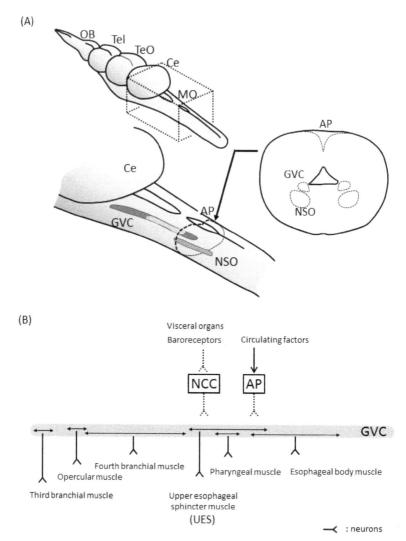

Figure 4. (A) Schematic representation of three-dimensional and planar cross-section views of the loci associated with drinking behavior. The spino-occipital motor nucleus (NSO) and the glossopharyngeal-vagal motor complex (GVC) are on both sides. Neurons in the area drawn by dots in the middle of the GVC column innervate the pharyngeal, upper esophageal sphincter (UES) muscle and esophageal body muscle. B) Distribution of neurons from (right line) and to (dotted line) the GVC column. Distribution of cell bodies in the GVC to each muscle is indicated by arrows. The nerve pathway *via* the NCC is suggested in species other than eels (Kanwal and Caprio 1987, Daiz-Regueria and Anadon 1992, Goehler and Finger 1992, Ito et al. 2006). AP, area postrema; Ce, cerebellum; MO, medulla oblongata; OB, olfactory bulb; Tel, telencephalon; TeO, optic tectum. After Mukuda and Ando (2003) and Mukuda et al. (2005).

have not been identified (Mukuda et al. 2005, Tsukada et al. 2007). The AP has a poor blood-brain barrier (BBB) as it is stained by Evans blue injected intraperitoneally (Mukuda et al. 2005). In fact, the AP is the site of action of circulating atrial natriuretic peptide (ANP) as evidenced by the following results: (1) the AP is an established site for the expression of ANP receptor, and (2) the antidipsogenic effect of ANP is diminished in AP-lesioned eels (APx eels) (Tsukada et al. 2007). In addition, AP-lesioning abolished the dipsogenic or antidipsogenic effects induced by peripheral injection of Ang II, isotocin, and ghrelin (Nobata and Takei 2011). Therefore, it is likely that these hormones also act on the AP, although the expression of their receptors at this site has not been demonstrated yet. In mammals, AP neurons have receptors for many hormones including Ang II, ANP and ghrelin, and are activated by peripheral injection of ghrelin (McKinley et al. 2003, Hashimoto et al. 2007). Thus, AP is included in the neural circuit for drinking regulation by these circulating hormones. On the other hand, the antidipsogenic effect of urotensin II (UII) is not abolished in APx eels (Nobata and Takei 2011). As UII is a potent vasopressor hormone in eels (Nobata et al. 2011), the antidipsogenic effect is probably induced by baroreflex and the AP is not responsible for pressure-regulated drinking as mentioned below.

The AP is physiologically important for body fluid homeostasis. When FW APx eels are transferred to SW, eels start drinking at almost the same rate as do sham controls (Fig. 2), suggesting that the AP is not involved in SW-induced drinking. However, plasma osmolality gradually increases in APx eels and some of APx eels die within 1 mon after the transfer even though all sham controls survive. In addition, APx eels do not respond to artificial dehydration by increased drinking (Fig. 5). In APx eels, drinking rate increases only slightly even though they are dehydrated to an extent similar to sham controls (Fig. 5). Thus, the AP is responsible for fine-tuning of body fluid balance by regulation of drinking in SW eels.

Swallowing Controlled by the GVC

Drinking behavior consists of (1) closing of the mouth and gill openings, (2) contraction of the pharynx, (3) relaxation of the sphincter muscle and the opening of the passageway between pharynx and esophagus, and (4) peristalsis in the esophageal body (Sibbing 1982, Bemis and Lauder 1986). Retrograde tracing showed that these drinking-related muscles are controlled by cholinergic neurons projecting from the NSO and GVC in the medulla oblongata (Mukuda and Ando 2003). Although environmental water is present in the oral cavity for respiration, it does not enter into the esophagus because the upper esophageal sphincter (UES), the last gate for introduction of water into the alimentary tract, is usually closed. However, when eels drink SW, the UES relaxes and water enters the esophagus

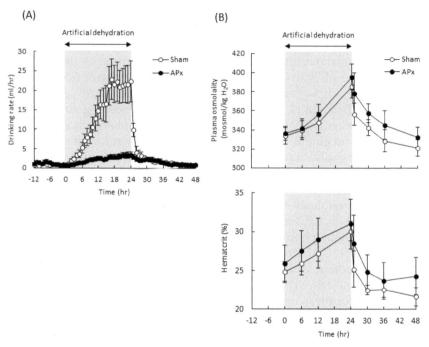

Figure 5. Effects of drainage of ingested water from the esophageal fistula (artificial dehydration) on water intake (A), plasma osmolality and hematocrit (B) in seawater (SW)-acclimated sham controls (open circle) and area postrema (AP)-lesioned eels (APx eels) (closed circle). Gray zones in each graph represent a period in which ingested water was eluted. Although both groups experience dehydration, induced drinking is more profound in sham controls than in APx eels. With reintroduction of 50% SW by the pulse injector, water intake, osmolality and hematocrit are returned to normal values (Nobata and Takei 2011).

(swallowing). The UES muscle is constricted by acetylcholine through the nicotinic ACh receptor, and is controlled by nerve terminals of cholinergic neurons with cell bodies located in a middle part of the GVC column (Fig. 4) (Kozaka and Ando 2003, Mukuda and Ando 2003).

Innervation to the GVC

In the hemisphere of eel brain studied *in vitro*, catecholamines (CAs) inhibit the activities of GVC neurons that innervate the drinking-related muscles. CA-responsive neurons are present at high density in the middle part of the GVC column, corresponding to the area that controls the UES muscle. As the inhibitory action of CAs is abolished by prazosin, an α_1-adrenergic receptor antagonist, it is likely that CAs act on α-like adrenoceptors. After bindings to their receptors, CAs inhibit neuronal activity in the GVC, thus relaxing the UES, and resulting in acceleration of swallowing (Ito et al. 2006).

Indeed, there are many catecholaminergic cell bodies and fibers in the medulla oblongata of the eel. Among these, nerve fibers from the AP and the commissural nucleus of Cajal (NCC) appear to project to the GVC as shown by immunohistochemistry for tyrosine hydroxylase (Ito et al. 2006). In the catfish, goldfish and mullet, the NCC is innervated by viscerosensory neurons (Diaz-Regueria and Anadon 1992, Goehler and Finger 1992, Kanwal and Caprio 1987), and is homologous to the nucleus tractus solitarius (NTS) that is innervated by the sensory neurons from the baroreceptors and visceral organs in mammals. Therefore, it is likely that circulating hormones regulate neural activity of the GVC through the AP, and the baroreflex response may regulate the GVC activity through the NCC, although this idea is still speculative.

Hormones

The dipsogenic and antidipsogenic effects in eels of many hormones have been demonstrated (Takei 2002, Kozaka et al. 2003), and most hormones fall into the vasopressor/antidipsogenic or vasodepressor/dipsogenic categories (Hirano and Hasegawa 1984). Such cooperative regulation of drinking with blood pressure is probably due to secondary neural reflex actions *via* the baroreceptor. However, there are exceptions such as the renin-angiotensin system and isotocin that are vasopressor/dipsogenic hormones, and natriuretic peptides that are vasodepressor/antidipsogenic hormones. These hormones may have direct actions on drinking that are independent of the baroreflex regulation.

Dipsogenic Hormones

Angiotensin

Angiotensin II (Ang II) is the best known dipsogenic hormone in vertebrates to be examined thus far. The Ang II effect has been examined in many fish species where, in most, Ang II increases water intake (Kobayashi et al. 1983, Perrott et al. 1992). Especially in eels, a unique time-course of water intake after a bolus injection is obtained using the system shown in Fig. 1. After injection, water intake transiently increases within 10 min. This is followed by a sustained decrease (secondary inhibition) (Fig. 6A). Peripheral injection of Ang II promotes water intake even in eels in which the telencephalon, diencephalon and a part of mesencephalon are removed (Takei et al. 1979). In addition, the dipsogenic effect of peripheral Ang II is abolished by AP-lesioning (Fig. 6A) (Nobata and Takei 2011). These findings indicate that circulating Ang II acts on the AP to elicit drinking. Although bolus injections show the biphasic effects on water intake as shown in Fig.

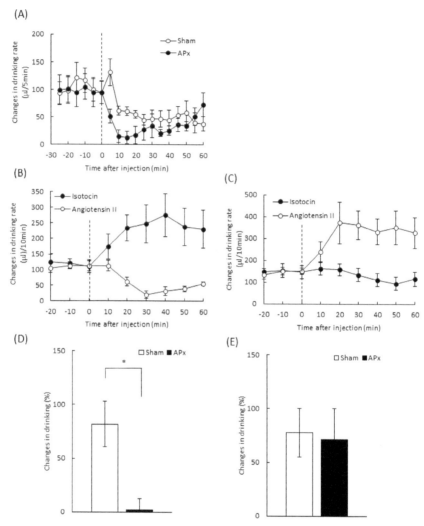

Figure 6. Time course of drinking rate changes after peripheral injection of angiotensin II (Ang II) at 5.0 nmol/kg in seawater (SW)-acclimated sham controls (open circle) and area postrema (AP)-lesioned eels (APx eels) (closed circle) (A). Effects of Ang II and isotocin (IT) injected peripherally at 1.0 nmol/kg (B) and centrally at 0.3 nmol/eel (C) on drinking rate. Changes in drinking after peripheral injection of IT (D) and central injection of Ang II (E) in sham controls and APx eels. In (D) and (E), changes in drinking (%) are represented as averaged drinking rate (μl/30 min) for 60 min after injection, which is normalized as a percentage to drinking rate for 30 min before injection. Vertical dotted line represents the time of each injection (A–C). Changes in drinking rates (%) were compared between sham controls and APx eels by Student's t test (*P<0.05). (Nobata and Takei 2011, unpublished data).

6A, infusion of Ang II induces only the dipsogenic effect (Tsuchida and Takei 1999). As infusion is more similar to the physiological state than bolus injection, circulating Ang II may act intrinsically as a dipsogenic hormone, whereas secondary inhibition may be induced by baroreceptors through the NCC-GVC axis (see above), and not *via* the AP. Consistently, the secondary inhibition of drinking by Ang II was emphasized by AP-lesioning (Fig. 6A).

Ang II injected centrally increases water intake without the secondary inhibition of drinking (Fig. 6C). Central Ang II is dipsogenic when injected into the third ventricle but not when delivered into the forth ventricle (Ogoshi et al. 2008), suggesting an action on the forebrain. As the dipsogenic effect of central Ang II is not abolished by AP-lesioning (Fig. 6E) (unpublished data), it may not be mediated through the AP.

The dipsogenic effect of Ang II in eels suggests its physiological role in chronic drinking of SW eels. However, the accumulated evidence does not support this idea. In Japanese eels, plasma renin activity and Ang II concentration do not differ between FW and SW eels although these parameters transiently increase after SW transfer (Sokabe et al. 1973, Okawara et al. 1987). However, plasma Ang II concentration is higher in European eels acclimated to SW than those to FW (Tierney et al. 1995). Infusion of captopril, an inhibitor of angiotensin-converting enzyme (ACE), reduces drinking rate with a decrease in plasma Ang II concentration in both Japanese and European eels (Tierney et al. 1995), but this reduction in drinking may be attributed to an increase in plasma bradykinin (BK), a potent antidipsogen in eels (Takei et al. 2001). ACE not only cleaves Ang II from Ang I but also inactivates BK. Additionally, infusion of anti-Ang II serum has no effect on drinking rate even if plasma Ang II concentration is decreased to undetectable levels (Takei and Tsuchida 2000). These findings suggest that circulating Ang II does not participate in chronic drinking in SW but in acute drinking caused by hypovolemia (see below).

Ang I, a precursor of Ang II, is also diposogenic with similar potency to Ang II, but this effect is completely blocked by pretreatment with captopril. Although captopril may induce an antidipsogenic effect via BK, the effect of Ang II was not blocked by pretreatment with captopril, indicating that the antidipsogenic effect of BK is negligible in this study (Ando et al. 2000). Therefore, drinking is induced by Ang I after conversion to Ang II by ACE. The N-terminal sequence seems to be important for the dipsogenic potency, with the order of potency being [Asn^1]-Ang II > [Asp^1]-Ang II > Ang III (Ando et al. 2000).

Isotocin

The role of isotocin (IT), teleostean homologue of oxytocin (OXT), in osmoregulation has not been examined in fishes, but the natriuretic effect of OXT has been shown in mammals (Chan 1965). IT relaxes the UES *in vitro* and it has a dipsogenic effect *in vivo* in eels (Watanabe et al. 2007). After peripheral injection of IT, drinking rate increased 2-to-3-fold even with profound elevation of blood pressure (Fig. 6B). The dipsogenic effect of IT is more potent and efficacious than that of Ang II when injected peripherally. Conversely, IT injected centrally has no effect on water intake, which is in contrast to the potent dipsogenic effect of central Ang II (Fig. 6C). The dipsogenic effect of peripheral IT is abolished by AP-lesioning, suggesting mediation by the AP (Fig. 6D). Therefore, it is likely that peripheral IT may inhibit the GVC through the AP. Since the GVC neurons are cholinergic (Mukuda and Ando 2003) and acetylcholine contracts the UES muscle (Kozaka and Ando 2003), the peripheral IT may inhibit acetylcholine release from the GVC. In addition to the central action, peripheral IT may relax the UES muscle directly, since the isolated UES muscle is relaxed by IT (Sakihara et al. 2007, Watanabe et al. 2007). The peripheral relaxant actions of IT may further accelerate drinking rate.

Other hormones

Adrenomedullins (AM1, AM2 and AM5) are also thought to be dipsogenic hormones. Among the three AMs, AM2 and AM5, but not AM1, promote drinking when injected peripherally. Although AM2 and AM5 are potent vasodepressor hormones, these increase water intake without vasodepressor effects during infusion (Ogoshi et al. 2008). Thus, AM2 and AM5 accelerate water intake at least in part independently of their effects on blood pressure. Peripheral injection of histamine, serotonin, acetylcholine and substance P also increased the water intake in eels (Ando et al. 2000).

Antidipsogenic Hormones

Natriuretic peptides

Natriuretic peptide (NP) family consists of seven NPs in teleosts: atrial NP (ANP), B-type NP (BNP), ventricular NP (VNP) and four C-type NPs (CNP1, 2, 3 and 4) (Inoue et al. 2003). In eels, NP cDNAs for all except CNP2 have been identified, and ANP, BNP and VNP, which are categorized as cardiac NPs, are expressed mainly in the heart with the three CNPs expressed in the brain or pituitary (Nobata et al. 2010). Cardiac NPs, even though they are vasodepressor, consistently depressed water intake after peripheral injections in the potency order of ANP ≥ VNP > BNP in both SW and FW eels. In addition, CNP3, an ancestral molecule of cardiac NPs, depresses water intake in SW eels but not in FW eels; CNP1 and CNP4 have no effect on water intake (Tsukada and Takei 2001, Miyanishi et al. 2011). As is the case with peripheral injection, infusion of ANP also decreased water intake in both FW and SW eels in parallel with reduction of plasma Ang II concentration (Tsuchida and Takei 1998). The antidipsogenic effect of ANP is not due to the reduction of plasma Ang II because drinking rate does not decrease even with removal of plasma Ang II using anti-Ang II serum (Takei and Tsuchida 2000). Little is known about the effect of NPs injected centrally on drinking. Although ANP injected into the fourth ventricle reduces water intake in SW eels (Kozaka et al. 2003), it remains unknown whether or not CNPs have antidipsogenic effects in the brain. The AP is the site of action of circulating ANP as mentioned above (Tsukada et al. 2007), and VNP and BNP may also act on the AP because these NPs share the same receptor (NPR-A) with ANP (Kashiwagi et al. 1999).

The physiological significance of the antidipsogenic effect of circulating NPs may be fine-tuning of drinking to avoid hypernatremia caused by excess drinking in SW eels. Inhibition of drinking following the burst of drinking just after SW exposure of FW eels (Fig. 2) occurs concomitantly with transient elevation of plasma ANP and VNP concentrations (Kaiya and Takei 1996b). In addition, infusion of ANP at 0.3 nmol/kg/min, a concentration that reduces water intake in SW eels, elevates plasma ANP concentration to around 200 fmol/mL, a value that is within the physiological range (Tsuchida and Takei 1998, Kaiya and Takei 1996a). These findings suggest that circulating ANP and VNP are physiologically involved in regulation of drinking. This is further supported by the following findings: 1) immunoneutralization of plasma ANP and VNP increases the drinking rate in SW eels (Tsukada and Takei 2006) and 2) hypertonic saline injected peripherally is a profound stimulus for ANP and VNP secretion from the heart (Kaiya and Takei 1996c) and reduces drinking through the AP as does ANP (Nobata and Takei 2011). As drinking is essential for survival in a hypertonic environment, inhibition of drinking may be disadvantageous for acclimation to SW. However, as excess drinking in SW induces severe hypernatremia, inhibition of drinking may moderate it to promote SW acclimation. In fact, the slower the drinking rate, the more efficient is desalination in the esophagus of SW eels (Ando and Nagashima 1996).

BNP is the only common NP throughout the vertebrate species examined so far (Takei et al. 2011). As BNP was only more recently found in eels compared with the discovery of ANP and VNP, its plasma concentration and response to osmotic stimulus have not yet been studied. More recently, however, the antidipsogenic effect of BNP was found to be weaker than that of ANP and VNP (Miyanishi et al. 2011), suggesting a supplementary role in regulation of drinking in eels.

Other hormones

The antidipsogenic effect of ghrelin is similar to that of ANP when injected both peripherally and centrally (Kozaka et al. 2003, unpublished data). Circulating ghrelin is known to exert its antidipsogenic effect through the AP (Nobata and Takei 2011). Peripheral ghrelin probably contributes to the regulation of drinking because ghrelin mRNA is detected in the stomach and intestine but not in the brain. Plasma ghrelin concentration transiently increases 6 hr after SW transfer of FW eels (Kaiya et al. 2006), corresponding to the time when the initial burst of drinking is reduced (see Fig. 2). Therefore, ghrelin might be involved in the transient reduction of drinking during SW acclimation. Although highly speculative, the antidipsogenic effect may be induced by ghrelin secreted by gastric distension or an increase in luminal osmolality of the stomach caused by the initial copious drinking. On the other hand, ghrelin is known to exhibit an orexigenic effect in vertebrates including fishes (Unniappan and Peter 2005). Likewise, neuropeptide Y (NPY) is antidipsogenic (M. Ando, unpublished data) and orexigenic (López-Patiño et al. 1999). The reason why ghrelin and NPY inhibit water intake despite stimulating food intake is not known.

Arginine vasotocin (AVT) reduces drinking rate when injected peripherally (Hirano and Hasegawa 1984, Ando et al. 2000, Watanabe et al. 2007). BK is also an antidipsogen in eels but a dipsogen in mammals (Takei et al. 2001). As AVT and BK are potently vasopressor, the antidipsogenic effect may be due to the baroreflex response.

Osmotic and Volemic Stimuli

Osmotic Stimulus (cellular dehydration)

Hyperosmotic stimulus by injection of membrane impermeable solutes such as NaCl and sucrose inhibits drinking in a concentration-dependent manner in SW and FW eels (Takei et al. 1988, Ando et al. 2000). Osmotic stimulus induces cellular dehydration (cell shrinkage).

FW eels

After intravenous injection of hypertonic solution into FW eels, hypertonicity continues for a long time, probably due to the poor ability to excrete excess salts (Takei et al. 1988). Even with continued plasma hyperosmolality, FW eels decrease water intake. This is completely opposite to the dipsogenic response in terrestrial animals. Hyperosmotic stimulus promotes hypervolemia caused by movement of water from the cytoplasm to the intravascular space and osmotic influx from environmental water across the gill in FW eels. Blood volume appears to be more strictly maintained than plasma osmolality in eels after intravascular injection of hypertonic solution, which contrasts with the cases in mammals and birds (Takei 2000). Consistently, eels can tolerate an increase in plasma Na^+ concentration by 80% (Takei et al. 1998), which is much greater than the tolerance in mammals. Therefore it is possible that eels are more sensitive to changes in blood volume than to changes in plasma osmolality for regulation of drinking. In fact, expansion of blood volume decreases drinking in eels (see below). Taken together, hyperosmotic stimulus is likely to inhibit drinking in part *via* hypervolemia in FW eels. It is more likely that the inhibition of drinking is due to the increased secretion of ANP and VNP after hyperosmotic stimulus (Kaiya and Takei 1996c).

SW eels

Unlike FW eels, SW eels are able to excrete ions across the gill immediately after injections of hypertonic solution (Takei et al. 1988). Nonetheless, they reduce water intake after intravascular hyperosmotic stimuli (Ando et al. 2000). As SW eels face a threat of hypernatremia, injections of hypertonic solution add further risk of cellular dehydration even with the high capability to excrete ions. Therefore this inhibition of drinking may be an anticipatory response to forestall such a premonition.

Hyperosmotic stimulus may also promote secretion of ANP and VNP from the heart that inhibit drinking *via* the AP. Indeed, the antidipsogenic effect of hyperosmotic stimulus is blocked by AP-lesioning (Nobata and Takei 2011), supporting AP mediation of hormonal inhibition of drinking. It is also possible that some sensors for extracellular osmolality or NaCl concentration might exist in the AP as suggested for mammals (Liedtke and Friedman 2003, Hiyama et al. 2004) to maintain optimal volume of all cells.

Hypovolemic Stimulus (extracellular dehydration)

When hypovolemia is induced by hemorrhage, water intake is elevated in both FW and SW eels (Hirano 1974, Ando et al. 2000). These findings indicate that it is important for eels to maintain blood volume because even FW eels that can obtain water by osmotic influx across the gill start drinking in response to hypovolemia. In SW eels, however, the dipsogenic effect is influenced by the basal drinking rate; hemorrhage of 1mL significantly elevates water intake in the eel that drinks at lower rates but not in the eel that drinks at rates higher than 0.9 mL/hr, probably to avoid hypernatremia by excess drinking (Nobata and Takei 2011). Hemorrhage elevates plasma Ang II concentration (Kobayashi and Takei 1996) and the hemorrhage-induced drinking is inhibited by pretreatment with captopril, indicating an involvement of Ang II (Ando et al. 2000). As mentioned above, however, the role of Ang II in extracellular dehydration is not fully understood, since Ang II may not be a strong regulator for drinking in SW eels and since captopril enhances antidipsogenic BK action (Takei et al. 2001).

Hypervolemic Stimulus (extracellular overhydration)

Infusion of 1 mL of isotonic 0.9% NaCl inhibits drinking in FW eels (Hirano 1974). Likewise, injection of 1mL 0.9% NaCl decreases drinking in SW eels (Nobata and Takei 2011). This reduction in drinking may serve to avoid excess volume loading. Although little is known about the volume receptor in eels, baroreflex may be one such mechanism in response to volume expansion. Blood volume expansion is known to stimulate secretion of ANP and VNP into the blood (Kaiya and Takei 1996c). Therefore, circulating ANP and VNP may also be responsible for the antidipsogenic effect induced by blood volume expansion, as AP-lesioning abolished drinking inhibition by volume expansion (Nobata and Takei 2011).

Absolute Dehydration

Under absolute dehydration, plasma osmolality increases (cellular dehydration), and blood volume decreases (extracellular dehydration). Eels suffer absolute dehydration after transfer from FW to SW (Takei et al. 1998). Although cellular dehydration reduces drinking after acute osmotic stimulus in eels, SW eels drink water more copiously than FW eels to compensate for osmotic water loss (extracellular dehydration).

Although eels are subjected to absolute dehydration in SW, the actual dehydration does not occur in the adapted state. However, absolute dehydration can be experimentally induced in SW eels by ligation of the esophagus (Hirano et al. 1972), or drainage of ingested water from the esophageal fistula using the system shown in Fig. 1 (Takei et al. 1998, Nobata and Takei 2011). When SW eels are subjected to artificial dehydration, osmotic water loss induces the elevation of plasma osmolality and reduction of blood volume, and ingestion of water increases more than 30-fold (Fig. 5). Enhanced ingestion of SW shows that improvement of hypovolemia is primary to that of hypernatremia under artificial dehydration. This phenomenon also supports the concept that blood volume is maintained more preferentially than plasma osmolality in fishes (Takei 2000).

Figure 5 shows that the artificial dehydration also enhances drinking rate through the AP. In contrast to sham controls, APx eels increased water intake only slightly after artificial dehydration. This is consistent with the fact that APx eels cannot respond to hypovolemia and Ang II (Nobata and Takei 2011), as these dipsogenic stimuli are enhanced after artificial dehydration. Central Ang II might not be involved in this drinking, since the dipsogenic effect of central Ang II did not change after AP-lesioning (Fig. 6E) (unpublished data).

Feedback from the Alimentary Tract

Gastric Distention

In SW eels, a cyclical pattern of drinking is often observed. This might be caused by a repeated rhythm of distension and emptying of the stomach. In fact, when the stomach is distended by isotonic mannitol solution, the drinking rate remarkably decreased in SW eels even after hemorrhage (Hirano 1974). Supporting this observation and conclusion, the eel under artificial dehydration lacks the cyclic drinking pattern (Takei et al. 1998), probably due to the continuously empty stomach.

Intestinal Content

Levels of Na^+ and Cl^- in luminal fluid of the intestine have an effect on drinking (Ando and Nagashima 1996). Cl^- inhibits and Na^+ enhances drinking, and the inhibitory effect of Cl^- is predominant compared with the effect of Na^+. This may also be important to avoid excess SW drinking and to desalt efficiently the ingested SW in the esophagus. Ingested SW must be desalted in the esophagus in order to achieve effective water absorption in the intestine, and thus Cl^- may indirectly regulate drinking by acting as a regulator of proper drinking and desalting. However, the ion sensors for this response have not been identified in the intestine yet.

Perspective

Many regulators in drinking have been found in eels by intensive and extensive studies, and future studies will elucidate whether they are physiologically significant for regulation of drinking in their life history. Among regulatory hormones, ANP and Ang II have been extensively studied. In the case of ANP, most of the results seem to confirm that this hormone is responsible for fine-tuning of drinking to avoid excessive intake. On the other hand, the physiological significance of peripheral Ang II is still controversial because Ang II injected peripherally is only weakly and transiently dipsogenic in eels, and is followed by a longer antidipsogenic effect (Fig. 6A). Centrally produced Ang II and peripheral IT are more reliable dipsogens in eels. During absolute dehydration, central Ang II and peripheral IT levels might be increased, with the former acting on the forebrain and the latter on the AP. Judging from artificial dehydration experiments, most of the enhanced water intake depends on the presence of the AP (Fig. 5A), suggesting the artificial dehydration stimulates secretion of dipsogens that act on the AP. IT might be such a dipsogen, but plasma IT levels have not been measured in such situations. Alternatively, IT secretion from the pituitary might be inhibited by AP-lesioning, since AP-lesioning is known to inhibit the secretion of oxytocin, a mammalian homologue of IT, from the pituitary in water-deprived rats (Huang et al. 2000). In either case, IT might be secreted from the pituitary and enhance drinking rate by acting on the AP. To test this hypothesis, plasma IT levels must be measured. In addition, some sensory machinery, which detects cellular and extracellular dehydration and transmit their signals to the pituitary, must be found.

Recently, the transient receptor potential vallinoid-type (TRPV) ion channel and Na_x channel were suggested as molecular candidates for the sensor of extracellular osmolality and Na^+ (Liedtke and Friedman 2003, Hiyama et al. 2004). Knockout of these molecules induced abnormal drinking behavior after osmotic challenges in mice. Thus these channels may also be candidates for the sensors in eels. It is highly probable that drinking may be regulated not only by hormones but also by a neural network linked to sensors of plasma osmolality or Na^+ located in the brain. In fact, TRPV2 and TRPV4 cDNAs are identified by molecular cloning in the eel brain (unpublished data). Such studies will enable deeper understanding of the mechanism that initiates migration between FW and SW during the life of eels. The involvement of these molecules in regulation of drinking and hormone release will be clarified by functional analysis or comparing gene expression between FW and SW, as shown in the tilapia (Watanabe et al. 2012). In addition, the information from eels may help to understand the general regulatory system of drinking in vertebrates.

Acknowledgement

We thank Dr. Yoshio Takei for the opportunity to write this chapter and for his comments, and Dr. Francesca Trischitta and Dr. Philippe Sebert for comments on the manuscript. We also thank Dr. Christopher Loretz for editing of the manuscript.

References

Ando, M., Y. Fujii, T. Kadota, T. Kozaka, T. Mukuda, I. Takase and A. Kawahara. 2000. Some factors affecting drinking behavior and their interactions in seawater acclimated eel, *Anguilla japonica*. Zool. Sci. 17: 171–178.

Ando, M. and K. Nagashima. 1996. Intestinal Na$^+$ and Cl$^-$ levels control drinking behavior in the seawater-adapted eel *Anguilla Japonica*. J. Exp. Biol. 199: 711–716.

Bemis, W.E. and G.V. Lauder. 1986. Morphology and function of the feeding apparatus of the lungfish, *Lepidosiren paradoxa* (Dipnoi). J. Morphol. 187: 81–108.

Chan, W.Y. 1965. Effects of neurohypophysial hormones and their deamino analogues on renal excretion of Na, K and water in rats. Endocrinology 77: 1097–1104.

Diaz-Regueria, S. and R. Anadon. 1992. Central projections of vagus nerve in *Chelon labrosus* Risso (Telestei, *O. Perciformes*). Brain Behav. Evol. 40: 297–310.

Evans, D.H. 1968. Measurements of drinking rates in fish. Comp. Biochem. Physiol. 25: 751–753.

Goehler, L.E. and T.E. Finger. 1992. Functional organization of vagal reflex systems in the brain stem of the goldfish, *Carassius auratus*. J. Comp. Neurol. 319: 463–478.

Hashimoto, H., H. Fujihara, M. Kawasaki, T. Saito, M. Shibata, H. Otsubo, Y. Takei and Y. Ueta. 2007. Centrally and peripherally administered ghrelin potently inhibits water intake in rats. Endocrinology 148: 1638–1647.

Hirano, T. 1974. Some factors regulating water intake by the eel, *Anguilla japonica*. J. Exp. Biol. 61: 737–747.

Hirano, T. and S. Hasegawa. 1984. Effects of angiotensins and other vasoactive substances on drinking in the eel, *Anguilla japonica*. Zool. Sci. 1: 106–113.

Hirano, T. and N. Mayer-Gostan. 1976. Eel esophagus as an osmoregulatory organ. Proc. Natl. Acad. Sci. USA 73: 1348–1350.

Hirano, T., M. Satou and S. Utida. 1972. Central nervous system control of osmoregulation in the eels (*Anguilla japonica*). Comp. Biochem. Physiol. 43A: 537–544.

Hiyama, T.Y., E. Watanabe, H. Okado and M. Noda. 2004. The subfornical organ is the primary locus of sodium-level sensing by Nax sodium channels for the control of salt-intake behavior. J. Neurosci. 24: 9272–9281.

Huang, W., A.F. Sved and E.M. Stricker. 2000. Vasopressin and oxytocin release evoked by NaCl loads are selectively blunted be area postrema lesions. Am. J. Physiol. Regul. Integr. Comp. Physiol. 278: R732–R740.

Inoue, K., K. Naruse, S. Yamagami, H. Mitani, N. Suzuki and Y. Takei. 2003. Four functionally distinct C-type natriuretic peptides found in fish reveal evolutionary history of natriuretic peptide system. Proc. Natl. Acad. Sci. USA 100: 10079–10084.

Ito, S., T. Mukuda and M. Ando. 2006. Catecholamines inhibit neuronal activity in the glossopharyngeal-vagal motor complex of the Japanese eel: Significance for controlling swallowing water. J. Exp. Zool. 305A: 499–506.

Kaiya, H. and Y. Takei. 1996a. Atrial and ventricular natriuretic peptide concentrations in plasma of freshwater and seawater-adapted eels. Gen. Comp. Endocrinol. 102: 183–190.

Kaiya, H. and Y. Takei. 1996b. Changes in plasma atrial and ventricular natriuretic peptide concentrations after transfer of eels from freshwater to seawater or vice versa. Gen. Comp. Endocrinol. 104: 337–345.

Kaiya, H. and Y. Takei. 1996c. Osmotic and volaemic regulation of atrial and ventricular natriuretic peptide secretion in conscious eels. J. Endocrinol. 149: 441–447.

Kaiya, H., T. Tsukada, S. Yuge, H. Mondo, K. Kangawa and Y. Takei. 2006. Identification of eel ghrelin in plasma and stomach by radioimmunoassay and histochemistry. Gen. Comp. Endocrinol. 148: 375–382.

Kamiya, M. and S. Utida. 1968. Changes in activity of sodium-potassium-activated adenosinetriphosphatase in gills during adaptation of the Japanese eel to sea water. Comp. Biochem. Physiol. 26: 675–685.

Kanwal, J.S. and J. Caprio. 1987. Central projections of the glossopharyngeal and vagal nerves the channel fish, *Ictalurus punstatus*: clues to the differential processing of the visceral inputs. J. Comp. Neurol. 264: 216–230.

Kashiwagi, M., K. Miyamoto, Y. Takei and S. Hirose. 1999. Cloning, properties and tissue distribution of natriuretic peptide receptor-A of euryhaline eel, *Anguilla japonica*. Eur. J. Biochem. 259: 204–211.

Kirsch, R. and N. Mayer-Gostan. 1973. Kinetics of water and chloride exchanges during adaptation of the European eel to sea water. J. Exp. Biol. 58: 105–121.

Kobayashi, H. and Y. Takei. 1996. Regulation of renin release. *In*: S.D. Bradshaw, W. Burggren, H.C. Heller, S. Ishii, H. Langer, G. Neuweiler and D.J. Randall [eds]. The Renin-angiotensin System. Springer, Berlin, Germany. pp. 53–75.

Kobayashi, H., H. Uemura, Y. Takei, N. Itatsu, M. Ozawa and K. Ichinohe. 1983. Drinking induced by angiotensin II in fishes. Gen. Comp. Endocrinol. 49: 295–306.

Kozaka, T. and M. Ando. 2003. Cholinergic innervation to the upper esophagus sphincter muscle in the eel, with special reference to drinking behavior. J. Comp. Physiol. 173: 135–140.

Kozaka, T., Y. Fujii and M. Ando. 2003. Central effects of various ligands on drinking behavior in eels acclimated to seawater. J. Exp. Biol. 206: 687–692.

Li, Y.Y. and Y. Takei. 2003. Ambient salinity-dependent effects of homologous natriuretic peptides (ANP, VNP and CNP) on plasma cortisol level in the eel. Gen. Comp. Endocrinol. 130: 317–323.

Liedtke, W. and J.M. Friedman. 2003. Abnormal osmotic regulation in trpv$^{-/-}$ mice. Proc. Natl. Acad. Sci. USA 100: 13698–13703.

López-Patiño, MA., AI. Guijarro, E. Isorna, MJ. Delqado, M. Alonso-Bdate and N. de Pedro. 1999. Neuropeptide Y has a stimulatory action on feeding behavior in goldfish (*Carassius auratus*). Eur. J. Pharmacol. 21: 147–153.

Marshall, W.S. and M. Grosell. 2006. Ion transport, osmoregulation, and acid-base balance. *In*: D.H. Evans and J.B. Claiborne [eds]. The Physiology of Fishes, 3rd ed. CRC Press, Boca Raton, FL, USA. pp. 177–230.

Mayer-Gostan, N. and T. Hirano. 1976. The effects of trasnsecting the IXth and Xth cranial nerves on hydromineral balance in the eel *Anguilla anguilla*. J. Exp. Biol. 64: 461–475.

McKinley, M.J. and A.K. Johnson. 2004. The physiological regulation of thirst and fluid intake. News Physiol. Sci. 19: 1–6.

McKinley, M.J., R.M. McAllen, P. Davern, M.E. Giles, J. Penschow, N. Sunn, A. Uschakov and B.J. Oldfield. 2003. Neurochemical aspects of sensory circumventricular organs. *In*: F. Beck, B. Christ, W. Kriz, W. Kummer, E. Marani, R. Putz, Y Sano, T.H. Schiebler and K. Zilles [eds]. The Sensory Circumventricular Organs of the Mammalian Brain. Springer, Berlin, Germany. pp. 35–54.

Miyanishi, H., S. Nobata and Y. Takei. 2011. Relative antidipsogenic potencies of six homologous natriuretic peptides in eels. Zool. Sci. 28: 719–726.

Mukuda, T. and M. Ando. 2003. Medullary motor neurons associated with drinking behavior of Japanese eels. J. Fish Biol. 62: 1–12.

Mukuda, T., Y. Matsunaga, K. Kawamoto, K. Yamaguchi and M. Ando. 2005. "Blood-contacting neurons" in the brain of the Japanese eel *Anguilla japonica*. J. Exp. Zool. 303A: 366–376.

Nobata, S., J.A. Donald, R.J. Balment and Y. Takei. 2011. Potent cardiovascular effects of homologous urotensin II (UII)-related peptide (URP) and UII in unanesthetized eels after peripheral and central injections. Am. J. Physiol. Regul. Integr. Comp. Physiol. 300: R437–R446.

Nobata, S. and Y. Takei. 2011. The area postrema in hindbrain is a central player for regulation of drinking behavior in Japanese eels. Am. J. Physiol. Regul. Integr. Comp. Physiol. 300: R1569–R1577.

Nobata, S., A. Ventura, H. Kaiya and Y. Takei. 2010. Diversified cardiovascular actions of six homologous natriuretic peptides (ANP, BNP, VNP, CNP1, CNP3, and CNP4) in conscious eels. Am. J. Physiol. Regul. Integr. Comp. Physiol. 298: R1549–R1559.

Ogoshi, M., S. Nobata and Y. Takei. 2008. Potent osmoregulatory actions of homologous adrenomedullins administered peripherally and centrally in eels. Am. J. Physiol. Regul. Integr. Comp. Physiol. 295: R2075–R2083.

Okawara, Y., T. Karakida, M. Aihara, K. Yamaguchi and H. Kobayashi. 1987. Involvement angiotensin II in water intake in the Japanese eel, *Anguilla japonica*. Zool. Sci. 4: 523–528.

Olson, K.R. and A.P. Farrell. 2006. The cardiovascular system. *In*: D.H. Evans and J.B. Claiborne [eds]. The Physiology of Fishes, 3rd ed. CRC Press, Boca Raton, USA. pp. 119–152.

Perrott, M.N., C.E. Grierson, N. Hazon and R.J. Balment. 1992. Drinking behavior in sea water and fresh water teleosts, the role of the renin-angiotensin system. Fish Physiol. Biochem. 10: 161–168.

Sakihara, T., Y. Watanabe, T. Mukuda and M. Ando. 2007. Post- and pre-synaptic action of isotocin in the upper esophagus sphincter muscle of the eel: its role in water drinking. J. Comp. Physiol. B. 177: 927–933.

Satou, M. and T. Hirano. 1969. Effects of decerebration on osmoregulation of the eel. Zool. Mag. Tokyo 78: 394.

Sibbing, P.A. 1982. Pharyngeal mastication and food transport in the carp (*Cyprinus carpio* L.): a cineradiographic and electromyographic study. J. Morphol. 172: 223–258.

Smith, H.W. 1930. The absorption and excretion of water and salts by marine teleosts. Am. J. Physiol. 93: 480–505.

Sokabe, H., H. Oide, M. Ogawa and S. Uchida. 1973. Plasma renin activity in Japanese eels (*Anguilla japonica*) adapted to seawater or in dehydration. Gen. Comp. Endocrinol. 21: 160–167.

Takei, Y. 2000. Comparative physiology of body fluid regulation in vertebrates with special reference to thirst regulation. Jpn. J. Physiol. 50: 171–186.

Takei, Y. 2002. Hormonal Control of deinking in eels: an evolutionary approach. *In*: N. Hazon and G. Flik [eds]. Osmoregulation and Drinking in Vertebrates. BIOS Sci. Publ. Ltd., Oxford, UK. pp. 61–82.

Takei, Y. and R.J. Balment. 2009. The neuroendocrine regulation of fluid intake and fluid balance. *In*: N.J. Bernier, G. Van Der Kraak, A.P. Farrell and C.J. Brauner [eds]. Fish Neuroendocrinology, Fish Physiology Vol. 28. Academic Press, San Diego, USA. pp. 366–421.

Takei, Y., T. Hirano and H. Kobayashi. 1979. Angiotensin and water intake in the Japanese eel, *Anguilla japonica*. Gen. Comp. Endocrinol. 38: 466–475.

Takei, Y., K. Inoue, S. Trajanovska and J.A. Donald. 2011. B-type natriuretic peptide (BNP), not ANP, is the principal cardiac natriuretic peptide in vertebrates as revealed by comparative studies. Gene. Comp. Endocrinol. 171: 258–266.

Takei, Y. and C.A. Loretz. 2011. The gastrointestinal tract as an endocrine/neuroendocrine/paracrine organ: Organization, chemical messengers and physiological targets. *In*: M. Grosell, A.P. Farrell and C.J. Brauner [eds]. The Multifunctional Gut of Fish, Fish Physiology Vol. 30. Academic Press, San Diego, USA. pp. 261–317.

Takei, Y., J. Okubo and K. Yamaguchi. 1988. Effects of cellular dehydration on drinking and plasma angiotensin II level in the eel, *Anguilla japonica*. Zool. Sci. 5: 43–51.

Takei, Y. and T. Tsuchida. 2000. Role of the renin-angiotensin system in drinking of seawater-adapted eels *Anguilla japonica*: a reevaluation. Am. J. Physiol. Regul. Integr. Comp. Physiol. 279: R1105–R1111.

Takei, Y., T. Tsuchida, Z. Li and M. Conlon. 2001. Antidipsogenic effects of eel bradykinins in the eel *Anguilla japonica*. Am. J. Physiol. Regul. Integr. Comp. Physiol. 281: R1090–R1096.

Takei, Y., T. Tsuchida and A. Tanakadate. 1998. Evaluation of water intake in seawater adaptation in eels using a synchronized drop counter and pulse injector system. Zool. Sci. 15: 677–682.

Tierney, M.L., G. Luke, G. Cramb and N. Hazon. 1995. The role of renin-angiotensin system in the control of blood pressure and drinking in the European eel, *Anguilla anguilla*. Gen. Comp. Endocrinol. 100: 39–48.

Tsuchida, T. and Y. Takei. 1998. Effects of homologous atrial natriuretic peptide on drinking and plasma ANG II level in eels. Am. J. Physiol. Regul. Integr. Comp. Physiol. 275: R1605–R1610.

Tsuchida, T. and Y. Takei. 1999. A potent dipsogenic action of homologous angiotensin II infused at physiological doses in eels. Zool. Sci. 16: 479–483.

Tsukada, T. and Y. Takei. 2001. Relative potency of three homologous natriuretic peptides (ANP, CNP and VNP) in eel osmoregulation. Zool. Sci. 18: 1253–1258.

Tsukada, T. and Y. Takei. 2006. Integrative approach to osmoregulatory action of atrial natriuretic peptide in seawater eels. Gen. Comp. Endocrinol. 147: 31–38.

Tsukada, T., S. Nobata, S. Hyodo and Y. Takei. 2007. Area postrema, a brain circumventricular organ, is the site of antidipsogenic action of circulating atrial natriuretic peptide in eels. J. Exp. Biol. 210: 3970–3978.

Unniappan, S. and R.E. Peter. 2005. Structure, distribution and physiological functions of ghrelin in fish. Comp. Biochem. Physiol. A. 140: 396–408.

Utida, S., M. Kamiya and N. Shirai. 1971. Relationship between the activity of Na^+-K^+-activated adenosinetriphosphatase and the number of chloride cells in eel gills with special references to sea-water adaptation. Comp. Biochem. Physiol. 38A: 443–447.

Watanabe, Y., T. Sakihara, T. Mukuda and M. Ando. 2007. Antagonistic effects of vasotocin and isotocin on the upper esophageal sphincter muscle of the eel acclimated to seawater. J. Comp. Physiol. B. 177: 867–73.

Watanabe, S., A.P. Seale, E. Gordon Grau and T. Kaneko. 2012. Stretch-activated cation channel TRPV4 mediates hyposmotically induced prolactin release from prolactin cells of Mozambique tilapia *Oreochromis mossambicus*. Am. J. Physiol. Regul. Integr. Comp. Physiol. 302: R1004–R1011.

Renal Sulfate Regulation

Akira Kato[1],* and Taro Watanabe[2]

Introduction

Since seawater (SW) is a hyperosmotic environment where various ions are dissolved, marine teleost fish cope with a threat of dehydration and invasion of excess ions. To compensate for osmotic loss of water, they drink copious SW and absorb most of the ingested water and monovalent ions via the intestine (Grosell 2011, Marshall and Grosell 2006, Smith 1930, Takei and Balment 2009). The excess Na^+ and Cl^- are excreted actively by mitochondria-rich cells in the gills (Evans et al. 2005). In contrast to the many studies on Na^+ and Cl^- regulation in fishes, studies on the regulation of divalent ions such as Mg^{2+}, Ca^{2+} and SO_4^{2-} are still limited. In terms of concentration difference between plasma and environmental SW, the ratio is 3–4 for Na^+ and Cl^- (e.g., Na^+, 170 mM vs. 450 mM; Cl^-, 120 mM vs. 500 mM), but it is 30–50 for Mg^{2+} and sulfate (SO_4^{2-}) (e.g., Mg^{2+}, 1 mM vs. 50 mM; SO_4^{2-}, ~1 mM vs. 30 mM). Therefore, ions unavoidably enter the body via the gills and digestive tracts across the concentration gradient (Hickman 1968). The excess divalent ions are excreted into the urine by the renal tubular secretion systems (Beyenbach 2004, Beyenbach et al. 1986, Renfro and Pritchard 1983).

[1]Department of Biological Sciences, Tokyo Institute of Technology, 4259-B19 Nagatsuta-cho, Midori-ku, Yokohama, Kanagawa 226-8501, Japan.
E-mail: akirkato@bio.titech.ac.jp
[2]Center for Cooperative Research Promotion, Atmosphere and Ocean Research Institute, The University of Tokyo, 5-1-5 Kashiwanoha, Kashiwa, Chiba 277-8564, Japan.
E-mail: wataro@aori.u-tokyo.ac.jp
*Corresponding author

SO_4^{2-} is the second most abundant anion in SW following Cl^-. Because of the relative difficulty of its measurement, research on SO_4^{2-} regulation has been largely neglected in marine teleost fish (Marshall and Grosell 2006). The gills and digestive tracts (lumen is an external environment) are possible sites of SO_4^{2-} influx. It is generally accepted that the absorption of SO_4^{2-} across the intestinal epithelium is limited (Marshall and Grosell 2006), but Hickman (Hickman 1968) estimated 11.3% of SO_4^{2-} derived from ingested SW is absorbed by the intestine of southern flounder (*Paralichthys lethostigma*). Concerning the gills, SO_4^{2-} permeability has been examined in freshwater (FW) teleosts and variable data were reported: significant permeability was found in the euryhaline guppy (*Poecillia reticula*) (Rosenthal 1961), while little permeability was detected in the stenohaline goldfish (*Carassius auratus*) (Romeu and Maetz 1964). However, detailed analyses of SO_4^{2-} budget between body fluids and media have not been reported in euryhaline fish that experience profound changes in SO_4^{2-} concentration in their natural habitats. More recently, we have shown that 85% of $^{35}SO_4^{2-}$ in environmental SW enters the body *via* the gills and skin and 15% from the digestive tract by comparing the uptake between eels with and without esophageal ligation (Watanabe and Takei 2012). The eels used in our experiments described in this chapter are cultured Japanese eels (*Anguilla japonica*) acclimated to SW for more than two weeks.

The molecular mechanism of how SO_4^{2-} is secreted has never been clarified. Recent studies from our laboratory demonstrated that eels maintain low plasma SO_4^{2-} (~1 mM) in SW by excreting SO_4^{2-}, while they accumulate high plasma SO_4^{2-} (6–19 mM) in FW by active reabsorption of SO_4^{2-} in the kidney (Nakada et al. 2005, Watanabe and Takei 2011a). In addition, molecular studies in eel and euryhaline pufferfish (*Takfiugu obscurus*) have identified molecules that mediate renal tubular SO_4^{2-} secretion and reabsorption (Kato et al. 2009, Nakada et al. 2005, Watanabe and Takei 2011b). Accordingly, the eel is now the best-studied species on the renal SO_4^{2-} regulation in teleost fishes. In the first half of this chapter, therefore, we will describe in some detail the basic morphology and physiology of the eel kidney. In the latter half of this chapter, we will attempt to introduce our works on the regulatory mechanism of SO_4^{2-} handling by the eel kidney.

1. Anatomy, Histology, and Physiology of the Eel Kidney

Anatomy of the Eel Kidney

The detailed anatomy of the eel kidney was described by Mott (1950) and Nakamura and Toyohiro (1983). In eels, the kidney lies along the backbone to the dorsal wall of the body cavity (Fig. 1A) (Naito 1990). The eel kidney is

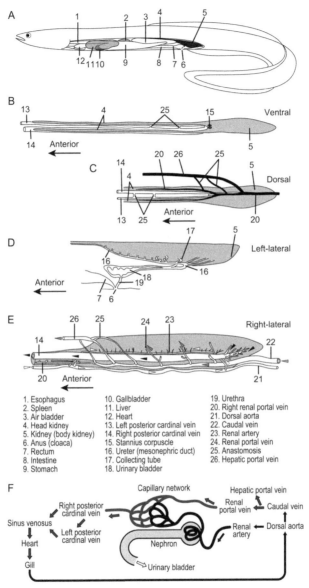

Figure 1. Anatomy of the eel kidney. (A) Schematic anatomical chart of eel. (B) Ventral view of the kidney. (C) Dorsal view of the kidney. (D) Lateral view of the kidney and urinary bladder. (E) Vascular distribution in the trunk kidney. White, gray, and black arrowheads indicate blood flows of renal artery, caudal vein/renal portal vein, and right posterior cardinal vein, respectively. (F) Diagram of expected blood flow in renal microcirculation. A, B, and D, Redrawn from Naito (1990) with permission of the Zoological Society of Japan; C, redrawn from Mott (1950) with permission of John Wiley & Sons Ltd.; E, redrawn from Nakamura and Toyohiro (1983) with permission of the Japanese Society of Fisheries Science.

long and thin, and consists of the head kidney (anterior part) and the trunk kidney (body kidney, posterior part) (Fig. 1B). Embryologically, the head kidney derives from pronephros, and the trunk kidney from mesonephros. There is no clear border between the head and trunk (body) kidney. The kidney weight of *A. japonica* is 0.65 ± 0.03 g/100g body weight in FW and is 0.68 ± 0.04 g/100g body weight in SW (Sokabe et al. 1973). The blood is supplied by (i) the renal artery which is derived from the dorsal aorta and (ii) the renal portal vein derived from the caudal vein, and the blood flows out to the right and left posterior cardinal vein (Mott 1950, Nakamura and Toyohiro 1983) (Fig. 1E). The renal artery seems to form a capillary network named the glomerular afferent artery (Fig. 1F). The renal artery and the renal portal vein are connected with the right posterior cardinal vein *via* capillary networks. The right posterior cardinal vein originates in posterior trunk kidney, and the left posterior cardinal vein originates from the anastomosis with the right posterior cardinal vein (Fig. 1C). Peripheral branches of the renal artery also connect the renal portal vein *via* anastomosis. The ureters are located ventrally along the kidney and mediate urine outflow from the renal collecting duct to the urinary bladder (Fig. 1D) (Nakamura and Toyohiro 1983). The nephrons (glomerulus and renal tubules) are dispersed in the trunk kidney. The head kidney and intertubular tissue of the trunk kidney are the hematopoietic tissue, which contains numerous reticular cells and capillaries.

Function of the Eel Kidney

The major function of the kidney is urine production. The urine contains inorganic ions and nitrogenous end products such as creatine, creatinine, and trimethylamine oxide (TMAO). Ammonia, urea, and monovalent electrolytes (Na^+, Cl^-, and K^+) are excreted mainly from the gill (Evans et al. 2005, Furukawa et al. 2012). The urine electrolyte composition and volume are different between eels in FW and SW (Tables 1 and 2). The kidney of FW eels excretes up to 10 times more urine than that of SW eels. In FW, the kidney produces urine hypotonic to body fluid and thereby excretes water. In SW, on the other hand, the kidney produces urine with high concentration of divalent ions such as SO_4^{2-} and Mg^{2+} to get rid of them. Therefore, the kidney is an important organ to maintain body fluid homeostasis in combination with the gill and intestine. The kidney, particularly head kidney, is a site of hematopoiesis in fishes, the detail of which has been recently described in the zebrafish, *Danio rerio* (de Jong and Zon 2005, Lin et al. 2005). The erythrocyte, thrombocyte, neutrophil, eosinophil, monocyte/macrophage, and lymphocyte have been shown to develop in the zebrafish kidney.

Table 1. Ionic composition and osmolarity of environmental water, plasma, serum and urine of eel.

Species (acclimation salinity)	Sample	mM								mOsM	Ref.
		Na	Cl	K	Ca	Mg	SO$_4$	PO$_4$	Urea	Osmolality	
Fresh water		1.0	0.5	0.5	0.3	0.5	0.3	-	-	<10	a
Seawater		450.0	525.0	12.5	10.0	50.0	30.0	-	-	1080	a
European eel (FW)	P	150.0	88.0	1.8	2.4	2.1	-	1.8	-	-	b
Japanese eel (FW)	S, P	139.4–165.3	78.7–114.1	4.2	1.2–1.8	0.6–1.6	6.2–18.8	-	1.9	293.3–295.2	c, d, e
American eel (FW)	P	133.6	76.2	2.7	2.6	0.9	-	-	-	263.2	f
Japanese eel (SW)	S, P	164.2–169.3	125.3–164.7	5.6	1.5–1.6	0.9–1.5	0.7–1.0	-	3.7	338.7–344.5	c, d, e
American eel (SW)	P	154.7	134.8	2.5	2.4	1.0	-	-	-	309.7	f
European eel (FW)	U	13.1	3.3	1.1	0.3	0.02	-	4.5	-	-	b
Japanese eel (FW)	U	54.2	-	-	0.3	0.3	2.7	-	-	-	e, g
American eel (FW)	U	11.7	6.0	2.1	1.0	0.54	-	-	-	44.5	f
European eel (SW)	U	64.4	128.0	1.6	8.2	28.7	-	0.6	-	-	b
Japanese eel (SW)	U	80.3	-	-	31.2	116.1	47.1	-	-	-	e, g
American eel (SW)	U	54.6	65.9–131.7	5.3	5.1–16.4	27.6–128.8	89.0	-	-	190.1	f, h

P, plasma; S, serum; U, urine
a) Watanabe and Takei 2011b; b) Jones et al. 1969; c) Mistry et al. 2005; d) Nakada et al. 2005; e) Watanabe and Takei 2012; f) Schmidt-Nielsen and Renfro 1975; g) Takei and Kaiya 1998; h) Smith 1930

Table 2. GFR and UFR of *Anguilla* species.

Species	Acclimation salinity	GFR (mL/kg/hr)	UFR (mL/kg/hr)	Ref.
European eel	FW	1.5–4.6	1.1–3.5	a, b
Japanese eel	FW	2.5–2.8	2.3	c, d
American eel	FW	2.2	1.5	e
European eel	SW	0.4–1.0	0.3–0.6	a, b
Japanese eel	SW	1.7–3.1	0.4 0.7	c, d
American eel	SW	2.1	0.8	e

a) Sharratt et al. 1964; b) Jones et al. 1969; c) Oide and Utida 1968; d) Sokabe et al. 1973; e) Schmidt-Nielsen and Renfro 1975

The kidney also acts as an endocrine organ. Cortisol is synthesized and secreted from the interrenal tissue in the head kidney. The key enzyme for corticosteroidogenesis (cytochrome P450 21-hydroxylase) is specifically expressed in the interrenal tissue and the expression is enhanced by ACTH (Li et al. 2003). Chromaffin cells are also present in the head kidney near the wall of posterior cardinal vein, which secrete catecholamines into the blood stream (Hathaway and Epple 1989). The afferent arterioles of the eel kidney contains granular epithelioid cells which are homologous with the renin-producing juxtaglomerular cells in mammals and are considered to be responsible for plasma renin activity (Krishnamurthy and Bern 1969, Nishimura et al. 1973, Sokabe and Ogawa 1974). The eel kidney contains immunoreactive erythropoietin (Epo) and suggested to be a major Epo-producing organ (Wickramasinghe et al. 1994). In pufferfish (*Takifugu rubripes*) and zebrafish, however, the heart and liver are the major organs that express Epo (Chou et al. 2004, Chu et al. 2007). Stanniocalcin is a glycoprotein hormone involved in the maintenance of calcium and phosphate homeostasis in fish, and secreted from paired corpuscle of Stannius at the anterior region of trunk kidney (Fig. 1B) (Hanssen et al. 1993).

Structure and Function of Eel Nephron

In both FW and SW eels, the nephron is present in the trunk kidney and basically consists of six parts, glomerulus, neck segment, proximal tubule segment I, proximal tubule segment II, distal tubule, and collecting tubule/duct (Fig. 2A). The basic architecture is common to those of FW and euryhaline fishes, but partially different from stenohaline marine fishes that lack distal tubule and/or glomerulus (Beyenbach 2004, Teranishi and Kaneko 2010).

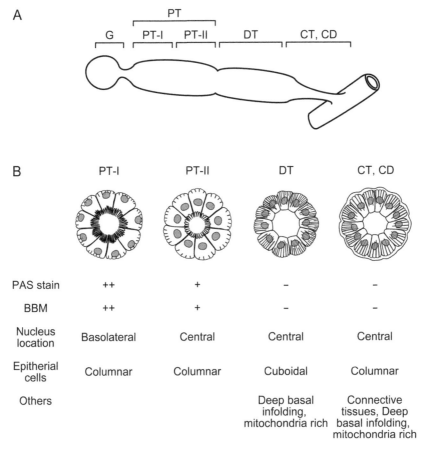

Figure 2. The eel nephron. (A) Schematic drawing of a nephron. (B) Cross sectional views of each segment. G, glomerulus; PT, proximal tubule; PT-I, first segment of proximal tubule; PT-II, second segment of proximal tubule; DT, distal tubule; CT collecting tubule; CD, collecting duct. Morphological characteristics of each segment are noted in (B).

Glomelurus

The glomelurus is the site of filtration of plasma consisting of capillary endothelial cells, mesangial cells, podocytes, and parietal epithelial cells, and is significantly smaller in SW fish than in FW fish (Colville et al. 1983). This difference may account for the reduced glomerular filtration rate (GFR), the volume of glomerular filtrate per unit time, of SW fishes than that of FW fishes. However, the mechanism of how the balance of contraction between afferent and efferent arterioles and the contraction of mesangial cells regulate different levels of GFR in FW and SW are largely unknown.

Proximal tubule

The proximal tubule is characterized by the presence of brush borders in the apical membrane, which is visualized by phalloidin, periodic acid-Schiff (PAS) stain, or WGA (Teranishi and Kaneko 2010). The brush border is thicker in the first segment of proximal tubule (PT-I) than in second segment (PT-II), and the nucleus locates basally in the PT-I cells and centrally in the PT-II cells (Fig. 2B). Proximal tubules express Na^+/K^+-ATPase on the basolateral membrane, which drives trans-epithelial transport of ions by the secondary active mechanism. In the epithelial cells, mitochondria are moderately developed in the cytoplasm, and the basal infolding continuous with the basolateral membrane is typically observed in the basal half of PT-I and PT-II. Analyses of a single tubule from the killifish (*Fundulus heteroclitus*) demonstrated that proximal tubules from either SW or FW fish exhibit low lumen-negative trans-epithelial voltage and low resistances typical of the proximal tubules of other vertebrates (Cliff and Beyenbach 1992). This segment consists of leaky epithelia with paracellular transport of Na^+ and Cl^-. Like in mammals, this segment is considered to be the dominant site for reabsorption of nutrients, phosphate, and HCO_3^-. In pronephros of zebrafish (Wingert et al. 2007) or rainbow trout (*Oncorhynchus mykiss*) kidney (Sugiura et al. 2003), expression of Na^+-glucose cotransporter SGLT1 and Na^+-phosphate cotransporter NaPi1 is observed in the proximal tubule. Apical Na^+/H^+ exchanger 3 (NHE3), apical H^+-ATPase (Ivanis et al. 2008), and basolateral electrogenic $Na^+/nHCO_3^-$ cotransporter NBCe1 (Sussman et al. 2009) are also expressed in the proximal tubule, suggesting that fishes acidify primitive urine and reabsorb CO_2/HCO_3^- by the mechanism similar to that of mammals. In contrast to mammalian proximal tubule that largely reduces the volume of primitive urine by reabsorption, the proximal tubule of marine teleosts is the site to secrete fluid that contains Na^+, Cl^-, SO_4^{2-}, Mg^{2+}, Ca^{2+}, etc. (Fig. 3) (Beyenbach 1982, 2004, Schmidt-Nielsen and Renfro 1975). The proximal tubular fluid secretion is considered to involve the following three steps (Beyenbach 2004): (i) an apical Cl^- channel (unidentified) secretes Cl^- by using negative membrane potential as a driving force, produces luminal-negative trans-epithelial potential that drives paracellular Na^+ secretion, and increases the osmotic pressure of the luminal fluid; (ii) high luminal NaCl drives the secretion of SO_4^{2-}, Mg^{2+}, and Ca^{2+} by anion-exchange and cation-exchange systems in the apical membrane; and (iii) water follows ion flux to maintain osmotic equilibrium. As in other leaky epithelia, movement of ions and water occurs in both directions, but net secretion of Cl^-, Na^+, SO_4^{2-}, Mg^{2+}, Ca^{2+}, and water are observed in proximal tubules. In the apical membrane of the proximal tubule, Cl^-/SO_4^{2-} exchanger (solute carrier family 26 member 6, Slc26a6) (Kato et al. 2009, Watanabe and Takei 2011b), Na^+/Mg^{2+} exchanger (unidentified) (Beyenbach 2004),

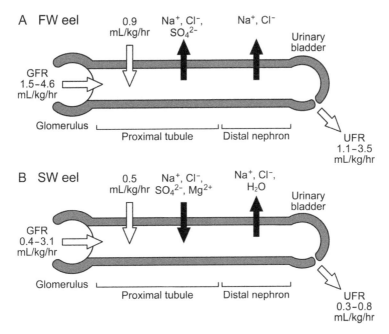

Figure 3. Ion and fluid movement reported to date in the nephron of SW (A) and FW (B) eels. GFR, glomerular filtration rate; UFR, urine flow rate.

and Na^+/Ca^{2+} exchanger (NCX2) (Islam et al. 2011) may be involved in the secretion of SO_4^{2-}, Mg^{2+}, and Ca^{2+}. This fluid secretion is essential for SW fishes to excrete divalent ions (Mg^{2+}, SO_4^{2-}, Ca^{2+}) from the kidney, and fluid secretion is also observed in FW fishes (see below). In FW eels, the proximal tubule is also the site of SO_4^{2-} reabsorption (Nakada et al. 2005) that will be discussed below.

Distal tubule

Distal tubule is known as a diluting segment and consists of cuboidal cells with a centrally located nucleus and scanty microvilli at the apical membrane (Fig. 2B) (Kato et al. 2011, Teranishi and Kaneko 2010). The distal tubule cells have a rich population of mitochondria. The basal infolding is well developed and invaginated much more deeply than in proximal tubules, and Na^+/K^+-ATPase exists at high levels on the basolateral membrane. Analyses of a single renal tubule from the rainbow trout showed that distal tubule has negligible water permeability, Na^+ and Cl^- absorptive activity, and lumen positive trans-epithelial potential that is reduced by addition of furosemide to the lumen, suggesting the presence of Na^+-K^+-$2Cl^-$ cotransporter (NKCC) in the apical membrane (Nishimura et al. 1983).

These characteristics are similar to those of thick ascending limb of Henle's loop (TAL) in the mammalian kidney (Greger 1985) and early distal tubule of the shark kidney (Friedman and Hebert 1990, Hebert and Friedman 1990). In zebrafish, euryhaline pufferfish and Mozambique tilapia (*Oreochromis mossambicus*), distal tubule cells have NKCC2 and K^+ channel (ROMK2) in the apical membrane, and Na^+/K^+-ATPase and Cl^- channel (CLC-K) in the basolateral membrane (Kato et al. 2011, Katoh et al. 2008, Miyazaki et al. 2002, Wingert et al. 2007). Eel has two paralogs for NKCC2, *nkcc2α* expressed in the kidney and *nkcc2β* expressed in the intestine (Cutler and Cramb 2008). In mammals, TAL is the major site of Mg^{2+} reabsorption where ~70% of filtrated Mg^{2+} is reabsorbed, and tight junction protein claudin-16 (paracellin-1) and claudin-19 mediate paracellular Mg^{2+} absorption which is driven by the luminal positive trans-epithelial potential (Hou et al. 2008). In fishes, the role of distal tubule in Mg^{2+} reabsorption is unknown. Distal tubule is present in the nephron of FW and euryhaline fishes but is absent in stenohaline SW fishes (Hickman and Trump 1969, Ogawa 1962). In FW eels, the immunoreactive Na^+/K^+-ATPase in the distal tubular cells is denser than in SW eels (Teranishi and Kaneko 2010), suggesting that urine dilution by the distal tubule is essential for water excretion in FW environment. In both FW and SW eel kidney, furosemide inhibits Na^+ and Cl^- reabsorption (Schmidt-Nielsen and Renfro 1975) and the expression levels of *nkcc2α* are similar (Cutler and Cramb 2008). Therefore, distal tubule of euryhaline fishes acclimated to SW may also have some role in Na^+ and Cl^- reabsorption.

Collecting duct

Collecting tubule/duct (CT/CD) is the last segment of the nephron, and the early part of this segment is also called as late distal tubule. CD consists of columnar epithelial cells with a centrally located nucleus, and is surrounded by connective tissues (Fig. 2B). CD cells lack brush border at the apical membrane, and have deep basal infolding that expresses Na^+/K^+-ATPase. In zebrafish or euryhaline pufferfish, CD or late distal cells express apical Na^+-Cl^- cotransporter (NCC, Slc12a3), basolateral Na^+/K^+-ATPase, and basolateral Cl^- channel CLC-K (Kato et al. 2011, Miyazaki et al. 2002, Wingert et al. 2007). Eel has two paralogs for NCC, *ncca* expressed in the kidney and *nccβ* expressed in the intestine. In the distal convoluted tubule (DCT) and the cortical collecting duct (CCD) of the mammalian kidney, NCC and epithelial Na^+ channel (ENaC) mediate Na^+ reabsorption (Garty and Palmer 1997, Hebert et al. 2004). As the presence of ENaC gene has been demonstrated in terrestrial vertebrates, Australian lungfish (*Neoceratodus forsteri*), and elephant shark (*Callorhinchus milii*), but not in bony fishes

(Venkatesh et al. 2007, Uchiyama et al. 2012), the collecting duct of FW or euryhaline teleosts may have been utilized NCC instead of ENaC. When eel or euryhaline pufferfish is transferred from SW to FW, the immunoreactive Na^+/K^+-ATPase and transcriptional and protein levels of NCC (or NCCα) in the CD cells all increased (Cutler and Cramb 2008, Kato et al. 2011, Teranishi and Kaneko 2010). These results suggest that CD also acts as a diluting segment for water excretion.

In contrast, the function of CD of SW eels is unclear. The urinary bladder, which is embryologically homologous to the collecting duct, of SW fish expresses NCC at high levels, and contributes to NaCl and water absorption, thereby concentrating divalent ions such as Mg^{2+} and SO_4^{2-} without water loss (Beyenbach 2004, Marshall and Grosell 2006). FW acclimation or prolactin injection reduces water permeability and increases NaCl absorption in the urinary bladder of euryhaline flounder (*Platichthys stellatus*) (Foster 1975, Hirano et al. 1971, 1973). Therefore, CD may switch between urine dilution (salt reabsorption) mode in FW and volume reduction (salt and water reabsorption) mode in SW by regulating NCC and aquaporin(s).

Urine Production by the Kidney of FW Eel

FW eels have plasma osmolality of 260–290 mOsm (Table 1), which is similar to that of human and is 100–1000 times higher than that of environmental FW. In FW eel, GFR is 1.5–4.6 mL/kg/hr (Table 2), which is similar to those in other FW teleosts and is ~20 and ~70 times smaller than that in human and rat, respectively. The urine flow rate (UFR) of FW eel is 1.1–3.5 mL/kg/hr (Table 2, Fig. 3A). Like other FW fishes, the UFR is 67–92% that of GFR, which is in contrast to human and rat where UFR is 0.4–1.8% that of GFR (Bird et al. 1988, McDonald et al. 1964). The difference indicates that renal tubular reabsorption is suppressed and/or the reabsorption is balanced by tubular fluid secretion in the kidney of FW eels, as tubular secretion is also observed in FW fishes. The calculated fluid secretion rate of American eel (*A. rostrata*) in FW is ~0.9 mL/kg/hr (Schmidt-Nielsen and Renfro 1975). In some, but not all, FW eels examined, UFR was greater than GFR possibly due to the excess of tubular fluid secretion. In FW killifish, fluid is secreted from the proximal tubule segments (Cliff and Beyenbach 1992). Tubular secretion is more prominent in SW fishes than in FW fishes, particularly in aglomerular SW fishes that produce no urine by glomerular filtration. Hyposmotic urine of FW fishes is considered to be produced in the distal tubule and collecting duct by reabsorption of Na^+ and Cl^- in these water-impermeable segments.

Urine Production by the Kidney of SW Eel

Eels have plasma osmolality of 310–340 mOsm in SW (Table 1), which is one third that of SW and is only ~15% higher than that of FW eels. Teleost fishes in SW produce relatively small volumes of isotonic urine with high concentrations of Mg^{2+} and SO_4^{2-} (Table 1). Urine of SW eel contains 28–130 mM Mg^{2+} and 48–89 mM SO_4^{2-}, which are much higher than those in plasma. Primitive urine is primarily produced by glomerular filtration and tubular fluid secretion. In SW eels, GFR is 0.4–3.1 mL/kg/hr, tubular fluid secretion rate is 0.5 mL/kg/hr, and UFR is 0.3–0.8 mL/kg/hr (Fig. 3B). Therefore, the UFR is 3–10 times smaller than the sum of GFR and tubular fluid secretion rate. The urine volume is considered to be reduced in the collecting ducts and urinary bladder by reabsorbing water together with Na^+ and Cl^-.

The ion composition of glomerular filtrate is similar to that of plasma, and the secreted fluid is rich in Na^+, Cl^-, and divalent ions. Isolated single renal tubule of winter flounder (*Pleuronectes americanus*), in which tubular fluid secretion was observed as a growing gap between two mineral oil droplets in the tubules, secretes fluid that contains 26 mM Mg^{2+}, 10 mM SO_4^{2-}, 152 mM Na^+, and 155 mM Cl^- (Beyenbach 1982, 2004). The concentrations of Mg^{2+} and SO_4^{2-} in the secreted fluid are much higher than those of primary urine, indicating that Mg^{2+} and SO_4^{2-} are secreted against concentration gradient. The Mg^{2+} and SO_4^{2-} in primitive urine are further concentrated by the collecting ducts and urinary bladder, thus the Mg^{2+} and SO_4^{2-} concentrations exceed 100 mM and 45 mM respectively in the final urine.

2. Eel Kidney as a Model of SO_4^{2-} Secretion and Reabsorption

Overview

SO_4^{2-} plays important roles in a variety of metabolic processes, including production of sulfated glycosaminoglycan, sulfated steroid, and sulfated lipid, detoxification of exogenous substances by sulfation, elimination of waste compounds by sulfoconjugation, and tyrosine sulfation of proteins and peptides. For sulfation, SO_4^{2-} is activated to form adenosine 3'-phospate 5'-sulfatophosphate (PAPS), which is used as a universal SO_4^{2-} donor by various sulfotransferases (Markovich 2001). Almost all vertebrate animals maintain plasma SO_4^{2-} concentration at 0.2–2 mM, except for the special case of FW eels which have plasma SO_4^{2-} concentration of 6–19 mM (Nakada et al. 2005, Watanabe and Takei 2012). The kidney of FW eels plays a major role in the SO_4^{2-} retention, and for this special property it is a good model to study the mechanism of renal SO_4^{2-} reabsorption. In contrast, plasma SO_4^{2-} concentration in SW eel is low (~ 1 mM), although SO_4^{2-} is abundant in environmental SW. To avoid hypersulfatemia, marine teleosts actively excrete SO_4^{2-} primarily *via* kidney (Kato et al. 2009, Watanabe and Takei 2011b). The details are described below.

Renal SO_4^{2-} excretion in SW Eel

As mentioned above, SO_4^{2-} influx occurs 84.5% *via* body surfaces (gill and skin) and 15.5% *via* digestive tracts in SW-acclimated eels (Fig. 4A) (Watanabe and Takei 2012). Therefore, the major site of SO_4^{2-} influx is the gill. The rate of SO_4^{2-} influx is 1.32 μmol/kg/hr from the gill and skin and 0.23 μmol/kg/hr from the digestive tract. The gills of marine teleosts have many mitochondria-rich cells or ionocytes, which are NaCl-extruding cells in the gill epithelia with high ion-transport activity and are surrounded by accessory cells. The intercellular junctions between mitochondria-rich cells and accessory cells are considered to be leaky to ions (Evans et al. 2005), thus SO_4^{2-} may enter the body through the leaky junction rather

A

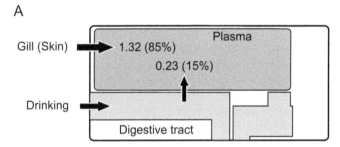

B

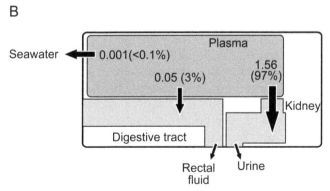

Figure 4. Schematic drawing of SO_4^{2-} influx (A) and efflux (B) in each osmoregulatory site of seawater-acclimated eels. Values are an average of six fish and are expressed in μmol/kg/hr. Relative values (%) among the sites are given in parentheses. From Watanabe and Takei (2012).

than the transcellular route. The influx rate of SO_4^{2-} is much lower than that of Na^+ (13.2 mmol/kg/hr) and Cl^- (0.48 mmol/kg/hr) in SW eels (Kirsch and Meister 1982, Tsukada et al. 2005), as in other marine teleosts (Marshall and Grosell 2006). Therefore, it seems that SO_4^{2-} transport across the body surfaces is suppressed in SW eels compared with that of Na^+ and Cl^- (Watanabe and Takei 2012). The drinking rate of SW eels is 2 ml/kg/hr (Tsuchida and Takei 1998), which contains 54 µmol/kg/hr of SO_4^{2-}. SO_4^{2-} transport activity is generally low in the intestine of marine teleosts. In SW eel, only 0.45% of ingested SO_4^{2-} is absorbed by the digestive tracts and the rest is rectally excreted. The SO_4^{2-} concentration of rectal fluid in SW eel is as high as 110 mM (Watanabe and Takei 2012).

Most SO_4^{2-} is excreted by the kidney. An efflux study after injection of $^{35}SO_4^{2-}$ into the blood revealed that 97% of SO_4^{2-} is removed *via* the kidney and 3% *via* digestive tracts (Fig. 4B) (Watanabe and Takei 2012). Urine radioactivity of SW eel increased rapidly within 1 h after $^{35}SO_4^{2-}$ injection, and a high level was maintained for 6 hr followed by a gradual decrease. When eels were transferred from FW to SW, they excreted SO_4^{2-} into the urine most abundantly 1 day after the transfer, and plasma SO_4^{2-} concentration slightly increased after 1 day but started to decrease thereafter probably due to the sustained excretion for 2 weeks (Fig. 5). Urine SO_4^{2-} concentration of FW eels was as low as 2 mM, and the peak after 1 day of SW transfer was 110 mM, and the level was maintained at ~40 mM from 3 to 14 days after the transfer.

SO_4^{2-} in plasma does not bind to proteins and is freely filtrated by the glomeruli. SW fishes exhibit low GFR (Table 2) and moderate tubular absorption of water, thus the high SO_4^{2-} concentration in the final urine of SW fishes may not due to high GFR and subsequent water absorption. Direct observations of renal tubules isolated from the eel, flounder, coho salmon, and killifish, all of which have glomerular kidney, demonstrated that the proximal tubule secretes fluid rich in Mg^{2+} and SO_4^{2-} (Beyenbach 1982, 2004, Schmidt-Nielsen and Renfro 1975). In the kidney of typical glomerular marine fish, the rate of tubular fluid secretion is 3.5 times that of GFR (Beyenbach 2004). As mentioned above, SO_4^{2-} is secreted into the primitive urine against the concentration gradient, the mechanism of which has been shown only recently.

Transcellular SO_4^{2-} secretion by the proximal tubule is mediated by basolateral SO_4^{2-} uptake from body fluids into the cytoplasm and apical SO_4^{2-} secretion from the cytosol into primitive urine. The SO_4^{2-} efflux of apical brush border membrane (BBM) vesicles isolated from the kidney of southern flounder is dependent on the presence of counter-flow anions such as Cl^- or HCO_3^-, and net SO_4^{2-} secretion by primary cultures of winter flounder renal proximal tubule epithelium required luminal Cl^- or HCO_3^-

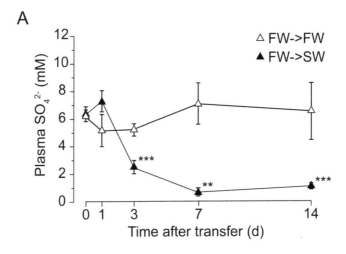

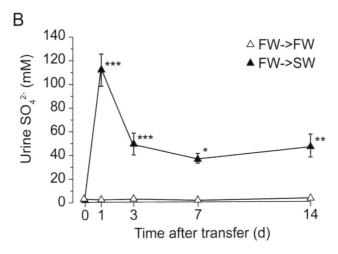

Figure 5. Time course of changes in SO$_4^{2-}$ concentration in plasma (A) and urine (B) after transfer of FW eels to FW or SW. *$P<0.05$, **$P<0.01$, ***$P<0.001$ compared with day 0. From Watanabe and Takei (2011b).

(Pelis et al. 2003, Renfro and Pritchard 1983). These studies suggest that the luminal SO$_4^{2-}$ secretion is mediated by Cl$^-$/SO$_4^{2-}$ or HCO$_3^-$/SO$_4^{2-}$ exchange. Studies using basolateral membrane (BLM) vesicles isolated from the kidney of southern flounder suggest that SO$_4^{2-}$ uptake is mediated by exchange of HCO$_3^-$/SO$_4^{2-}$ or OH$^-$/SO$_4^{2-}$.

Molecules Involved in Apical SO$_4$$^{2-}$ Secretion by Proximal Tubules of SW Eel

There has been accumulating evidence showing that fish-specific paralogs for Slc26a6 provide major routes for apical SO$_4$$^{2-}$ secretion (Figs. 6 and 7, Table 3) (Kato et al. 2009, Watanabe and Takei 2011b). Slc26 is a family of

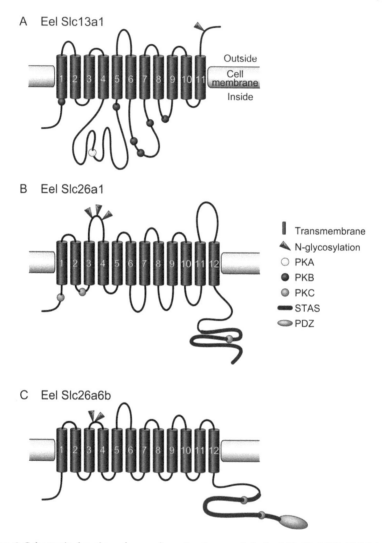

Figure 6. Schematic drawing of secondary structure model of eel Slc13a1 (A), Slc26a1 (B), and Slc26a6b (C). Putative transmembrane domain and phosphorylation sites of Slc26a6b are predicted by comparison with orthologs of other species (Kurita et al. 2008, Lohi et al. 2003).

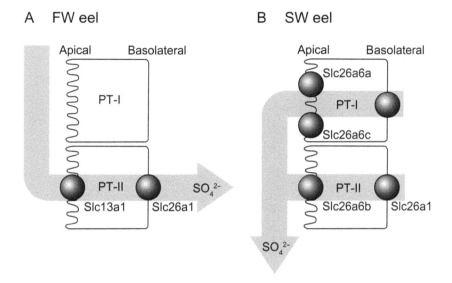

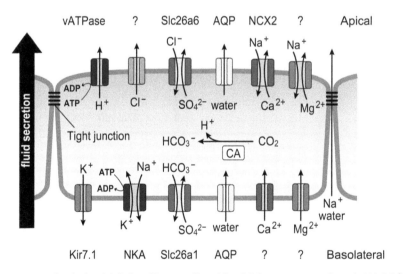

Figure 7. Renal tubular SO$_4^{2-}$ handling mediated by SO$_4^{2-}$ transporters in eel. (A) SO$_4^{2-}$ reabsorption by Slc13a1 and Slc26a1 at proximal tubule of FW eel. (B) SO$_4^{2-}$ secretion by Slc26a6 and Slc26a1 in SW eel. (C) Localization of transporters in proximal tubule of SW eel. AQP, aquaporin; NCX, Na$^+$/Ca^{2+} exchanger; NKA, Na$^+$/K$^+$-ATPase; CA, carbonic anhydrase; PT-I, first segment of proximal tubule; PT-II, second segment of proximal tubule.

Table 3. Slc26 family.

	Protein name	Substrates	Transporting type	Tissue distribution in mammals	Tissue distribution in eel[a,b]	Associated disease(s)	Reports within teleost
Slc26a1	Sat1, Sat-1	SO_4^{2-}, Oxalate	Exchanger	Liver, Kidney	Kidney, Rectum, Urinary bladder	-	Eel[a-d], Pufferfish[e], Rainbow trout[f], Zebrafish[g]
Slc26a2	DTDST	SO_4^{2-}, Cl^-	Exchanger	Widespread	?	Chondrodysplasias	-
Slc26a3	DRA, CLD	SO_4^{2-}, Cl^-, HCO_3^-, OH^-, Oxalate	Exchanger	Intestine, Sweat gland, Pancreas, Prostate	Intestine, Rectum	Congenital chloride-losing diarrhea	Zebrafish[b,i]
Slc26a4	Pendrin	Cl^-, HCO_3^-, I^-, Formate	Exchanger	Inner ear, Kidney, Thyroid	?	Pendrated syndrome, deafness (DFNB4)	Zebrafish[b,i]
Slc26a5	Prestin	?	?	Inner ear	?	Deafness	Zebrafish
Slc26a6	CFEX, PAT-1	SO_4^{2-}, Cl^-, HCO_3^-, OH^-, Oxalate, Formate	Exchanger	Widespread	Kidney, Intestine, Rectum	-	Eel[a-d], Pufferfish[e,k], Zebrafish[b,i]
Slc26a7	SUT-1	SO_4^{2-}, Cl^-, Oxalate	Exchanger	Kidney	?	-	-
Slc26a8	Tat-1	SO_4^{2-}, Cl^-, Oxalate	Exchanger	Sperm, Brain	?	-	-
Slc26a9	-	SO_4^{2-}, Cl^-, Oxalate	Exchanger	Lung, Stomach	?	-	-
Slc26a10	-	Pseudogene	-	-	?	-	-
Slc26a11	-	SO_4^{2-}	exchanger	Widespread	?	-	-

a) Nakada et al. 2005; b) Watanabe and Takei 2011a; c) Watanabe and Takei 2011b; d) Watanabe and Takei 2012; e) Kato et al. 2006; f) Katoh et al. 2006; g) Hong et al. 2010; h) Bayaa et al. 2009; i) Perry et al. 2009; j) Schaechinger and Oliver 2007; k) Kurita et al. 2008; others, Mount and Romero 2004

anion exchangers that transport Cl^-, SO_4^{2-}, HCO_3^-, I^-, OH^-, formate, oxalate, etc, across the plasma membrane. In mammals, they are involved in skeletal development, synthesis of thyroid hormone, facilitation of trans-epithelial Na^+-Cl^- transport, and bicarbonate excretion by the distal nephron and the exocrine pancreas (Mount and Romero 2004). Slc26 family consists of 11 members (Slc26a1–11) in mammals (Table 3). Pufferfish and zebrafish genome contain 9 and 11 Slc26 gene members, respectively, of which some are fish-specific paralogs. Slc26a8, 9, and 10 are not present in the genome databases of pufferfish and zebrafish, and Slc26a4 is absent in pufferfish genome database (Bayaa et al. 2009, Kato et al. 2009).

Expression analyses of fish Slc26 family indicate that *slc26a1* (*sat1*) and paralogs for *slc26a6* (*pat-1*, *cfex*) are highly expressed in the kidney (Table 3) (Kato et al. 2009, Watanabe and Takei 2011b). When Japanese eel or euryhaline pufferfish are acclimated in SW, the expression of three paralogs of *slc26a6* (*slc26a6a*, *b*, and *c*) are up-regulated in the kidney. Immunohistochemical analyes demonstrated that fish Slc26a6 paralogs are localized to the apical (brush-border) region of the proximal tubules, and no other Slc26 transporters have been demonstrated in the apical membrane. Electrophysiological analyses of *Xenopus* oocytes expressing Slc26a6 members demonstrated that fish paralogs of Slc26a6 have strong electrogenic Cl^-/SO_4^{2-} exchange activity in euryhaline pufferfish (Kato et al. 2009). This activity is ideal for luminal SO_4^{2-} secretion against the gradient because the epithelial cells can utilize high Cl^- concentration in primitive urine and negative membrane potential as driving forces. These results indicate that renal tubular SO_4^{2-} secretion is mediated by apical Slc26a6.

The presence of three Slc26a6 paralogs is commonly observed in eel and pufferfish, and is also observed in genome databases of other teleost species. The amino-acid sequences of the paralogs share ~55% identities. Mammals have one Slc26a6 gene. Some differences have been observed among the fish Slc26a6 paralogs at the levels of transcriptional regulation and protein function. The *slc26a6a* and *slc26a6b* genes are expressed in the kidney and intestine, whereas *slc26a6c* is expressed only in the kidney (Kato et al. 2009, Kurita et al. 2008, Watanabe and Takei 2011b). The renal expression of *slc26a6a* is not observed in FW but is strongly induced in SW. In contrast, *slc26a6b* and *slc26a6c* are expressed in the kidney of both FW and SW eels. Both Slc26a6a and Slc26a6c are expressed in PT-I segment of proximal tubule, while slc26a6b is expressed only in the PT-II segment (Fig. 7) (Watanabe and Takei 2011b). When the genes are expressed in *Xenopus* oocytes, Slc26a6a exhibits 50–200-fold higher SO_4^{2-} current than Slc26a6b in the pufferfish, though the current elicited by oxalate or HCO_3^- are similar (Kato et al. 2009). Therefore, Slc26a6a may drastically increase SO_4^{2-} conductance at the apical membrane of proximal tubule. The function of Slc26a6c has not been reported possibly due to the failure of surface expression in *Xenopus* oocyte.

Molecules Involved in Basolateral SO_4^{2-} Uptake by Proximal Tubules of SW Eel

Expression and immunohistochemical analyses of SW-acclimated eel kidney demonstrated that *slc26a1* is highly expressed in the kidney and is localized at the basolateral membrane of proximal tubule (Figs. 6 and 7) (Nakada et al. 2005, Watanabe and Takei 2011b). No other SO_4^{2-} transporters have been identified in the basolateral membrane of proximal tubule. Therefore, it is likely that Slc26a1 is the anion exchanger that mediates basolateral SO_4^{2-} uptake. SO_4^{2-} transporting activity of eel Slc26a1 was shown by $^{35}SO_4^{2-}$ uptake in *Xenopus* oocytes (Nakada et al. 2005). Although Slc26a1 is assumed to be an electro-neutral $2OH^-/SO_4^{2-}$ or $2HCO_3^-/SO_4^{2-}$ exchanger, there is still no direct evidence for OH^- or HCO_3^- transport in the recombinant Slc26a1. Considering the low HCO_3^- concentrations in plasma and urine of fishes, cellular substrate for HCO_3^- secretion must be increased by hydration of CO_2 in the proximal tubular cells (Grosell 2009). CO_2 hydration is mediated by carbonic anhydrase (CA) (Pelis et al. 2003) and HCO_3^- secretion must be accompanied by H^+ extrusion mediated by apical NHE3 and H^+-ATPase (Ivanis et al. 2008). Immunohistochamical studies showed that Slc26a1 and Slc26a6b are colocalized in the same epithelial cells in the PT-II. However, *slc26a1* is not expressed in PT-I that expresses *slc26a6a* and *slc26a6c*. The SO_4^{2-} transporter that mediates basolateral SO_4^{2-} uptake by PT-I remains to be determined (Fig. 7) (Watanabe and Takei 2011b).

Renal SO_4^{2-} Reabsorption in FW Eel

Renal SO_4^{2-} retention was initially studied by the kidney of terrestrial vertebrates, where net renal SO_4^{2-} reabsorption was observed (Bastlein and Burckhardt 1986, Renfro et al. 1989). BBM vesicles isolated from the kidney of mammals and chicken have Na^+-dependent $^{35}SO_4^{2-}$ transport activity. This is contrastive to the Na^+-independent $^{35}SO_4^{2-}$ transport activity of BBM vesicles isolated from the kidney of SW fishes that normally display only net SO_4^{2-} secretion (Renfro and Pritchard 1982). Using the expression cloning technique, Na^+-SO_4^{2-} cotransporter (Slc13a1 or NaS1) was cloned from the rat kidney (Markovich et al. 1993). Though SO_4^{2-} metabolism have not been well studied in FW fishes, recent analyses showed that similar SO_4^{2-} retention system is present in FW eels that allows them to accumulate SO_4^{2-} in plasma (Nakada et al. 2005, Watanabe and Takei 2011a, 2012).

FW eels show a marked increase in plasma SO_4^{2-} concentration and a concomitant decrease in Cl^- concentration. The reported SO_4^{2-} concentrations in plasma range from 18.8 mM (Nakada et al. 2005) to 6.2 mM (Watanabe and Takei 2011a, 2012). Plasma SO_4^{2-} concentrations of other fishes are normally maintained at 0.2–2 mM, suggesting that this high accumulation is specific to eels (Watanabe and Takei 2012). When eels are transferred from FW to SW, plasma SO_4^{2-} concentration decreases gradually after 3 days (Fig. 5A). Urine SO_4^{2-} concentration of FW eel is ~3 mM, indicating active SO_4^{2-} reabsorption by the kidney (Watanabe and Takei 2011a, 2012).

Tracer experiments showed that the influx rate of SO_4^{2-} was 0.09 µmol/kg/hr in FW eels (Watanabe and Takei 2012). Although this value is much smaller than the influx rate of SO_4^{2-} in SW eel, FW eels seem to absorb SO_4^{2-} actively from the environment that contains SO_4^{2-} at µM concentration.

Molecules Involved in Apical SO_4^{2-} Reabsorption by Proximal Tubules of FW Eel

FW eels have a system for renal SO_4^{2-} reabsorption which is mediated by Slc13a1, an electrogenic Na^+-SO_4^{2-} co-transporter (Nakada et al. 2005, Watanabe and Takei 2011b). The eel Slc13a1 is specifically expressed in the kidney of FW eel (Figs. 6 and 7, Table 4). The expression was reduced to undetectable level in 1–3 days after transfer from FW to SW, and no expression was observed in the kidney of SW eel. In FW eels, Slc13a1 is localized on the apical membrane of PT-II that expresses Slc26a1 (Fig. 7) (Nakada et al. 2005, Watanabe and Takei 2011b). Zebrafish kidney also expresses Slc13a1 in the second segment of proximal tubule (Wingert et al. 2007). The *slc13a1* gene is expressed in the kidney and intestine of mammals and zebrafish (Markovich et al. 1993, 2008), but not in eel intestine (Nakada et al. 2005, Watanabe and Takei 2011b). Rat, eel, and zebrafish Slc13a1 expressed in *Xenopus* oocytes have Na^+ dependent SO_4^{2-} transport activity, which is advantageous for SO_4^{2-} absorption rather than secretion (Markovich et al. 2008, Nakada et al. 2005).

In addition to the Na^+-SO_4^{2-} co-transport system, SO_4^{2-} can be transported by anion exchange mechanism. However, the role of the apical anion exchangers in FW fishes has not been well investigated. As mentioned above, Slc26a6 paralogs are involved in renal SO_4^{2-} secretion of SW fishes. Although the expression of *slc26a6a* is strongly suppressed in the FW eel kidney, *slc26a6b* and *slc26a6c* are expressed in the kidney of FW eel (Watanabe and Takei 2011b). The role of Slc26a6 paralogs can be predicted from studies of mammalian kidney. In mammals, addition of oxalate to the perfuste markedly stimulated proximal tubule NaCl absorption (Markovich and Aronson 2007, Wang et al. 1992, 1996). This oxalate stimulated NaCl

Table 4. Slc13 family.

	Protein name	substrates	Transporting type	Tissue distribution in mammals	Tissue distribution in eel[a,c]	Associated disease(s)	Reports within teleost
Slc13a1	Nas1, NaSi-1	SO_4^{2-}, Selenate, Thiosulfate	Cotransporter	Kidney	Kidney	-	Eel[a-c], Pufferfish[d], Zebrafish[e]
Slc13a2	NaC1, NaDC-1, SDCT1, NaDC-2	Succinate, Citrate, a-Ketoglutarate	Cotransporter	Kidney, Intestine	?	-	-
Slc13a3	NaC2, NaDC-3, SDCT2	Succinate, Citrate, a-Ketoglutarate	Cotransporter	Kidney, Intestine, Liver, Pancreas, Brain, Placenta	?	-	Zebrafish[f]
Slc13a4	NaS2, SUT-1	SO_4^{2-}	Cotransporter	Placenta, Tonsillar high endothelial venules, Testis, Heart	?	-	-
Slc13a5	NaC2, NaCT	Citrate	Cotransporter	Liver, Brain, Testis	?	-	-

a) Nakada et al. 2005; b) Watanabe and Takei 2011a; c) Watanabe and Takei 2011b; d) Kato et al. 2009; e) Markovich et al. 2008; f) Strungaru et al. 2011; others, Markovich and Murer 2004

transport was abolished when SO_4^{2-} is removed from the perfusates, and was not observed in *Slc26a6*-null mice (Wang et al. 2005). These findings indicate that NaCl absorption across the apical membrane of proximal tubule cells takes place by Cl^-/oxalate exchange in parallel with SO_4^{2-}/oxalate exchange and Na^+-SO_4^{2-} co-transport. Similarly, Slc26a6b and Slc26a6c in FW eel kidney may be involved in the SO_4^{2-} homeostasis.

Molecules Involved in Basolateral SO_4^{2-} Secretion by Proximal Tubules of FW Eel

In BLM vesicles isolated from the kidneys of fishes and mammals, SO_4^{2-} is transported by an anion exchange mechanism (Pritchard and Renfro 1983). Only the expression of Slc26a1 has been reported in the proximal tubule of FW eels (Nakada et al. 2005). Therefore, Slc26a1 is considered to be the major pathway that transports SO_4^{2-} across the basolateral membrane. In eel, FW acclimation induced the expression of Slc26a1 gene, suggesting more active functioning of Slc26a1 in FW eels (Nakada et al. 2005).

It is interesting to consider how Slc26a1 mediates both absorption (FW) and secretion (SW) of SO_4^{2-}. The direction of transporters is determined by electrochemical gradients of the substrates across the plasma membrane. In FW eel, induced expression of Slc13a1 and high SO_4^{2-} concentration in increased primary urine may together maintain a high intracellular SO_4^{2-} concentration, which drives SO_4^{2-} efflux into blood via Slc26a1 at the basolateral side. When plasma SO_4^{2-} is increased, SO_4^{2-} in primary urine is also increased, which is advantageous to maintain high intracellular SO_4^{2-} in FW eels.

Lessons from Mammalian Model

The internal environment of terrestrial mammals is rich in SO_4^{2-} relative to the external environment, which results in the loss of SO_4^{2-}, especially via the kidney. This situation is similar to fishes in FW. To avoid the loss, terrestrial mammals obtain metabolic SO_4^{2-} from sulfur-containing amino acids in food, and reabsorb as much SO_4^{2-} as possible across the kidney (Goudsmit et al. 1939, Swan et al. 1956). Proximal tubule is the major site of renal SO_4^{2-} reabsorption (Hierholzer et al. 1960). When plasma SO_4^{2-} concentration increases, the filtered load of SO_4^{2-} exceeds the capacity of tubular reabsorption and the remaining SO_4^{2-} is excreted into urine (Lin and Levy 1983, Mudge et al. 1973).

In mammalian kidney, Slc13a1 and Slc26a1 are involved in SO_4^{2-} reabsorption (Markovich 2012, Markovich and Aronson 2007). Slc13a1 was first cloned from rat kidney by expression cloning as a Na^+-SO_4^{2-}

cotransporter (Markovich et al. 1993), and then the orthologs were isolated from other mammals (Beck and Markovich 2000, Lee et al. 2000). Rat Slc13a1 functions as an electrogenic $3Na^+$-SO_4^{2-} cotransporter with a K_m of 93 µM for SO_4^{2-} and 16–24 mM for Na^+. Slc13a1 also transport thiosulfate (K_m = 84 µM) and selenate (K_m = 580 µM). The SO_4^{2-}-transport activity is inhibited by molybdate, selenate, tungstate, thiosulfate, succinate, and citrate, but not by oxalate and DIDS. The Slc13a1 protein is located on the apical membrane of the epithelial cells lining the renal proximal tubule and of the small and large intestine of mammals, and is thought to be a major regulator of serum SO_4^{2-} concentrations (Besseghir and Roch-Ramel 1987, Markovich 2001). The *Slc13a1*-null mice suffer from reduced blood SO_4^{2-} levels (hyposulfatemia), increased urine SO_4^{2-} levels (hypersulfaturia), growth retardation, reduced fertility, spontaneous seizures, reduced circulating steroid levels and increased urinary glucocorticoid excretion, altered hepatic lipid and cholesterol metabolism, reduced serum levels of dehydroepiandrosterone-sulfate, reduced intestinal mucin sulfonation and impaired intestinal barrier function (Markovich 2012).

Meanwhile, Slc26a1 was first identified as a Na^+-independent SO_4^{2-}-anion exchanger by expression cloning from the rat liver (Bissig et al. 1994). Slc26a1 transport SO_4^{2-} with a K_m of 136 µM and the SO_4^{2-}-transport activity is inhibited by oxalate, molybdate, selenate, tungstate, thiosulfate, phenol red, probenecid, and DIDS (Markovich 2012). Slc26a1 protein is located on the basolateral membrane of epithelial cells lining the proximal tubule. In the proximal tubular cells, Slc26a1 and Slc13a1 are co-localized (Karniski et al. 1998). *Slc26a1*-null mice also show hyposulfatemia and hypersulfaturia. These are reverse-genetic evidences showing that Slc13a1 and Slc26a1 mediate renal SO_4^{2-} reabsorption. *Slc26a1*-null mice also have increased blood and urine oxalate levels and calcium oxalate urolithiasis (Markovich 2012).

Slc26a6 is the third SO_4^{2-} transporter in the mammalian kidney, but its role on SO_4^{2-} handling has not been established yet. Slc26a6 was first identified in human through database mining (Everett and Green 1999), and the anion transport activity was later established (Knauf et al. 2001, Ko et al. 2002, Xie et al. 2002). The *Slc26a6* gene is expressed in the small intestine, kidney, pancreas, heart, and placenta. Slc26a6 is localized at the apical membrane of the renal proximal tubule, gastric parietal cells, and duodenal enterocytes. *Slc26a6*-null mice develop hyperoxalemia, hyperoxaluria, and calcium-oxalate stones as a result of a defect in intestinal oxalate secretion (Jiang et al. 2006), and show reduced luminal Cl^-/HCO_3^- exchange activity and thereby reduced NaCl absorption in the small intestine (Seidler et al. 2008). SO_4^{2-}/HCO_3^- exchange was eliminated in the duodenal villous epithelium of *Slc26a6*-null mice (Simpson et al. 2007), but urine SO_4^{2-} concentration was normal. In contrast to fish in which Slc26a6 mediates active SO_4^{2-} efflux in the kidney, function of mammalian Slc26a6 in SO_4^{2-} regulation is not clear yet.

Regulation of Renal SO_4^{2-} Handling in Eel

Eels are unique in that they maintain much higher plasma SO_4^{2-} concentration in SO_4^{2-}-poor FW than in SO_4^{2-}-rich SW, showing drastic changes in SO_4^{2-} regulating system between FW and SW. When FW eels are transferred to SW, renal expression of *slc13a1* is reduced to undetectable levels and renal *slc26a6a* expression appears in 1–3 days with concomitant decrease in plasma SO_4^{2-} concentration (Watanabe and Takei 2011a). These changes also occur in eels in SO_4^{2-}-free SW and solutions that contain Cl^-, but not in solutions that contain SO_4^{2-}, Mg^{2+}, Ca^{2+} or Na^+ (Watanabe and Takei 2011a). Injection of solutions that contain Cl^- alone into the circulation of FW eel also triggers the switching the transporters from FW-type to SW-type. Therefore, increased Cl^- in media followed by increased plasma Cl^- concentration may be a key regulator of renal SO_4^{2-} regulation (Fig. 8). Na^+-Cl^- cotransporter (NCC) is involved in the Cl^- uptake in FW eels, because a NCC inhibitor hydrochlorochiazide (HCTZ) impaired the switching of renal SO_4^{2-} transporters. The pathway from Cl^- detection to SO_4^{2-} regulation is yet unknown. High SO_4^{2-} diet induced a decrease in *slc13a1* expression in the renal proximal tubule of mammals (Markovich et al. 1998, Pena and Neiberger 1997) and injection of Na_2SO_4 into the circulation increased *slc26a1* transcripts in the kidney of rainbow trout (Katoh et al. 2006), indicating a role of internal SO_4^{2-} in the transporter regulation. However, injection of Na_2SO_4 in eels did not alter renal expression of *slc13a1* and *slc26a6*. These results suggest that the ions responsible for renal SO_4^{2-} regulation are different among genes and species.

In proximal tubule epithelium primary culture of winter flounder, addition of cortisol to medium increased CA activity and CA-dependent basolateral Slc26a1 activity for SO_4^{2-} secretion (Pelis et al. 2003). Glucocorticoid (dexamethasone) did not alter SO_4^{2-}-transport by BLM vesicles isolated from chicken kidney, but significantly reduced Na^+-dependent SO_4^{2-}-transport by the BBM vesicles (Renfro et al. 1989). In eel, elevated plasma cortisol promotes SW adaptation (Mayer et al. 1967). Thus it should be examined whether cortisol stimulates SO_4^{2-} excretion by increasing CA activity and reducing Slc13a1 in the eel kidney.

In mammals, tunicamycin (an *N*-glycosylation inhibitor) treatment or site-directed mutagenesis of the putative *N*-glycosylation sites reduced SO_4^{2-} transport activity of Slc13a1 and Slc26a1 in *Xenopus* oocytes (Li and Pajor 2003, Markovich 2001, Markovich and Aronson 2007). Pharmacological activators of protein kinase A and C completely inhibit rat Slc13a1 in *Xenopus* oocytes (Markovich and Aronson 2007). Phorbol ester PMA inhibits human

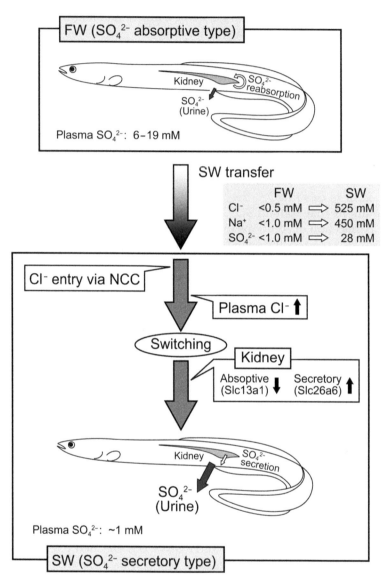

Figure 8. A schematic summary of how renal SO$_4^{2-}$ regulation is altered after transfer of eels from FW to SW. For details, see text. From Watanabe and Takei (2011a).

SLC26A6 in *Xenopus* oocytes, and protein kinase Cδ (PKCδ) mediates the inhibition and internalization (Hassan et al. 2007). The effect of the inhibitors on renal SO$_4^{2-}$ handling in eels should be examined in the near future.

Acknowledgments

We wish to express our sincere gratitude to Prof. Yoshio Takei (University of Tokyo, Japan), who gave this opportunity to summarize our work, for critically reviewing this manuscript. We also thank the members of the Prof. Yoshio Takei, Prof. Shigehisa Hirose (Tokyo Institute of Technology, Japan), and Prof. Michael F. Romero (Mayo Clinic, USA) groups for discussions and contribution to the original research discussed here. We are grateful to the following authors and publishers for permission to reproduce the following copyright material: Prof. Nobuko Naito and The Zoological Society of Japan for Figs. 1A, 1B, and 1D; Prof. Kaworu Nakamura and The Japanese Society of Fisheries Science for Fig. 1E; and John Wiley & Sons Ltd. for Fig. 1C.

References

Bastlein, C. and G. Burckhardt. 1986. Sensitivity of rat renal luminal and contraluminal sulfate transport systems to DIDS. Am. J. Physiol. 250: F226–234.

Bayaa, M., B. Vulesevic, A. Esbaugh, M. Braun, M.E. Ekker, M. Grosell and S.F. Perry. 2009. The involvement of SLC26 anion transporters in chloride uptake in zebrafish (*Danio rerio*) larvae. J. Exp. Biol. 212: 3283–3295.

Beck, L. and D. Markovich. 2000. The mouse Na$^+$-sulfate cotransporter gene Nas1. Cloning, tissue distribution, gene structure, chromosomal assignment, and transcriptional regulation by vitamin D. J. Biol. Chem. 275: 11880–11890.

Besseghir, K. and F. Roch-Ramel. 1987. Renal excretion of drugs and other xenobiotics. Ren. Physiol. 10: 221–241.

Beyenbach, K.W. 1982. Direct demonstration of fluid secretion by glomerular renal tubules in a marine teleost. Nature 299: 54–56.

Beyenbach, K.W. 2004. Kidneys sans glomeruli. Am. J. Physiol. Renal Physiol. 286: F811–827.

Beyenbach, K.W., D.H. Petzel and W.H. Cliff. 1986. Renal proximal tubule of flounder. I. Physiological properties. Am. J. Physiol. 250: R608–615.

Bird, J.E., K. Milhoan, C.B. Wilson, S.G. Young, C.A. Mundy, S. Parthasarathy and R.C. Blantz. 1988. Ischemic acute renal failure and antioxidant therapy in the rat. The relation between glomerular and tubular dysfunction. J. Clin. Invest. 81: 1630–1638.

Bissig, M., B. Hagenbuch, B. Stieger, T. Koller and P.J. Meier. 1994. Functional expression cloning of the canalicular sulfate transport system of rat hepatocytes. J. Biol. Chem. 269: 3017–3021.

Chou, C.F., S. Tohari, S. Brenner and B. Venkatesh. 2004. Erythropoietin gene from a teleost fish, *Fugu rubripes*. Blood 104: 1498–1503.

Chu, C.Y., C.H. Cheng, G.D. Chen, Y.C. Chen, C.C. Hung, K.Y. Huang and C.J. Huang. 2007. The zebrafish erythropoietin: functional identification and biochemical characterization. FEBS Lett. 581: 4265–4271.

Cliff, W.H. and K.W. Beyenbach. 1992. Secretory renal proximal tubules in seawater- and freshwater-adapted killifish. Am. J. Physiol. 262: F108–116.

Colville, T.P., R.H. Richards and J.W. Dobbie. 1983. Variations in renal corpuscular morphology with adaptation to sea water in the rainbow trout, *Salmo gairdneri* Richardson. J. Fish. Biol. 23: 451–456.

Cutler, C.P. and G. Cramb. 2008. Differential expression of absorptive cation-chloride-cotransporters in the intestinal and renal tissues of the European eel (*Anguilla anguilla*). Comp. Biochem. Physiol. B. Biochem. Mol. Biol. 149: 63–73.

de Jong, J.L. and L.I. Zon. 2005. Use of the zebrafish system to study primitive and definitive hematopoiesis. Annu. Rev. Genet. 39: 481–501.

Evans, D.H., P.M. Piermarini and K.P. Choe. 2005. The multifunctional fish gill: dominant site of gas exchange, osmoregulation, acid-base regulation, and excretion of nitrogenous waste. Physiol. Rev. 85: 97–177.

Everett, L.A. and E.D. Green. 1999. A family of mammalian anion transporters and their involvement in human genetic diseases. Hum. Mol. Genet. 8: 1883–1891.

Foster, R.C. 1975. Changes in urinary bladder and kidney function in the starry flounder (*Platichtys stellatus*) in response to prolactin and to freshwater transfer. Gen. Comp. Endocrinol. 27: 153–161.

Friedman, P.A. and S.C. Hebert. 1990. Diluting segment in kidney of dogfish shark. I. Localization and characterization of chloride absorption. Am. J. Physiol. 258: R398–408.

Furukawa, F., S. Watanabe, S. Kimura and T. Kaneko. 2012. Potassium excretion through ROMK potassium channel expressed in gill mitochondrion-rich cells of Mozambique tilapia. Am. J. Physiol. Regul. Integr. Comp. Physiol. 302: R568–576.

Garty, H. and L.G. Palmer. 1997. Epithelial sodium channels: function, structure, and regulation. Physiol. Rev. 77: 359–396.

Goudsmit, A., Jr., M.H. Power and J.L. Bollman. 1939. The excretion of sulphates by the dog. Am. J. Physiol. 125: 506–520.

Greger, R. 1985. Ion transport mechanisms in thick ascending limb of Henle's loop of mammalian nephron. Physiol. Rev. 65: 760–797.

Grosell, M. 2009. Editorial Focus: Using phenotypic plasticity: focus on "Identification of renal transporters involved in sulfate excretion in marine teleost fish". Am. J. Physiol. Regul. Integr. Comp. Physiol. 297: R1645–1646.

Grosell, M. 2011. The role of the gastrointestinal tract in salt and water balance. *In*: M. Grosell, P. Farrell and C.J. Brauner [eds]. The Multifunctional Gut of Fish. Academic Press, San Diego, USA.

Hanssen, R.G.J.M., N. Mayer-Gostan, G. Flik and S.E. Wendelaar Bonga. 1993. Stanniocalcin kinetics in freshwater and seawater european eel (*Anguilla anguilla*) Fish Physiol. Biochem. 10: 491–496.

Hassan, H.A., S. Mentone, L.P. Karniski, V.M. Rajendran and P.S. Aronson. 2007. Regulation of anion exchanger Slc26a6 by protein kinase C. Am. J. Physiol. Cell Physiol. 292: C1485–1492.

Hathaway, C.B. and A. Epple. 1989. The sources of plasma catecholamines in the American eel, *Anguilla rostrata*. Gen. Comp. Endocrinol. 74: 418–430.

Hebert, S.C. and P.A. Friedman. 1990. Diluting segment in kidney of dogfish shark. II. Electrophysiology of apical membranes and cellular resistances. Am. J. Physiol. 258: R409–417.

Hebert, S.C., D.B. Mount and G. Gamba. 2004. Molecular physiology of cation-coupled C$^-$ cotransport: the SLC12 family. Pflugers Arch. 447: 580–593.

Hickman, C.P., Jr. 1968. Urine composition and kidney tubular function in southern flounder, *Paralichthys lethostigma*, in seawater. Can. J. Zool. 46: 439–455.

Hickman, C.P. and B.F. Trump. 1969. The Kidney. *In*: S.H. William and J.R. David [eds]. Fish Physiology. Academic Press, New York, USA. pp. 91–239.

Hierholzer, K., R. Cade, R. Gurd, R. Kessler and R. Pitts. 1960. Stop-flow analysis of renal reabsorption and excretion of sulfate in the dog. Am. J. Physiol. 198: 833–837.

Hirano, T., D.W. Johnson and H.A. Bern. 1971. Control of water movement in flounder urinary bladder by prolactin. Nature 230: 469–470.

Hirano, T., D.W. Johnson, H.A. Bern and S. Utida. 1973. Studies on water and ion movements in the isolated urinary bladder of selected freshwater, marine and euryhaline teleosts. Comp. Biochem. Physiol. A Comp. Physiol. 45: 529–540.

Hong, S.K., C.S. Levin, J.L. Brown, H. Wan, B.T. Sherman, W. Huang da, R.A. Lempicki and B. Feldman. 2010. Pre-gastrula expression of zebrafish extraembryonic genes. BMC Dev. Biol. 10: 42.

Hou, J., A. Renigunta, M. Konrad, A.S. Gomes, E.E. Schneeberger, D.L. Paul, S. Waldegger and D.A. Goodenough. 2008. Claudin-16 and claudin-19 interact and form a cation-selective tight junction complex. J. Clin. Invest. 118: 619–628.

Islam, Z., A. Kato, M.F. Romero and S. Hirose. 2011. Identification and apical membrane localization of an electrogenic Na^+/Ca^{2+} exchanger NCX2a likely to be involved in renal Ca^{2+} excretion by seawater fish. Am. J. Physiol. Regul. Integr. Comp. Physiol. 301: R1427–1439.

Ivanis, G., M. Braun and S.F. Perry. 2008. Renal expression and localization of SLC9A3 sodium/hydrogen exchanger and its possible role in acid-base regulation in freshwater rainbow trout (*Oncorhynchus mykiss*). Am. J. Physiol. Regul. Integr. Comp. Physiol. 295: R971–978.

Jiang, Z., J.R. Asplin, A.P. Evan, V.M. Rajendran, H. Velazquez, T.P. Nottoli, H.J. Binder and P.S. Aronson. 2006. Calcium oxalate urolithiasis in mice lacking anion transporter Slc26a6. Nat. Genet. 38: 474–478.

Jones, I.C., D.K. Chan and J.C. Rankin. 1969. Renal function in the European eel (*Anguilla anguilla* L.): effects of the caudal neurosecretory system, corpuscles of Stannius, neurohypophysial peptides and vasoactive substances. J. Endocrinol. 43: 21–31.

Karniski, L.P., M. Lotscher, M. Fucentese, H. Hilfiker, J. Biber and H. Murer. 1998. Immunolocalization of sat-1 sulfate/oxalate/bicarbonate anion exchanger in the rat kidney. Am. J. Physiol. 275: F79–87.

Kato, A., M.H. Chang, Y. Kurita, T. Nakada, M. Ogoshi, T. Nakazato, H. Doi, S. Hirose and M.F. Romero. 2009. Identification of renal transporters involved in sulfate excretion in marine teleost fish. Am. J. Physiol. Regul. Integr. Comp. Physiol. 297: R1647–1659.

Kato, A., T. Muro, Y. Kimura, S. Li, Z. Islam, M. Ogoshi, H. Doi and S. Hirose. 2011. Differential expression of Na^+-Cl^- cotransporter and Na^+-K^+-Cl^- cotransporter 2 in the distal nephrons of euryhaline and seawater pufferfishes. Am. J. Physiol. Regul. Integr. Comp. Physiol. 300: R284–297.

Katoh, F., R.R. Cozzi, W.S. Marshall and G.G. Goss. 2008. Distinct $Na^+/K^+/2Cl^-$ cotransporter localization in kidneys and gills of two euryhaline species, rainbow trout and killifish. Cell Tissue Res. 334: 265–281.

Katoh, F., M. Tresguerres, K.M. Lee, T. Kaneko, K. Aida and G.G. Goss. 2006. Cloning of rainbow trout SLC26A1: involvement in renal sulfate secretion. Am. J. Physiol. Regul. Integr. Comp. Physiol. 290: R1468–1478.

Kirsch, R. and M.F. Meister. 1982. Progressive processing of ingested water in the gut of sea-water teleosts. J. Exp. Biol. 98: 67–81.

Knauf, F., C.L. Yang, R.B. Thomson, S.A. Mentone, G. Giebisch and P.S. Aronson. 2001. Identification of a chloride-formate exchanger expressed on the brush border membrane of renal proximal tubule cells. Proc. Natl. Acad. Sci. USA 98: 9425–9430.

Ko, S.B., N. Shcheynikov, J.Y. Choi, X. Luo, K. Ishibashi, P.J. Thomas, J.Y. Kim, K.H. Kim, M.G. Lee, S. Naruse and S. Muallem. 2002. A molecular mechanism for aberrant CFTR-dependent HCO_3^- transport in cystic fibrosis. EMBO J. 21: 5662–5672.

Krishnamurthy, V.G. and H.A. Bern. 1969. Correlative histologic study of the corpuscles of Stannius and the juxtaglomerular cells of teleost fishes. Gen. Comp. Endocrinol. 13: 313–335.

Kurita, Y., T. Nakada, A. Kato, H. Doi, A.C. Mistry, M.H. Chang, M.F. Romero and S. Hirose. 2008. Identification of intestinal bicarbonate transporters involved in formation of carbonate precipitates to stimulate water absorption in marine teleost fish. Am. J. Physiol. Regul. Integr. Comp. Physiol. 294: R1402–1412.

Lee, A., L. Beck and D. Markovich. 2000. The human renal sodium sulfate cotransporter (SLC13A1; hNaSi-1) cDNA and gene: organization, chromosomal localization, and functional characterization. Genomics 70: 354–363.

Li, H. and A.M. Pajor. 2003. Mutagenesis of the *N*-glycosylation site of hNaSi-1 reduces transport activity. Am. J. Physiol. Cell Physiol. 285: C1188–1196.

Li, Y.Y., K. Inoue and Y. Takei. 2003. Interrenal steroid 21-hydroxylase in eels: primary structure, progesterone-specific activity and enhanced expression by ACTH. J. Mol. Endocrinol. 31: 327–340.

Lin, H.F., D. Traver, H. Zhu, K. Dooley, B.H. Paw, L.I. Zon and R.I. Handin. 2005. Analysis of thrombocyte development in CD41-GFP transgenic zebrafish. Blood 106: 3803–3810.

Lin, J.H. and G. Levy. 1983. Renal clearance of inorganic sulfate in rats: effect of acetaminophen-induced depletion of endogenous sulfate. J. Pharm. Sci. 72: 213–217.

Lohi, H., G. Lamprecht, D. Markovich, A. Heil, M. Kujala, U. Seidler and J. Kere. 2003. Isoforms of SLC26A6 mediate anion transport and have functional PDZ interaction domains. Am. J. Physiol. Cell Physiol. 284: C769–779.

Markovich, D. 2001. Physiological roles and regulation of mammalian sulfate transporters. Physiol. Rev. 81: 1499–1533.

Markovich, D. 2012. Slc13a1 and Slc26a1 KO models reveal physiological roles of anion transporters. Physiology (Bethesda) 27: 7–14.

Markovich, D. and P.S. Aronson. 2007. Specificity and regulation of renal sulfate transporters. Annu. Rev. Physiol. 69: 361–375.

Markovich, D., J. Forgo, G. Stange, J. Biber and H. Murer. 1993. Expression cloning of rat renal Na$^+$/SO$_4$$^{2-}$ cotransport. Proc. Natl. Acad. Sci. USA 90: 8073–8077.

Markovich, D. and H. Murer. 2004. The SLC13 gene family of sodium sulphate/carboxylate cotransporters. Pflugers Arch. 447: 594–602.

Markovich, D., H. Murer, J. Biber, K. Sakhaee, C. Pak and M. Levi. 1998. Dietary sulfate regulates the expression of the renal brush border Na/Si cotransporter NaSi-1. J. Am. Soc. Nephrol. 9: 1568–1573.

Markovich, D., A. Romano, C. Storelli and T. Verri. 2008. Functional and structural characterization of the zebrafish Na$^+$-sulfate cotransporter 1 (NaS1) cDNA and gene (slc13a1). Physiol. Genomics 34: 256–264.

Marshall, W.S. and M. Grosell. 2006. Ion transport, osmoregulation, and acid-base balance. *In*: D.H. Evans and J.B. Claiborne [eds]. The Physiology of Fishes. CRC, New York, USA. pp. 177–224.

Mayer, N., J. Maetz, D.K. Chan, M. Forster and I.C. Jones. 1967. Cortisol, a sodium excreting factor in the eel (*Anguilla anguilla* L.) adapted to sea water. Nature 214: 1118–1120.

McDonald, R.H., Jr., L.I. Goldberg, J.L. McNay and E.P. Tuttle, Jr. 1964. Effect of Dopamine in Man: Augmentation of Sodium Excretion, Glomerular Filtration Rate, and Renal Plasma Flow. J. Clin. Invest. 43: 1116–1124.

Mistry, A.C., G. Chen, A. Kato, K. Nag, J.M. Sands and S. Hirose. 2005. A novel type of urea transporter, UT-C, is highly expressed in proximal tubule of seawater eel kidney. Am. J. Physiol. Renal Physiol. 288: F455–465.

Miyazaki, H., T. Kaneko, S. Uchida, S. Sasaki and Y. Takei. 2002. Kidney-specific chloride channel, OmClC-K, predominantly expressed in the diluting segment of freshwater-adapted tilapia kidney. Proc. Natl. Acad. Sci. USA 99: 15782–15787.

Mott, J.C. 1950. The gross anatomy of the blood vascular system of the Eel *Anguilla anguilla*. Proc. Zool. Soc. London 120: 503–518.

Mount, D.B. and M.F. Romero. 2004. The SLC26 gene family of multifunctional anion exchangers. Pflugers Arch. 447: 710–721.

Mudge, G.H., W. Berndt and H. Valtin. 1973. Tubular transport of urea, phosphate, uric acid, sulfate and thiosulfate. *In*: J. Orloff and R.W. Berliner [eds]. Handbook of Physiology. American Physiological Society, Washington DC, USA. pp. 587–652.

Naito, N. 1990. Eel. *In*: E. Kobayashi, S. Kawashima, C. Oguro and Y. Sasayama [eds]. Animal Anatomy. The Zoological Society of Japan, Tokyo, Japan. pp. 59–63.

Nakada, T., K. Zandi-Nejad, Y. Kurita, H. Kudo, V. Broumand, C.Y. Kwon, A. Mercado, D.B. Mount and S. Hirose. 2005. Roles of Slc13a1 and Slc26a1 sulfate transporters of eel kidney in sulfate homeostasis and osmoregulation in freshwater. Am. J. Physiol. Regul. Integr. Comp. Physiol. 289: R575–R585.

Nakamura, K. and M. Toyohiro. 1983. Distribution patterns of urinary tracts and renal blood vascular systems of the eel *Anguilla japonica*. Bull. Japan. Soc. Sci. Fisheries. 49: 1803–1808.

Nishimura, H., M. Imai and M. Ogawa. 1983. Sodium chloride and water transport in the renal distal tubule of the rainbow trout. Am. J. Physiol. 244: F247–254.

Nishimura, H., M. Ogawa and W.H. Sawyer. 1973. Renin-angiotensin system in primitive bony fishes and a holocephalian. Am. J. Physiol. 224: 950–956.

Ogawa, M. 1962. Comparative study on the internal structure of the teleostean kidney. Sci. Rept. Saitama Univ. 4: 107–129.

Oide, H. and S. Utida. 1968. Changes in intestinal absorption and renal excretion of water during adaptation to sea-water in the Japanese eel. Marine Biol. 1: 172–177.

Pelis, R.M., J.E. Goldmeyer, J. Crivello and J.L. Renfro. 2003. Cortisol alters carbonic anhydrase-mediated renal sulfate secretion. Am. J. Physiol. Regul. Integr. Comp. Physiol. 285: R1430–1438.

Pena, D.R. and R.E. Neiberger. 1997. Renal brush border sodium-sulfate cotransport in guinea pig: effect of age and diet. Pediatr. Nephrol. 11: 724–727.

Perry, S.F., B. Vulesevic, M. Grosell and M. Bayaa. 2009. Evidence that SLC26 anion transporters mediate branchial chloride uptake in adult zebrafish (*Danio rerio*). Am. J. Physiol. Regul. Integr. Comp. Physiol. 297: R988–997.

Pritchard, J.B. and J.L. Renfro. 1983. Renal sulfate transport at the basolateral membrane is mediated by anion exchange. Proc. Natl. Acad. Sci. USA 80: 2603–2607.

Renfro, J.L. 1999. Recent developments in teleost renal transport. J. Exp. Zool. 283: 653–661.

Renfro, J.L., N.B. Clark, R.E. Metts and M.A. Lynch. 1989. Glucocorticoid inhibition of Na-SO$_4$ transport by chick renal brush-border membranes. Am. J. Physiol. 256: R1176–1183.

Renfro, J.L. and J.B. Pritchard. 1982. H$^+$-dependent sulfate secretion in the marine teleost renal tubule. Am. J. Physiol. 243: F150–159.

Renfro, J.L. and J.B. Pritchard. 1983. Sulfate transport by flounder renal tubule brush border: presence of anion exchange. Am. J. Physiol. 244: F488–496.

Romeu, G.F. and J. Maetz. 1964. The mechanism of sodium and chloride uptake by the gills of a fresh-water fish, carassius auratus. I. evidence for an independent uptake of sodium and chloride ions. J. Gen. Physiol. 47: 1195–1207.

Rosenthal, H.L. 1961. The uptake and turnover of S^{35} sulfate by Lebistes. Biol. Bull. 120: 183–191.

Schaechinger, T.J. and D. Oliver. 2007. Nonmammalian orthologs of prestin (SLC26A5) are electrogenic divalent/chloride anion exchangers. Proc. Natl. Acad. Sci. USA 104: 7693–7698.

Schmidt-Nielsen, B. and J.L. Renfro. 1975. Kidney function of the American eel *Anguilla rostrata*. Am. J. Physiol. 228: 420–431.

Seidler, U., J. Rottinghaus, J. Hillesheim, M. Chen, B. Riederer, A. Krabbenhoft, R. Engelhardt, M. Wiemann, Z. Wang, S. Barone, M.P. Manns and M. Soleimani. 2008. Sodium and chloride absorptive defects in the small intestine in Slc26a6 null mice. Pflugers Arch. 455: 757–766.

Sharratt, B.M., D. Bellamy and I.C. Jones. 1964. Adaptation of the Silver Eel (*Anguilla Anguilla* L.) to sea water and to artificial media together with observations on the role of the gut. Comp. Biochem. Physiol. 11: 19–30.

Simpson, J.E., C.W. Schweinfest, G.E. Shull, L.R. Gawenis, N.M. Walker, K.T. Boyle, M. Soleimani and L.L. Clarke. 2007. PAT-1 (Slc26a6) is the predominant apical membrane Cl$^-$/HCO$_3^-$ exchanger in the upper villous epithelium of the murine duodenum. Am. J. Physiol. Gastrointest. Liver Physiol. 292: G1079–1088.

Smith, H.W. 1930. The absorption and excretion of water and salts by marine teleosts. Am. J. Physiol. 93: 480–505.

Sokabe, H. and M. Ogawa. 1974. Comparative studies of the juxtaglomerular apparatus. Int. Rev. Cytol. 37: 271–327.

Sokabe, H., H. Oide, M. Ogawa and S. Utida. 1973. Plasma renin activity in Japanese eels (*Anguilla japonica*) adapted to seawater or in dehydration. Gen. Comp. Endocrinol. 21: 160–167.

Strungaru, M.H., T. Footz, Y. Liu, F.B. Berry, P. Belleau, E.V. Semina, V. Raymond and M.A. Walter. 2011. PITX2 is involved in stress response in cultured human trabecular meshwork cells through regulation of SLC13A3. Invest. Ophthalmol. Vis. Sci. 52: 7625–7633.

Sugiura, S.H., N.K. McDaniel and R.P. Ferraris. 2003. *In vivo* fractional P(i) absorption and NaPi-II mRNA expression in rainbow trout are upregulated by dietary P restriction. Am. J. Physiol. Regul. Integr. Comp. Physiol. 285: R770–781.

Sussman, C.R., J. Zhao, C. Plata, J. Lu, C. Daly, N. Angle, J. DiPiero, I.A. Drummond, J.O. Liang, W.F. Boron, M.F. Romero and M.H. Chang. 2009. Cloning, localization, and functional expression of the electrogenic Na^+ bicarbonate cotransporter (NBCe1) from zebrafish. Am. J. Physiol. Cell Physiol. 297: C865–875.

Swan, R.C., H.M. Feinstein and H. Madisso. 1956. Distribution of sulfate ion across semi-permeable membranes. J. Clin. Invest. 35: 607–610.

Takei, Y. and R.J. Balment. 2009. The neuroendocrine regulation of fluid intake and fluid balance. *In*: N.J. Bernier, G.V.D. Kraak, A.P. Farrell and C.J. Brauner [eds]. Fish Neuroendocrinology. Academic Press, San Diego, USA.

Takei, Y. and H. Kaiya. 1998. Antidiuretic Effect of Eel ANP Infused at Physiological Doses in Conscious, Seawater-Adapted Eels, *Anguilla japonica*. Zoolog. Sci. 15: 399–404.

Teranishi, K. and T. Kaneko. 2010. Spatial, cellular, and intracellular localization of Na^+/K^+-ATPase in the sterically disposed renal tubules of Japanese eel. J. Histochem. Cytochem. 58: 707–719.

Tsuchida, T. and Y. Takei. 1998. Effects of homologous atrial natriuretic peptide on drinking and plasma ANG II level in eels. Am. J. Physiol. 275: R1605–1610.

Tsukada, T., J.C. Rankin and Y. Takei. 2005. Involvement of drinking and intestinal sodium absorption in hyponatremic effect of atrial natriuretic peptide in seawater eels. Zoolog. Sci. 22: 77–85.

Uchiyama, M., S. Maejima, S. Yoshie, Y. Kubo, N. Konno and J.M. Joss. 2012. The epithelial sodium channel in the Australian lungfish, *Neoceratodus forsteri* (Osteichthyes: Dipnoi). Proc. Biol. Sci. (in press).

Venkatesh, B., E.F. Kirkness, Y.H. Loh, A.L. Halpern, A.P. Lee, J. Johnson, N. Dandona, L.D. Viswanathan, A. Tay, J.C. Venter, R.L. Strausberg and S. Brenner. 2007. Survey sequencing and comparative analysis of the elephant shark (*Callorhinchus milii*) genome. PLoS Biol. 5: e101.

Wang, T., A.L. Egbert, Jr., T. Abbiati, P.S. Aronson and G. Giebisch. 1996. Mechanisms of stimulation of proximal tubule chloride transport by formate and oxalate. Am. J. Physiol. 271: F446–450.

Wang, T., G. Giebisch and P.S. Aronson. 1992. Effects of formate and oxalate on volume absorption in rat proximal tubule. Am. J. Physiol. 263: F37–42.

Wang, Z., T. Wang, S. Petrovic, B. Tuo, B. Riederer, S. Barone, J.N. Lorenz, U. Seidler, P.S. Aronson and M. Soleimani. 2005. Renal and intestinal transport defects in Slc26a6-null mice. Am. J. Physiol. Cell Physiol. 288: C957–965.

Watanabe, T. and Y. Takei. 2011a. Environmental factors responsible for switching on the SO_4^{2-} excretory system in the kidney of seawater eels. Am. J. Physiol. Regul. Integr. Comp. Physiol. 301: R402–411.

Watanabe, T. and Y. Takei. 2011b. Molecular physiology and functional morphology of SO_4^{2-} excretion by the kidney of seawater-adapted eels. J. Exp. Biol. 214: 1783–1790.

Watanabe, T. and Y. Takei. 2012. Vigorous SO_4^{2-} influx via the gills is balanced by enhanced SO_4^{2-} excretion by the kidney in eels after seawater adaptation. J. Exp. Biol. 215: 1775–1781.

Wickramasinghe, S.N., S. Shiels and P.S. Wickramasinghe. 1994. Immunoreactive erythropoietin in teleosts, amphibians, reptiles, birds. Evidence that the teleost kidney is both an erythropoietic and erythropoietin-producing organ. Ann. N. Y. Acad. Sci. 718: 366–370.

Wingert, R.A., R. Selleck, J. Yu, H.D. Song, Z. Chen, A. Song, Y. Zhou, B. Thisse, C. Thisse, A.P. McMahon and A.J. Davidson. 2007. The cdx genes and retinoic acid control the positioning and segmentation of the zebrafish pronephros. PLoS Genet. 3: 1922–1938.

Xie, Q., R. Welch, A. Mercado, M.F. Romero and D.B. Mount. 2002. Molecular characterization of the murine Slc26a6 anion exchanger: functional comparison with Slc26a1. Am. J. Physiol. Renal Physiol. 283: F826–838.

Cell Volume Regulation in the Eel Intestine

Maria Giulia Lionetto,[a] Maria Elena Giordano[b] and Trifone Schettino[c],*

Introduction

Control of cell volume is a fundamental and highly conserved physiological mechanism, essential for survival of the cells under varying environmental and metabolic conditions (Wehner et al. 2003, Hoffmann et al. 2009, Pedersen et al. 2011).

In general, the plasma membrane of animal cells is highly permeable to water. Therefore, cell water content and cell volume are determined by the intracellular content of osmotic active compounds and by the extracellular tonicity (Hoffmann et al. 2009). Under isotonic steady-state conditions cells are constantly threatened by colloid osmotic cell swelling due to entrance of diffusible ions and water into the cell where impermeable polyvalent anionic macromolecules are present. Cell swelling and lysis are avoided because of the low Na^+ permeability of the plasma membrane and because of active Na^+ extrusion via the Na^+-K^+-ATPase pump which makes the plasma membrane effectively impermeable to Na^+. Under anisotonic condition

Dept. of Biological and Environmental Sciences and Technologies, University of Salento, Via prov.le Lecce-Monteroni-73100 Lecce (Italy).
[a]E-mail: giulia.lionetto@unisalento.it
[b]E-mail: elena.giordano@unisalento.it
[c]E-mail: trifone.schettino@unisalento.it
*Corresponding author

rapid osmotic water flux occurs across the plasma membrane also thanks to the presence of aquaporins which increase the water permeability of the membrane. If intracellular osmolarity exceeds extracellular osmolarity water enters into the cell and cell swelling occurs. Conversely, if extracellular osmolarity exceeds intracellular osmolarity water exits from the cells and cell shrinkage takes place.

Changes in cell volume can alter the hydration of cytosolic proteins which in turn is reflected in alteration of several cellular functions. Alterations of cell volume can also impair the integrity of the cell membrane and in turn can compromise the survival of the cell itself. Therefore, to survive the cells need to promptly counteract the changes in their volume.

The mechanisms for the control of cell volume after osmotic stress are highly conserved and principally similar in cells from various tissues as well as between evolutionary distant species (Lang et al. 1998). Following a hypotonic cell swelling, they consist in the release of osmolytes from the cell followed by loss of osmotically obliged water. This response is termed Regulatory Volume Decrease (RVD) (Hoffmann et al. 2009). Following hypertonic cell shrinkage the cell increases the intracellular osmolarity by intake of osmolytes, and hence influx of water. This response is termed Regulatory Volume Increase (RVI). The most rapid mechanism to change intracellular osmolarity and accomplish cell volume regulation is ion transport across cell membrane. In the RVD response, the cells release KCl by the activation of K^+ and Cl^- efflux through independent K^+ and anion channels (Hoffmann and Dunham 1995, Fürst et al. 2000, Hoffmann 2000, Niemeyer et al. 2000, Nilius et al. 2000, Valverde et al. 2000). Electroneutral K^+-Cl^- cotransporter is an alternative system contributing to RVD in some cell types (Lauf and Adragna 2000). In the RVI response, the cells generally initiate a net gain of ions by the activation of Na^+-K^+-$2Cl^-$ cotransport, Na^+/H^+ exchange, and/or nonselective cation channels (Hoffman et al. 2009).

In the slightly longer term, changes in non ionic osmolyte catabolism/metabolism occur contributing to osmoregulation. This is accounted by altered transcription of a number of osmoregulatory genes most of which are involved in the catabolism/metabolism of methylamines, polyalcohols, aminoacids and their derivatives (Hoffmann et al. 2009).

A wide array of factors can modify intra- or extra-cellular osmolarity. Changes in intracellular osmolarity can occur as a result of ion and nutrient transport across the plasma membrane or accumulation of metabolic waste products in the cell (Hoffman et al. 2009). Changes in extracellular osmolarity are less frequent under normal physiological conditions in vertebrates, since the osmolarity of the extracellular fluid is kept constant by body fluid homeostasis. Some examples of anisosmotic extracellular condition are provided in mammals by intestinal epithelial cells and blood cells in intestinal capillaries, which are exposed to low extracellular osmolarity after water or hypotonic food intake, or by kidney medullar cells and blood cells in kidney medulla capillaries which are exposed to very high extracellular osmolarity during antidiuresis (Hoffmann et al. 2009).

However, euryhaline fish such as the eel, that lives in fresh waters in the adult stage of its life cycle but migrates back to the ocean to breed, can experience dramatic changes in the osmolarity of the external environment from close to zero mOsm in fresh water to nearly 1100 mOsm in seawater. During the transfer between fresh water and sea water, transitory changes of plasma osmolarity occur before the acclimation process in the new osmotic environment is completed in a period of few weeks.

Most of the studies that are the basis of our present insight on cell volume regulation have been carried out on mammalian cell models. In comparison, relatively few studies have addressed volume regulation in non-mammalian cells (Chara et al. 2011).

In this review, we summarize the present insight about cell volume regulation in the European eel (*Anguilla anguilla*), highlighting findings on the intestinal epithelium, which serves as one of the main osmoregulatory organs in the fish. This tissue is a useful model system for functional studies of epithelia that perform near-isosmotic fluid absorption and has specifically been used as a physiological model for the study of cell volume regulation in epithelia (Lionetto et al. 2001, 2002, 2005, 2008).

The Eel as a Model for the Study of Cell Volume Regulation

When eels move from freshwater to seawater and *vice versa*, their cells are exposed to osmotic stress due to the dramatic change in plasma osmolarity. During the transfer from fresh water to seawater, plasma osmolarity increases by about 50% (Lionetto et al. 2001), mostly due to the loss of water from all the permeable body surfaces (Fig. 1).

This is slowly followed by a return to the normal value, with the fish still in seawater. Recovery is due to the action of osmoregulatory mechanisms in the acclimatization process, involving ion movements mainly at the gill and intestine level. 30 h after the transfer of eels from fresh water to sea water, an increase in salt excretion through the gills occurs, leadings to a decrease in plasma osmolarity. It takes several days before the intestine fully adapts to process sea water under cortisol stimulation, which has been demonstrated to increase the expression of the Na^+-K^+-ATPase and of the Na^+-K^+-$2Cl^-$ cotransporter (Ando and Hara 1994). The animal reaches a steady state with respect to Na^+ balance and drinking becomes coordinated with the absorptive capacity of the intestine, thereby allowing the quantity of intestinally absorbed salt to match the amount excreted through the gills. On the other hand, during the transition from seawater to fresh water, plasma osmolarity decreases by about 20% because of the osmotic intake of water

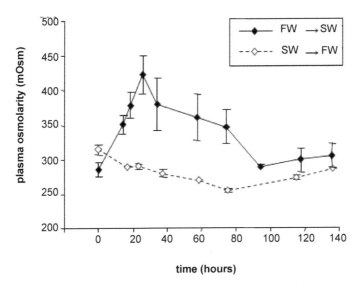

Figure 1. Plasma osmolarity changes in the European eel during the transfer from sea-water to fresh-water (SW→FW) and *vice versa* (FW→SW). Redrawn from Lionetto et al. (2001, 2005).

(Fig. 1) (Lionetto et al. 2005). However, over a period of 4–5 days it slowly tends to return to the initial level, thanks to the activation of osmoregulatory processes mainly at the gill and kidney level (Lionetto et al. 2005).

Therefore, when the fish transfers from fresh water to seawater and *vice versa*, cells must face dramatic changes in plasma osmolarity, which in turn are responsible for abrupt cell volume changes. This phenomenon gives rise to some interesting questions: what mechanisms enable the cells to overcome the high osmotic gradient during the initial phase of the transfer, before long term acclimatization can take place? How cells can counteract the cell volume changes that jeopardize the constancy of their intracellular milieu?

The ability of the eel to adapt to varying osmolarity of the external environment provides researches with good physiological models for studying cell volume regulation in anisosmotic conditions directly in native tissues.

Most of the studies on cell volume regulation in the eel have focused on intestine (Lionetto et al. 2001, 2002, 2005, 2008). The intestinal epithelium is one of the main interfaces of fish with the environment and, therefore, it is physiologically exposed to osmotic stress. When the eel moves from freshwater to seawater, the intestine is physiologically exposed to anisotonicity from either the basolateral side, because of the increase in the plasma osmolarity, or the luminal side, because of the ingested medium. In fact, the fish drinks seawater and although the medium is diluted and desalinated during its course from mouth to intestine, because of the uptake of ions at the oesophagus level (Nagashima and Ando 1994) it is still hypertonic to blood when it reaches the intestinal lumen (about 100 mOsm greater than plasma, whose osmolarity is about 290 mOsm) (Skadhauge 1967). When the eel moves from seawater to freshwater the intestine is exposed to hypotonicity on the basolateral side, because of the decrease in plasma osmolarity (about a 19% decrease) observed in the first hours after hypotonic exposure (Lionetto et al. 2005). Moreover, if one considers that the European eel does drink also in freshwater, although substantially less than in seawater (the drinking rate in freshwater is about 10% of that in sea water, as demonstrated by Tierney et al. (1995), it is reasonable to think that the intestinal epithelium is exposed to hypotonicity also from the apical side because of the ingested hypotonic fluid.

Cell Volume Regulation in Eel Intestine: RVI and RVD

The intestine is one of the main osmoregulatory organs of the eel. Thanks to its absorptive capacities it is widely utilized as model system for functional studies of epithelia that perform near-isosmotic fluid absorption.

The eel intestine operates a transepithelial Cl^- absorption. As shown in electrophysiological studies on Ussing chamber mounted intestine (Trischitta et al. 1992a, b, Marvão et al. 1994), Cl^- absorption occurs *via* bumetanide sensitive Na^+-K^+-$2Cl^-$ cotransport, localized on the luminal membrane (Fig. 2).

It operates in series with a basolateral Cl^- conductance and presumably a basolateral electroneutral KCl cotransport, and in parallel with a luminal K^+ conductance, which permits recycling of K^+ into the lumen and contributes to the operation of the luminal Na^+-K^+-$2Cl^-$ cotransporter. The basolaterally located Na^+–K^+–ATPase, by generating an inwardly directed electrochemical gradient for Na^+, provides the driving force for the active intracellular accumulation of Cl^- by the electroneutral luminal Na^+-K^+-$2Cl^-$ cotransporter (Fig. 2) (Ando and Utida 1986, Trischitta et al. 1992a, b, Marvão et al. 1994, Schettino and Lionetto 2003).

Cl^- absorption generates a transepithelial electrical potential of several millivolts, which is negative on the serosal side, and brings about a measurable short circuit current. Therefore, in the eel intestine the absorptive Na^+–K^+–$2Cl^-$ cotransporter can be functionally detected as bumetanide-sensitive short-circuit current or transepithelial potential; it represents about 90–95% of the overall short-circuit current of the epithelium. The ion transport model described for eel intestine, based on the operation of the luminal absorptive Na^+–K^+–$2Cl^-$ cotransporter, is basically the same as the model that has been proposed for the cortical thick ascending limb (cTAL)

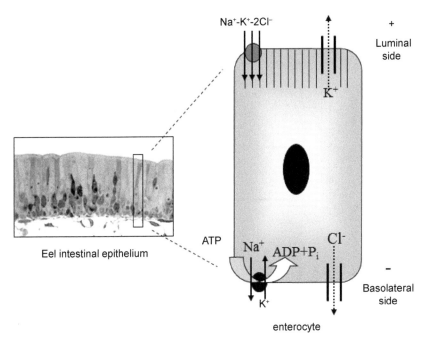

Figure 2. On the left: representative image of semithin (0.5 μm) sections of fixed eel intestine cut along planes perpendicular to the luminal epithelium surface. Sections were stained with 1% toluidine blue and observed by optical microscopy (100X magnification). On the right: model of ion transport mechanisms in eel enterocyte. Redrawn from Lionetto and Schettino 2006.

of the mammalian renal cortex (Greger 1985). Intestinal ion absorption is described in detail in Chapter 6 of this volume.

As previously demonstrated both by morphometrical analysis and electrophysiological measurements, the eel intestinal epithelium is sensitive to changes in the osmolarity of the external medium, either hypertonic (Lionetto et al. 2001) or hypotonic stress (Lionetto et al. 2005).

When the isolated eel intestine is experimentally exposed to hypertonic stress by 25% increase of the osmolarity of the perfusion solution by mannitol or sucrose addition, it initially shrinks, as indicated by the decrease in epithelium height (about 32% decrease after 5 min exposure), observed on fixed semithin (0.5 μm) sections of the intestine by optical microscopy (Fig. 3).

After about 45 min, the epithelium height increases again in the continued presence of the osmotic stress, suggesting that the eel intestine is able to perform a regulatory volume increase (RVI) following hypertonic shrinkage (Lionetto et al. 2001).

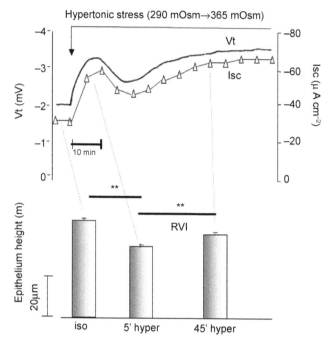

Figure 3. Cell volume changes and electrophysiological responses to hypertonic stress in the intestine of fresh-water acclimated eel. Cell volume changes were measured as epithelium height variation measured on semithin section (0.5 μm) of fixed intestine cut along planes perpendicular to the luminal epithelium surface. Vt: transepithelial potential (expressed as mV); Isc: short-circuit current (expressed as μA·cm²). Redrawn from Lionetto et al. 2001 and Lionetto and Schettino 2006.

When the isolated eel intestine is experimentally exposed to hypotonic stress (performed by reducing Ringer osmolarity by about 45% by decreasing the NaCl concentration) the epithelium swells as indicated by the increase in epithelium height after 5 min exposure (Fig. 4).

After about 45 min, the epithelium height starts decreasing again towards the initial value. These results indicate that the eel intestinal epithelium is able to regulate its volume after hypotonic swelling (Lionetto et al. 2005) through a RVD response. Hypertonic stress and hypotonic stress exposure experiments were performed on freshwater and seawater acclimated yellow eels, respectively (at least 15 days acclimation period) in order to experimentally reproduce the osmotic stress that the animals face in nature when they move from freshwater to seawater and *vice versa* (Lionetto et al. 2001, 2005). However, an RVI response was also observed for freshwater acclimated eels and RVD was observed for seawater acclimated eels (M.G. Lionetto, unpublished results) respectively, suggesting a remarkable 'osmotic plasticity' in these organisms.

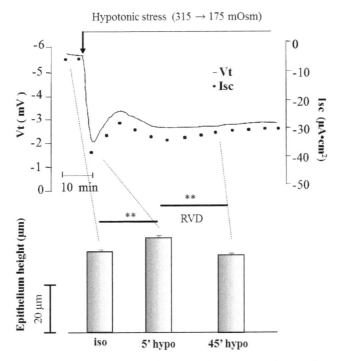

Figure 4. Cell volume changes and electrophysiological responses to hypotonic stress in the intestine of sea-water acclimated eel. Vt: transepithelial potential (expressed as mV); Isc: short-circuit current (expressed as $\mu A \cdot cm^2$). Details as Fig. 3. Redrawn from Lionetto et al. 2005 and Lionetto and Schettino 2006.

The volumetric responses of the epithelium to an osmotic challenge are paralleled by corresponding changes in the electrophysiological parameters (transepithelial potential, Vt, and short-circuit current, Isc,) (Lionetto et al. 2001, 2005) which reflect the ion transport mechanisms operating in the intestine. Thus, following hypertonic stress (Ringer osmolarity: 290 mOsm → 365 mOsm with mannitol), the Ussing chamber mounted epithelium shows a biphasic increase in Vt and Isc: an initial transient phase, which lasts about 10 min and a second sustained phase which reaches its plateau in about 30–45 min (Lionetto et al. 2001). Combining electrophysiological and morphometric analyses together, it is possible to observe that the initial transient phase correlates with the cellular shrinkage, while the second phase correlates with the RVI (Lionetto et al. 2001). On the other hand, following hypotonic stress (315 mOsm → 175 mOsm, by decreasing the NaCl content of the Ringer solution) the eel intestine shows a biphasic decrease in Vt and Isc (Fig. 3) (Lionetto et al. 2005). The first transient phase correlates with cell swelling measured by the morphometrical measurement of the epithelium height, while the second phase correlates with an RVD response.

Ion Transport Mechanisms Involved in Eel Intestine RVI

The use of specific ion transport inhibitors and ion substitution experiments allowed the clarification of the ionic nature of the RVI response in Ussing chamber mounted eel intestine. In the intestine, the first transient phase of the Vt and Isc response to hypertonic stress is completely inhibited by the serosal application of NPPB (5-nitro-2(3-phenyl-propylamino)-benzoate) or DPC (diphenylamine-2-carboxylate), known blockers of the basolateral Cl⁻ channels. Although the intracellular Cl⁻ concentration following hypertonic stress was not measured, these results could be tentatively attributed to Cl⁻ exit via basolateral channels resulting from the shrinkage induced increase in intracellular ion concentration (Lionetto et al. 2001). The second sustained phase is completely obliterated by 10 μM bumetanide (added in the luminal bathing solution), revealing that the second phase is due to the hypertonicity induced stimulation of the cotransporter following hypertonic shrinkage (Lionetto et al. 2001).

Luminal bumetanide was also able to completely abolish the epithelium RVI following hypertonic stress as demonstrated by morphometrical analysis, confirming that eel intestine Na^+-K^+-$2Cl^-$ transporter is not only a volume-sensitive transport mechanism, but its stimulation represents the physiological volume recovery mechanism after hypertonic shrinkage in this tissue. Furthermore, these data show that like the renal NKCC2 (Gamba 2005) the eel intestine luminal Na^+–K^+–$2Cl^-$ is active in basal condition and increases its activity following hypertonic stress.

Although the electrophysiological experiments with luminal bumetanide seem conclusive in identifying stimulation of the luminal Na^+–K^+–$2Cl^-$ cotransporter as the main mechanism responsible of RVI, the possible participation of Na^+/H^+ exchange was also tested in Ussing chamber mounted eel intestine. Na^+/H^+ exchange is known to be involved in hypertonic response in many cell types (Zadunaisky et al. 1995, Hoffmann et al. 2009), including gill pavement cells of Japanese eel (Tse et al. 2007). Functional Na^+/H^+ exchange has been previously demonstrated to occur only on the basolateral membrane of the eel enterocyte under isotonic conditions, but not on the apical membrane (Cassano et al. 1986, Vilella et al. 1995, Lionetto et al. 2001). The addition of 100 μM dimetylamiloride (DMA), a specific inhibitor of Na^+/H^+ exchange, to the serosal bathing solution does not alter the response of Isc to hypertonic challenge, hence excluding a role of the Na^+/H^+ exchanger in the RVI response in the eel intestine under the conditions studied.

The actin-based cytoskeleton has been proposed to play a key role in Na^+-K^+-$2Cl^-$ stimulation by cell shrinkage. The eel enterocytes exhibit a robust cortical F-actin skeleton, with especially prominent F-actin labelling in the apical brush border membrane, reflecting the presence of multiple microvilli in this region (Lionetto et al. 2002). Following hypertonic shrinkage, pre-treatment with 20 µM cytochalasin D or 0.5 µM latrunculin A, known to induce F-actin depolymerization by different molecular mechanisms, almost completely obliterates the bumetanide sensitive Vt and Isc response, suggesting that the Na^+-K^+-$2Cl^-$ cotransporter response to hypertonic stress is dependent on the integrity of the F-actin cytoskeleton (Lionetto et al. 2002). In several cell types the functional involvement of the cytoskeleton in response to osmotic stress has been shown to be associated with changes in the polymerization state of F-actin or with F-actin rearrangement (Pedersen et al. 2001). However, in eel intestine, no detectable alteration in F-actin organization after hypertonic stress exposure was observed by confocal imaging (Lionetto et al. 2002) and no significant change in F-actin content was measured by using a quantitative F-actin assay (Lionetto et al. 2002). Therefore, in eel intestine the role of actin cytoskeleton in the Na^+-K^+-$2Cl^-$ cotransporter stimulation following hypertonic stress requires the presence of an intact and organized microfilament system, but does not involve detectable changes in the polymerization state of F-actin or F-actin remodelling (Lionetto et al. 2002).

To further investigate the possible mechanism by which hypertonicity induced stimulation of Na^+-K^+-$2Cl^-$ is influenced by the cytoskeleton in eel intestine, Lionetto et al. (2002) investigated the possible involvement of protein phosphorylation processes known to interact with actin cytoskeleton and to be involved in the activation of volume-regulatory ion transport, on the bumetanide-sensitive Isc increase following hypertonic stress.

The Protein Kinase C (PKC) inhibitor chelerythrine completely abolished the bumetanide-sensitive Isc increase, suggesting that PKC is required for Na^+-K^+-$2Cl^-$ stimulation by hypertonicity in eel intestine (Lionetto et al. 2002). The Myosin Light Chain Kinase (MLCK) inhibitor ML-7 (20 µM) also completely blocked the bumetanide-sensitive Isc and Vt response to hypertonic stress in eel intestine in agreement with previous reports that ML-7 prevents the shrinkage induced activation of the Na^+-K^+-$2Cl^-$ cotransporter in bovine aortic endothelial cells (Klein and O'Neill 1995, O'Donnell et al. 1995). It is known that MLCK phosphorylates myosin light chain (MLC), allowing myosin II to interact with F-actin (Kohama et al. 1996). Given the above described important role of F-actin in Na^+-K^+-$2Cl^-$ regulation, in conjunction with the fact that MLC is the only known substrate of MLCK, and the known direct interaction of MLCK with actin (Hatch et al. 2001), the obtained results suggest a possible causal link between MLCK/MLC phosphorylation and the hypertonicity response. However, further studies are needed to clarify the precise roles of MLCK and MLC in the stimulation of Na^+-K^+-2Cl cotransporter in eel intestine.

Ion Transport Mechanisms Involved in Eel Intestine RVD

The ionic nature of the RVD response in eel intestine has been investigated by the use of specific ion transport inhibitors on Ussing chamber mounted intestine. As demonstrated by Lionetto et al. (2005) the decrease of Vt and Isc induced by the hypotonic stress reflects an electrogenic response of the tissue arising from the ion movements across the enterocyte cell membranes subsequent to their swelling-induced conductance changes. The response was reduced by the serosal application of iberiotoxin and apamin, specific inhibitors of high conductance (BK) and small conductance (SK2 and SK3) Ca^{2+} activated K^+ channels respectively, but was enhanced by the luminal application of iberiotoxin and slightly increased by luminal apamin. These results can be explained by the "well type" model of the transepithelial potential profile with the apical membrane potential in series to the basolateral one and Vt corresponding to their algebraic sum (Lionetto et al. 2005). According to this model, the K^+ efflux through the basolateral membrane will produce a depolarization of Vt (therefore its inhibition is associated with a decrease in the hypotonicity induced depolarization of Vt), while the K^+ efflux through the apical membrane is consistent with a hyperpolarization of Vt (therefore its inhibition is associated with an increase of the hypotonicity induced depolarization of Vt). Hence, the results are consistent with the hypotonicity induced activation of BK and SK Ca^{2+} activated K^+ channels on the apical and the basolateral membranes of the eel intestine.

Lionetto et al. (2008) demonstrated the molecular expression of the BK channel in the intestine of the European eel. Reverse transcription of mRNA extracted from eel intestinal mucosa cells and subsequent PCR amplification led to a sequence of 696 bp cDNA showing an 83% of similarity to human, mouse and rat BK channels. This sequence should correspond to a part of the pore-forming motif consisting of the P-loop, which bears the receptor for the binding of the pore blocker iberiotoxin. Functional studies on Ussing chamber mounted eel intestine indicate that BK channels are inactive in unstimulated intestinal cells but are activated when an increase in the free, intracellular Ca^{2+} concentration ($[Ca^{2+}]i$) occurs. In eel intestine Lionetto et al. (2005) and Trischitta et al. (2005) clearly demonstrated a strong dependence of the V_t and I_{sc} response to hypotonic stress on the extracellular Ca^{2+} concentration. Considering that BK channels are generally not regulated by cell volume changes *per se* (Stutzin and Hoffmann 2006, Hoffmann et al. 2009) it is, therefore, reasonable to assume that in eel intestine swelling activated influx of Ca^{2+} increases $[Ca^{2+}]i$, thereby activating BK channels on both membranes. This event in turn leads to the corrective K^+ efflux and the swelling induced hyperpolarization in either apical or basolateral membrane

On each membrane the K$^+$ efflux is accompanied by a parallel anion efflux through volume activated anion channel, as suggested by the inhibitory effect of 0.5 mM DIDS (applied either on the apical or basolateral membrane) on the electrophysiological response to hypotonic stress.

On the basis of these results Lionetto et al. (2005) concluded that in the eel intestinal epithelium, hypotonic stress activates separate K$^+$ and anion conductances on the basolateral and the apical membranes of the eel intestine. The electrogenic nature of the response can be explained by the K$^+$ and anion conductances increasing in a ratio different from 1:1 on the two membranes. Since the electrogenic response is a depolarization of Vt, the authors suggest that on the basolateral membrane the increase in the K$^+$ conductance exceeds the increase in the Cl$^-$ conductance, whereas on the apical membrane, the increase in the anion conductance exceeds the increase in the K$^+$ conductance. Together the various changes in conductances on the two membranes result in the measured decrease in the potential difference across the epithelium (normal polarity: serosal side negative) and thus also in the measured decrease in Isc. The typical time course of the electrogenic response could result from the sum of the different time-courses of activation of these separate conductive pathways on the two membranes.

Cell Volume Regulation and Apoptotic Volume Decrease in Eel Intestine

Eel enterocytes freshly isolated and maintained in suspension in a serum free medium undergo apoptosis following the detachment from the extracellular matrix. This phenomenon typical of epithelial and endothelial cells is known as anoikis (Meredith et al. 1993). Anoikis maintains homeostasis in tissues with high cyclic renewal rates (Fouquet et al. 2004). In the intestine, this process is thought to be involved in the normal homeostatic shedding of enterocytes at the tip of intestinal villi (Gordon and Hermiston 1994, Mayhew et al. 1999, Grossmann et al. 2001).

Suspended eel enterocytes showed a multistep decrease of their volume during the progression of anoikis: an early normotonic volume decrease, and a further cell volume decrease observed in permeable dead cells (Lionetto et al. 2010). While the last cell volume reduction can be attributed to cytoplasmic condensation due to blebbing of the plasma membrane and apoptotic body formation, the first cell isotonic volume reduction can be ascribed to Apoptotic Volume Decrease (AVD), an early and obligatory event in the apoptotic process. In isolated eel enterocytes, AVD was shown to precede key hallmarks of apoptosis such as externalization of membrane phosphatidylserine, suggesting that AVD is one of the earliest events in the anoikis program in eel enterocytes (Lionetto et al. 2010).

When the cells were resuspended in high K^+ saline solution or treated with iberiotoxin, the characteristic AVD was completely abolished, strongly suggesting that K^+ efflux through BK channels is a crucial event for the AVD process in eel enterocytes (Lionetto et al. 2010).

Involvement of K^+ efflux in apoptosis has been previously demonstrated in neurons (Nadeau et al. 2000, Pal et al. 2003), lymphocytes (Elliott and Higgins 2003), smooth muscle cells (Krick et al. 2001), cardiac cells (Ekhterae et al. 2003), Ehrlich ascites tumour cells (Poulsen et al. 2010), and various other cell lines (Yu 2003, Maeno et al. 2000), including IEC-6 cells, a line of cultured intestinal epithelial cells (Grishin et al. 2005).

Lionetto et al. (2010) demonstrated for the first time in native intestinal cells, by using suspended eel enterocytes, that the AVD process depends on BK channel activation and that the same type of K^+ channels responsible for cell volume regulation is also used for the AVD process. These results suggest that K^+ efflux through BK channels is a crucial event for the AVD activation and anoikis process in eel enterocytes. Although the signal transduction mechanisms underlying AVD activation in the eel intestine need to be further investigated, it is possible to argue that the signalling pathways responsible for AVD and RVD could share common downstream signals in BK channel activation. In turn, BK channel activation could represent a key event in the complex mechanisms of cell volume regulation and cell survival. This is of particular relevance for a tissue like the intestinal epithelium where cell volume regulation and apoptosis are processes central to the normal cellular physiology.

Conclusions

Cell volume regulation represents a key aspect of eel adaptative physiology contributing to the ability of the eel to cope with varying osmolarity of the external environment. This intriguing aspect of eel physiology provides researchers with good physiological models for studying cell volume regulation in anisosmotic conditions directly in native tissues. In particular the eel intestinal epithelium, because it is exposed to changes in extracellular osmolarity on both the apical and the basolateral side under physiological conditions, constitutes a good experimental model for cellular volume regulation, permitting the study of ion transport activation in response to osmotic stress directly in a native epithelium. Tissue responses can be studied by using freshly isolated epithelial cell sheets with intact tight junctions retaining the functional characteristics of the native cells in vivo. This opens interesting perspectives for the study of the signalling pathways underlying cell volume regulation processes in epithelia performing near-isosmotic fluid absorption such as the intestine.

Aknowledgement

The authors thank Prof. S.F. Pedersen for reviewing the chapter and providing very helpful comments.

References

Ando, M. and I. Hara. 1994. Alteration of sensitivity to various regulators in the intestine of the eel following seawater acclimation. Comp. Biochem. Physiol. (A) 109: 447–453.

Ando, M. and S. Utida. 1986. Effects of diuretics on sodium, potassium, chloride and water transport across the seawater eel intestine. Zool. Sci. 3: 605–612.

Cassano, G., S. Vilella, V. Zonno and C. Storelli. 1986. Na/H exchange activity is not present on the eel intestinal brush border membrane vesicles. *In*: F. Alvarado and C.H. Van Os [eds]. Ion Gradient-Coupled Transport. Elsevier Science Publisher. Amsterdam, Netherlands. pp. 282–286.

Chara, O., M.V. Espelt, G. Krumschnabel and P.J. Schwarzbaum. 2011. Regulatory volume decrease and P receptor signaling in fish cells: mechanisms, physiology, and modelling approaches. J. Exp. Zool. 315: 175–202.

Ekhterae, D., O. Platoshyn, S. Zhang, C.V. Remillard and J.X. Yuan. 2003. Apoptosis repressor with caspase domain inhibits cardiomyocyte apoptosis by reducing K^+ currents. Am. J. Physiol. 284: C1405– C1410.

Elliott, J.I and C.F. Higgins. 2003. IKCa1 activity is required for cell shrinkage, phosphatidylserine translocation and death in T lymphocyte apoptosis. EMBO 4: 189–194.

Fouquet, S., V.H. Lugo-Martinez, A.M. Faussat, F. Renaud, P. Cardot, J. Chambaz, M. Pincon-Raymond and S. Thenet. 2004. Early loss of e-cadherin from cell–cell contacts is involved in the onset of anoikis in enterocytes. J. Biol. Chem. 279: 43061–43069.

Fürst, J., M. Jakab, M. Köning, M. Ritter, M. Gschwentner, J. Rudzki, J. Danzl, M. Mayer, C.M. Burtscher, J. Schirmer, B. Maier, M. Nairz, S. Chwatal and M. Paulmichl. 2000. Structure and function of the ion channel ICln. Cell. Physiol. Biochem. 10: 329–334.

Gamba, G. 2005. Molecular physiology and pathophysiology of electroneutral cation-chloride cotrasporters. Physiol. Rev. 85: 423–493.

Gordon, J.I. and M.L. Hermiston. 1994. Differentiation and self-renewal in the mouse gastrointestinal epithelium. Curr. Opin. Cell. Bio. l6: 795–803.

Greger, R. 1985. Ion transport mechanisms in thick ascending limb of Henle's loop of mammalian nephron. Physiol. Rev. 65: 760–797.

Grishin, A., H. Ford, J. Wang, H. Li, V. Salvador-Recatala, E.S. Levitan and E. Zaks-Makhina. 2005. Attenuation of apoptosis in enterocytes by blockade of potassium channels. Am. J. Physiol. 289: G815–G821.

Grossmann, J., K. Walther, M. Artinger, S. Kiessling and J. Scholmerich. 2001. Apoptotic signaling during initiation of detachment-induced apoptosis ("anoikis") of primary human intestinal epithelial cells. Cell Growth Differ. 12: 147–155.

Hatch, V., G. Zhi, L. Smith, J.T. Stull, R. Craig and W. Lehman. 2001. Myosin light chain kinase binding to a unique site of F-actin revealed by three-dimensional image reconstruction. J. Cell. Biol. 154: 611–617.

Hoffmann, E.K. and P.B. Dunham. 1995. Membrane mechanisms and intracellular signaling in cell volume regulation. Int. Rev. Cytol. 161: 173–262.

Hoffmann, E.K. 2000. Intracellular signalling involved in volume regulatory decrease. Cell. Physiol. Biochem. 10: 273–288.

Hoffmann, E.K., I.H. Lambert and S.F. Pedersen. 2009. Physiology of cell volume regulation in vertebrates. Physiol. Rev. 89: 193–277.

Klein, J. and W. O'Neill. 1995. Volume-sensitive myosin phosphorylation in vascular endothelial cells: correlation with Na-K-2Cl cotransport. Am. J. Physiol. 269: C1524–C1531.

Kohama, K., L.H. Ye, K. Hayakawa and T. Okagaki. 1996. Myosin light chain kinase: an actin-binding protein that regulates an ATP-dependent interaction with myosin. Trends Pharmacol. Sci. 17: 284–287.

Krick, S., O. Platoshyn, M. Sweeney, H. Kim and J.X. Yuan. 2001. Activation of K^+ channels induces apoptosis in vascular smooth muscle cells. Am. J. Physiol. 280: C970–C979.

Lang, F., G.L. Busch and H. Volkl. 1998. The diversity of volume regulatory mechanisms. Cell. Physiol. Biochem. 8: 1–45.

Lauf, P.K. and N.C. Adragna. 2000. K^+-Cl^- cotransport: properties and molecular mechanims. Cell. Physiol. Biochem. 10: 341–354.

Lionetto, M.G., M.E. Giordano, G. Nicolardi and T. Schettino. 2001. Hypertonicity stimulates Cl^- transport in the intestine of freshwater acclimated eel, *Anguilla anguilla*. Cell. Physiol. Biochem. 11: 41–54.

Lionetto, M.G., S.F. Pedersen, E.K. Hoffmann, M.E. Giordano and T. Schettino. 2002. Roles of the cytoskeleton and of protein phosphorilation events in the osmotic stress response in eel intestinal epithelium. Cell. Physiol. Biochem. 12: 163–178.

Lionetto, M.G., M.E. Giordano, F. De Nuccio, G. Nicolardi, E.K. Hoffmann and T. Schettino. 2005. Hypotonicity induced K^+ and anion conductive pathways activation in eel intestinal epithelium. J. Exp. Biol. 208: 749–760.

Lionetto, M.G. and T. Schettino. 2006. The Na^+-K^+-$2Cl^-$ cotransporter and the osmotic stress response in a model salt transport epithelium. Acta Physiol. 187: 115–124.

Lionetto, M.G., A. Rizzello, M.E. Giordano, M. Maffia, F. De Nuccio, G. Nicolardi, E.K. Hoffman and T. Schettino. 2008. Molecular and functional expression of high conductance Ca^{2+} activated K^+ channels in the eel intestinal epithelium. Cell. Physiol. Biochem. 21: 361–37.

Lionetto, M.G., M.E. Giordano, A. Calisi, R. Caricato, E.K. Hoffman and T. Schettino. 2010. Role of BK channels in the Apoptotic Volume Decrease in native eel intestinal cells. Cell. Physiol. Biochem. 25: 733–744.

Maeno, E., Y. Ishizaki, T. Kanaseki, A. Hazama and Y. Okada. 2000. Normotonic cell shrinkage because of disordered volume regulation is an early prerequisite to apoptosis. Proc. Natl. Acad Sci. 97: 9487–9492.

Marvão, P., M.G. Emílio, K. Gil. Ferreira, P.L. Fernandes and H.G. Ferreira. 1994. Ion transport in the intestine of *Anguilla anguilla*: gradients and translocators. J. Exp. Biol. 193: 97–117.

Mayhew, T.M., R. Myklebust, A. Whybrow and R. Jenkins. 1999. Epithelial integrity, cell death and cell loss in mammalian small intestine. Histol. Histopathol. 14: 257–267.

Meredith, J. Jr., B. Fazeli and M.A. Schwartz. 1993. The extracellular matrix as a survival factor. Mol. Biol. Cell. 4: 953–961.

Nadeau, H., S. McKinney, D.J. Anderson and H.A. Lester. 2000. ROMK1 (Kir1.1) causes apoptosis and chronic silencing of hippocampal neurons. J. Neurophysiol. 84: 1062–1075.

Nagashima, K. and M. Ando. 1994. Characterization of esophageal desalination in the seawater eel *Anguilla japonica*. J. Comp. Physiol. B. 164: 47–54.

Niemeyer, M.I., C. Hougaard, E.K. Hoffmann, F. Jørgensen, A. Stutzin and F.V. Sepulveda. 2000. Characterisation of a cell swelling-activated K^+-selective conductance of Ehrlich mouse ascites tumour cells. J. Physiol. 524: 757–767.

Nilius, B., J. Eggermont and G. Droogmans. 2000. The endothelial volume-regulated anion channel, VRAC. Cell. Physiol. Biochem. 1: 313–320.

O'Donnell, M.E., A. Martinez and D. Sun. 1995. Endothelial Na^+-K^+-Cl^- cotransport regulation by tonicity and hormones: phosphorylation of cotransport protein. Am. J. Physiol. 269: C1513–C1523.

Pal, S., K.A. Hartnett, J.M. Nerbonne, E.S. Levitan and E. Aizenman. 2003. Mediation of neuronal apoptosis by Kv2.1-encoded potassium channels. J. Neurosi. 23: 4798–4802.

Pedersen, S.F., E.K. Hoffmann and J.W. Mills. 2001. The cytoskeleton and cell volume regulation. Comp. Biochem. Physiol. 130: 385–399.

Pedersen, S.F., A. Kapus and E.K. Hoffmann. 2011. Osmosensory Mechanisms in Cellular and Systemic Volume Regulation. J. Am. Soc. Nephrol. 22: 1587–1597.

Poulsen, K.A., E.C. Andersen, T.K. Klausen, C. Hougaard, I.H. Lambert and E.K. Hoffmann. 2010. Deregulation of Apoptotic Volume Decrease and ionic movements in Multidrug Resistant Tumor Cells: Role of the chloride permeability. Am. J. Physiol. 298: C14–25.

Schettino, T. and M.G. Lionetto. 2003. Cl⁻ absorption in European eel intestine and its regulation. J. Exp. Zool. 300A: 63–68.

Skadhauge, E. 1967. The mechanism of salt and water absorption in the intestine of the eel (*Anguilla anguilla*) adapted to waters of various salinities. J. Physiol. 204: 125–158.

Stutzin, A. and E.K. Hoffmann. 2006. Swelling activated ion channels: functional regulation in cell-swelling, proliferation and apoptosis. Acta Physiol. 187: 27–42.

Tierney, M.L., G. Luke, G. Cramb and N. Hazon. 1995. The role of the renin-angiotensin system in the control of blood pressure and drinking in the European eel, *Anguilla anguilla*. Gen. Comp. Endocrinol. 100: 39–48.

Trischitta, F., M.G. Denaro, C. Faggio and T. Schettino. 1992a. Comparison of Cl⁻ absorption in the intestine of the seawater and freshwater adapted eel, *Anguilla anguilla*: evidence for the presence of Na⁺-K⁺-2Cl⁻ cotransport system on the luminal membrane of the enterocyte. J. Exp. Zool. 263: 245–253.

Trischitta, F., M.G. Denaro, C. Faggio, T. Schettino. 1992b. An attempt to determine the mechanisms of Cl⁻-exit across the basolateral membrane of eel intetsine: use of different Cl⁻transport pathway inhibitors. J. Exp. Zool. 264: 11–18.

Trischitta, F., M.G. Denaro and C. Faggio. 2005. Cell volume regulation following hypotonic stress in the intestine of the eel, *Anguilla anguilla*, is Ca²⁺-dependent. Comp. Biochem. Physiol. B. 140: 359–367.

Tse, W.K.F., D.W.T. Au, C.K.C. Wong. 2007. Effect of osmotic shrinkage and hormones on the expression of NA⁺/H⁺ exchanger-1, Na⁺/K⁺/2Q⁻ cotransporter and Na⁺/K⁺-ATPase in gill pavement cells of fresh water adapted japanese eel, *Anguilla Japonice*. J. Exp. Biol. 210: 2113–2120.

Valverde, M.A., E. Vázquez, F.J. Muñoz, M. Nobles, S.J. Delaney, B.J. Wainwright, W.H. Colledge and D.N. Sheppard. 2000. Murine CFTR channel and its role in regulatory volume decrease of small intestine crypts. Cell. Physiol. Biochem. 10: 321–328.

Vilella, S., V. Zonno, M. Lapadula, T. Verri and C. Storelli. 1995. Characterization of plasma membrane Na⁺-H⁺ exchange in eel *Anguilla anguilla* intestinal epithelial cells. J. Exp. Zool. 271: 18–26.

Wehner, F., H. Olsen, H. Tinel, E. Kinne Saffran and R.K.H. Kinne. 2003. Cell volume regulation: osmolytes, osmolyte transport, and signal transduction. Rev. Physiol. Biochem. Pharmacol. 148: 1–80.

Yu, S.P. 2003. Regulation and critical role of potassium homeostasis in apoptosis. Prog. Neurobiol. 70: 363–386.

Zadunaisky, J.A., S. Cardona, L. Au, D.M. Roberts, E. Fisher, B. Lowestain, E.J. Cragoe and K.R Spring. 1995. Chloride transport activation by plasma osmolarity during rapid adaptation to high salinity of *Fundulus heteroclitus*. J. Membr. Biol. 143: 207–217.

The Immune System

Michael Engelbrecht Nielsen,[1,a,]* Joanna J. Miest[2] and
Jacob Günther Schmidt[1]

Introduction

The immune system is an intricate complex of interrelated cellular, molecular and genetic components that provide the host with a defence against foreign organisms such as bacteria, viruses, parasites and fungi or from other foreign substances, but also from native cells, e.g., cell fragments from dead cells.

The immune system can be divided into innate and adaptive branches. Innate (also called non-specific) immune responses can be found in all organisms down to even primitive multicellular ones. It is the innate immune system as well as the physical properties of skin and mucus that provides the first barrier for invading pathogens.

Due to the physical nature of water not only viruses and bacteria but also parasites inhabit this environment which is why fish essentially can be said to swim in a soup of potential pathogens. As protection against the threats from the surrounding environment fish have developed a strong first line defence from the live skin and the mucus layer covering it. This defence is

[1]The Technical University of Denmark, Biological Quality Group, Denmark.
[a]E-mail: mice@food.dtu.dk
[2]Keele University, School of Life Sciences, U.K.
*Corresponding author

not least important in intensive aquaculture where high stocking densities are common and can lead to not only horizontal transfer of pathogens, but also be stressful to the fish, which eventually can lead to immune suppression (Siwicki and Robak 2011) and thus a reduced protection and a greater susceptibility to infection.

Adaptive (also called acquired or specific) immunity, on the other hand, is the hallmark for vertebrates, and it arose at least twice during early vertebrate evolution. Two basic, convergent adaptive immune systems are recognized today; the type found in jawless fishes such as hagfish and lampreys where the variable domain consists of leucine-rich repeats the type found in jawed fishes which consist of immunoglobulin modules (Litman et al. 2010).

The eels are some of the most ancient teleost fishes, and eels thus hold an evolutionarily interesting position. At the same time, eels are highly esteemed eating fish, which are produced in large numbers. In spite of this there has been surprisingly little research into the anguillid immune system. Thus most of our "knowledge" on the immune system of eels is inferred from other fish species or from functional studies indicating the likely presence of certain molecules or molecular pathways. The eel immune system was reviewed in an paper by Nielsen and Esteve-Gassent (2006), and since then relatively little new information has been published on the subject. However, if the current attempts to close the Japanese and European eel life-cycles are fruitful and will result in propagation at production scale, investigations into eel immunology is likely to increase, to avoid disease in these valuable, selectively bred fish.

The Innate Immune System

Humoral Factors

Humoral factors are cellular secretions found in the body fluids such as blood and mucus. The humoral innate factors provide an important first barrier against pathogens. Among the humoral factors are also antibodies, which are discussed in the section on the adaptive immune system.

Acute Phase Proteins

Complement

Complement is a complex of factors that upon activation takes part in the cascade leading to the assembly of the so-called membrane attack complex (MAC). The MAC essentially creates pores in the membrane of cells, thereby

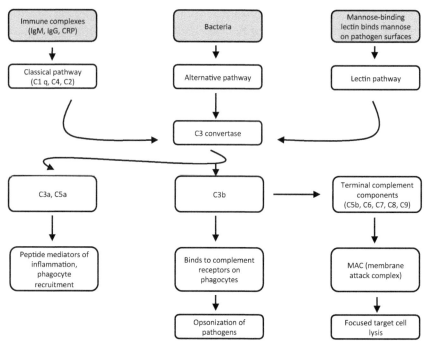

Figure 1. Overview of possible activation pathways of the complement system in eels, based on present knowledge from eels and other fish species.

upsetting the osmotic balance and ultimately leading to cell death. The complement system is an important defence mechanism against bacteria (Boshra et al. 2006). Cleavage products from this cascade additionally serve as opsonising and chemotactic factors for phagocytes (Dunkelberger and Song 2010).

The complement system can be activated in three ways; the classical complement pathway, the alternative complement pathway and the mannan-binding lectin (MBL) pathway. The classical complement pathway is initiated by antibody molecules and thus bridges the innate and the adaptive immune systems. Homologs to many of the factors involved in the well-described mammalian complement system are found in fish, and—as in mammals—complement is considered an important defence mechanism in this taxon (Nakao et al. 2011).

A way of studying the effect of complement—or rather the effect of the absence of it—is by heat inactivation, since some components of the complement pathway are heat labile. However, inactivation of eel complement cannot be obtained by the same methodology used for many other fish species as, e.g., rainbow trout. This may be explained by an evolutionary adaptation of the eel complement system to warm water, as the methodology used for other warm water species such as carp can be applied (Sakai 1981). Iida and Wakabayashi (1988) suggested complement C3 and C5 activities in eel through functional studies, since neutrophils showed reduced migration towards heat-treated eel serum compared to non-treated. Wang et al. (2007) found an activation effect of glycan immunomodulators on complement in Japanese eel serum. However, none of the involved factors have been described directly in the anguillids.

Lectins

Lectins are a group of proteins that bind to cell surface carbohydrates and play important roles in innate immunity. Several types of lectins have through the last few decades been described from eel serum (Baldus et al. 1996, Honda et al. 2000). The functionality of eel serum lectins is well known, but structures have only been determined for a few such as the fucolectins, which have a highly specific binding to only fucose. Eel lectin-producing cells have suggested to originate from liver and to be present in mucosal surfaces (Honda et al. 2000). The production of lectins by eel hepatocytes have been shown to increase by lipopolysaccharide (a structure found in gram-negative bacteria outer membrane, and which consists of lipids and saccharides such as fucose) stimulation, suggesting a role for fucolectins in host defense against bacteria. Another lectin, important for the activation of one of the complement pathways, is Mannan-binding lectin which has been described in *A. anguilla* as well as the conger, *Conger myriaster* (Gercken and Renwrantz 1994, Tsutsui et al. 2007a).

Anti-microbial Peptides and Proteins

Anti-microbial peptides (AMPs) are small, secreted proteins with bacteriostatic properties. They can exert their effect through various mechanisms such as pore formation or interfering with protein synthesis. AMPs are recognized as a critical first line of defence against intruding pathogens. AMPs have been found in different tissues as well as in numerous taxa including mammals, amphibians, fishes, and insects (Jenssen et al. 2006). AMPs have been described from mucus of several species of fish, including eel (Ebran et al. 1999, Ebran et al. 2000, Cho et al. 2002, Liang et al. 2011). Liang and co-workers (Liang et al. 2011) recently discovered a novel anti-microbial peptide (AJN-10) in the Japanese eel after intra-peritoneal injection with the important fish pathogen *Aeromonas hydrophila*. This AMP showed little similarity to other known proteins in a BLAST search.

In one study, bactericidal activity in mucus from elvers after vaccination did not correlate with lysozyme activity nor antibody levels (Esteve-Gassent et al. 2004b, Esteve-Gassent et al. 2004a), and the authors hypothesized that AMPs could account for the bactericidal activity. While this possibility cannot be excluded, a novel piscine antibody isotype called IgT/IgZ was discovered in trout and zebrafish (and subsequently several other teleost species) shortly after this paper was published (Hansen et al. 2005, Danilova et al. 2005), and could have accounted for the antibody response. Studies have since elucidated that this isotype is important at mucosal surfaces and is possibly more important than IgM during early ontogeny (Zhang et al. 2011).

A group of AMPs named piscidins have only been discovered in the most evolutionary advanced order of fish, the perciformes (Mulero et al. 2008) and have not been found in the anguilliformes.

Lysozyme

The name lysozyme was coined by Alexander Fleming. Fish lysozyme has a broader activity than mammalian lysozyme (Demers and Bayne 1997). Lysozyme lyses Gram-positive bacteria by cleaving the glycosidic bonds in the cell wall peptidoglycans. Gram-negative bacteria are not affected by lysozyme until other factors (such as complement) have exposed the inner peptidoglycan layer (Saurabh and Sahoo 2008). Lysozyme can also hydrolyse chitin and thus have a deleterious effect on fungi and some parasites (Saurabh and Sahoo 2008). The effect of lysozyme is not strictly lytic it can also promote phagocytosis. Lysozyme has been found in the eggs, mucus and blood of several fish species, but only indirectly so in European eels (Esteve-Gassent et al. 2004a, Esteve-Gassent et al. 2004b). Eel serum lysozyme levels are intermediate when compared to other fish species (Ren et al. 2007, Saurabh and Sahoo 2008).

Lysozyme levels were relatively unaffected by starvation during the experimental period (2 months) in an investigation by Caruso and co-workers (Caruso et al. 2010). Thus, starvation could seem to be a fairly natural and unstressful event in the eel.

Proteases

The cysteine proteinases cathepsin L and B have been described from Japanese eel epidermal secretory cells (Aranishi et al. 1998). These proteases have also been described from European and American eel where they accounted for bactericidal activity against several Gram-positive as well as Gram-negative eel pathogenic bacteria such as *Edwardsiella tarda*, *Vibrio anguillarum* and *Flavobacterium columnare* (Aranishi et al. 1998, Aranishi 1999b, Aranishi 1999a, Aranichi et al. 1999, Aranishi 2000).

Cellular Immunity

Phagocytic immune cells such as macrophages and neutrophils are relatively well described from fish, albeit once again only sporadically from eel. Neutrophils are the first of the two mentioned cell types to arrive at the scene of infection or injury, with the more efficiently phagocytic macrophages arriving in significant numbers usually within a few days. Phagocytes engulf pathogens and kill them in vacuoles into which lysozyme and/or reactive oxygen species are released. Phagocytes express receptors for conserved microbe-specific molecules and structures—the so-called pathogen-associated molecular patterns (PAMPs). The receptors are called pattern recognition receptors (PRRs). The PRRs include toll-like receptors (TLRs), mannose and C-type lectin receptors and NOD-like receptors (NLRs). Several members of these families have been discovered in fish (see Boltaña et al. 2011 for a review of teleost PRRs). Activation of PRRs can lead to the expression and/or activation of pro-inflammatory cytokines, phagocytosis and complement activation.

The Adaptive Immune System

The innate immune system is considered the most important branch of the piscine immune system. However, in an aquaculture setting certain pathogenic bacteria such as for example *Vibrio vulnificus* may become a serious threat and require vaccination of the fish for sufficient resistance. An adaptive immune system is a prerequisite for vaccination. Vaccination leads to the development of pathogen-specific antibodies as well as immunological memory so that a specific, secondary antibody response can be quickly triggered upon contact with the pathogen.

However, the adaptive immune response is not very efficient in fish as compared with mammals. Fish respond relatively poorly to immunization: affinity maturation is limited and the secondary response not very prominent (Kaattari et al. 2002, Pilström et al. 2005). The Atlantic cod even seems incapable of producing specific antibodies altogether (Dacanay et al. 2006, Star et al. 2011). In addition, Atlantic cod has very little lysozyme activity. However, the Atlantic cod has high serum titres of so-called natural antibodies, which can be considered in a way innate antibodies. However, due to the mainly tetrameric nature, IgM has many antigen binding sites, and thus can have a high avidity in spite of a low affinity for a given antigen. Antigen-bound IgM is a potent activator of complement through the classical pathway, and also functions as an opsonin for phagocytes.

In spite of a relatively poor piscine adaptive immune system, vaccination does improve resistance and is common practice in many aquaculture settings. A vaccine against *Vibrio vulnificus* serovar E, which is the cause of vibriosis, has been shown to induce protection in eels when administred by immersion, injection or orally (Fouz et al. 2001, Esteve-Gassent et al. 2003, Esteve-Gassent et al. 2004a, Esteve-Gassent et al. 2004b). Esteve-Gassent et al. (2003) found a faster IgM (this was before the discovery of IgT) antibody response to vaccination in the skin than in the blood, with specific antibody titres peaking at day four in skin mucus compared to day seven in blood. The authors also observed a stronger antibody response after challenge than after vaccination.

Edwardsiella tarda is a Gram-negative bacterium and is the cause of edwardsiellosis in a number of commercially important fish species including eel. Attempts to produce an efficient vaccine has been going on for years (Salati and Kusuda 1985, Gutierrez and Miyazaki 1994, Hossain et al. 2011) with varying results. Depending on the method of inactivation of the *E. tarda* used as bacterin in the vaccine survival rates one month post-vaccination varied between ~35% and ~85%. This correlated well with the measured specific antibody titres. Additional challenge experiments for up to six months post-vaccination showed a persisting protection from vaccine with pressure-killed and formalin-killed bacteria, but no lasting protection with electric current-killed bacteria (Hossain et al. 2011, Hossain and Kawai 2009).

Humoral Factors

Immunoglobulins

Immunoglobulins (antibodies) account for the humoral adaptive immune system. The basic design of immunoblobulins is similar across the vertebrate taxa, but there are isotypic differences. Humans have five immunoglobulin classes; IgA, IgD, IgE, IgG and IgM, which are further subdivided. In teleost fish three classes have been described; IgD, IgM and IgT. The three isotypes probably have very different effector functions, but the elucidations of these have only begun in recent years (Zhang et al. 2011). IgM is the most investigated piscine isotype, and was described almost half a century ago (Feinstein and Munn 1969, Acton et al. 1971). IgM is the most abundant immunoglobulin in the blood and IgM has been described from the European as well as the Japanese eel (Buchmann et al. 1992, Hung et al. 1996, Uchida et al. 2000, Feng et al. 2009).

In teleosts the anterior part of the kidney is a lymphoid organ which has been compared immunologically to the bone marrow of higher vertebrates (Zapata 1979). Feng et al. (2009) found that the kidney had the highest level of expression of IgM of all investigated organs. It is not known if eel possess all the three isotypes of antibodies found in other species (or even different isotypes altogether), but in most investigated teleost species all three isotypes have been discovered.

IgD was first reported in a teleost in 1997 (Wilson et al. 1997), but has not been subject of much investigation since then. However, there has been some research relating to IgD across the animal kingdom (Chen et al. 2009) and recently also work on fish IgD function has been published (Edholm et al. 2010, Edholm et al. 2011). In humans IgD is found to be co-expressed with IgM on B cells, and initial work on rainbow trout (*Oncorhynchus mykiss*) indicates a similar scenario in fish (Li et al. 2006). In the channel catfish (*Ictalurus punctatus*) IgM$^-$/IgD$^+$ subpopulations of B cells have since been found (Edholm et al. 2010) although for the present no reports relating to eels.

IgT was discovered almost simultaneously in zebrafish (published as IgZ) and rainbow trout in 2005 (Danilova et al. 2005, Hansen et al. 2005). Since then this isotype has been discovered in a number of fish species, although in the channel catfish it remains elusive (Bengtén et al. 2006, Zhang et al. 2011). It has not been reported from any anguilliform species. The function of IgT in rainbow trout has been under relatively intense investigations, and results so far show that IgT is an important antibody at mucosal surfaces (Salinas et al. 2011, Zhang et al. 2011), and thus seems to be functionally related to mammalian IgA and amphibian IgX. However, these teleost, amphibian and mammalian mucosal immunoglubolins arose independently during vertebrate evolution.

IgT could thus be speculated to be the most important isotype under normal conditions, with IgM being more important once the mucosal barrier has been breached. This could result from injuries from predators or parasites, and in relation to aquaculture abrasion or conspecific biting could be the cause.

Cellular Immunity

Antigen-presenting cells

Some of the phagocytic cells present peptides from the engulfed particles in MHC II molecules on their surface. These so-called professional antigen-presenting cells (APCs) include dendritic cells (which are very peripherally described in fish), macrophages and B cells. B cells were until recently not thought to be capable of phagocytosis, but this was demonstrated recently for rainbow trout and African clawed frog B cells (Li et al. 2006). The peptides are presented to T cells. These cell types or cells with similar effects are most likely to be found also in the anguillids.

Lymphocytes

The existence of functional B and T lymphocytes, and at least some subtypes of these, has been known for some time from functional studies. Fish have thus been shown to be capable of delayed-type hypersensitivity (Bartos and Sommer 1981), graft rejection and graft-versus-host reaction (Hildemann and Haas 1960, Qin et al. 2002, Nakanishi et al. 2011) and proliferation in a mixed leukocyte reaction (Etlinger et al. 1977, Kusuda and Xia 1992). Anti-IgM antibodies have been available for several fish species for some time, and thus the isolation and investigation of IgM$^+$ B cells has been possible. However, the tools to positively detect, isolate and study T lymphocytes are still very limited at the present time, although genomic and expressed sequence tag (EST) databases have confirmed the presence of key T cell molecules such as CD3, CD4, CD8, CD28, CTLA4 in several species, thus further indicating the presence of T cell subtypes such as cytotoxic T cells and T helper cells type 1, 2 and 17 (Castro et al. 2010). The lack of tools thus results in very rudimentary information of piscine T cells. Malaczewska and colleagues (Małaczewska et al. 2010) investigated the effect of the immunomodulator bovine lactoferrin (BL) on rainbow trout, wels catfish and European eel leukocytes. It was found that BL enhanced the proliferation of lymphocytes when Concanavalin A (a T-cell mitogen) was added. However, no additional verification of the nature of the proliferated cells could be provided.

In a development study of thymus in Japanese eel expression of lymphocyte-specific protein tyrosine kinase (*lck*) was found as early as 3 days post-fertilization (artificial fertilization). This is indicative of the presence of T cells as the *lck* is most commonly found in these cells where it interacts with the cytoplasmic domains of CD8 and CD4 (Kawabe et al. 2012). Contaminants in the aquatic environment also affect the cells of the immune system of the resident fish. Carlson and Zelikoff (2008) found that cadmium and mercury affected blood and kidney lymphocytes.

In mammals, natural killer (NK) cells are part of the lymphoid lineage. They can be considered intermediate between the adaptive and innate immune system and are important cells in controlling virus infection and tumour growth. In channel catfish, cells with more ore less similarity to mammalian NK cells have been described, and are described as NK-like and non-specific cytotoxic cells (NCCs) respectively (Shen et al. 2004, Yoder 2004).

In recent years several novel leukocyte receptor families have been discovered in fish and are likely to also be described in eels. These include the novel immunoglobulin-like transcripts (NILTs), leukocyte immune-type receptors (LITRs) and novel immune-type receptors (NITRs) (Yoder et al. 2002), reviewed by Montgomery et al. (2011) and Yoder and Litman (2011).

They are expressed on phagocytes and NCCs, however, the functional significance of these receptors are still being investigated.

Cytokines

Cytokines are small, secreted signalling molecules that orchestrate the immune response. Cytokines include chemokines, interleukins and interferons, which have a variety of functions such as maturation, activation and recruitment of leukocytes.

A large number of different cytokines have been discovered in fish during the last decade (see, e.g., Alejo and Tafalla 2011, Secombes et al. 2011). The first cytokine sequences within the order anguilliformes (transforming growth factor (TGF)-β1 and -β3) were reported in 1999 from *A. anguilla* (Laing et al. 1999) followed in 2007 by IL-1β in *Conger myriaster* (Tsutsui et al. 2007b). Since then IL-10 has been reported from *A. anguilla* (van Beurden et al. 2011). Cytokines evolve more rapidly than most other immune genes which makes it more difficult to identify them in nucleotide databases based on homology searches (Alejo and Tafalla 2011). More elaborate approaches such as synteny (genes are presumed to be arranged similar relative to each other in different species) may give better results (Bird et al. 2005). The large differences in cytokine structure even within the investigated fish species, makes extrapolation to cytokine function in the eel dubious in with the present methodology.

Immune Responses to Some Pathogens

Virus infection

Several viruses affecting eel have been described. These include EV2 virus (causing Cauliflower disease), eel virus American (EVA), eel virus European (EVE), eel virus European X (EVEX) and Japanese eel endothelial cell-infecting virus (JEECV) (van Beurden et al. 2012, van Beurden et al. 2011, van Ginneken et al. 2005, Mizutani et al. 2011). But probably *Herpesvirus anguillae* (HVA) has received the most attention. *Herpesvirus anguillae* can cause mortality in farmed and in wild populations of both European and Japanese eels (van Beurden et al. 2010), and is considered a significant viral threat to the eel. HVA has been detected in various regions including Europe and Asia (Davidse et al. 1999, Chang et al. 2002, Haenen et al. 2002, Jakob et al. 2009, Kim et al. 2012) and HVA all stages of eels (Lee et al. 1999, Hangalapura et al. 2007). The clinical signs of HVA are a loss of appetite and dermal haemorrhagic lesions, which affect mainly the pectoral fin, the opercular regions and the abdominal body surface. However the major pathological changes occur in the gills, where extensive necrosis occurs in the connective tissue of the gill filaments. Necrotic regions are highly infiltrated by erythrocytes and inflammatory cells as macrophages, lymphocytes and neutrophils. Internally HVA causes swelling of the kidney and the epithelial cells of renal tubules probably as an immunological reaction. The liver can be congested and hepatocytes are swollen and display atrophy (Lee et al. 1999). A common characteristic of herpes viruses is the development of latency in the host and reactivation of the virus due to stress-inducing factors, which reduce immune competence. Stress induced outbreak of HVA have been observed in European eels possibly caused by a compromised immune competence (Nieuwstadt et al. 2001). Polymerase chain reaction (PCR) detection methods have been developed to detect HVA in asymptomatic eels (Shih 2004 and Rijsewijk et al. 2005).

The immunological response to HVA has been very poorly described, but anti-viral measures in other fish species usually start with an innate response induced through the group of cytokines called interferons. These in turn activate a plethora of anti-viral genes (Verrier et al. 2011, Workenhe et al. 2010). Although not described from eel, the ubiquitous presence of interferon genes in other fish species indicates that these are probably also present in eels. Later in the immune response cytotoxic T cells and NK cells (or NCCs) are important for clearing of virus-infected cells (Verrier et al. 2011, Nakanishi et al. 2011). Virus-specific antibodies are also produced by the fish host, and several anti-viral vaccines have been developed for fish (Verrier et al. 2011, Lorenzen and La Patra 2005). In Channel catfish with latent *Ictalurid herpesvirus 1* (CCV), recrudescence was induced with the immunosuppressant dexamethasone and an increased serum CCV-neutralizing activity was found after administration of dexamethasone (Arnizaut and Hanson 2011).

Bacterial infections

Bacterial infections elicit both innate and adaptive reactions. Esteve-Gassent et al. (2004c) described mucosal antibody reactions against *Vibrio vulnificus* indicating a specific first line of defence in the mucosal surface of European eels. *Aeromonas* spp. is gram-negative bacteria that infect a wide variety of fish including eels. *Aeromonas hydrophila* is considered the most common disease agent in all fresh and brackish water environments (Nielsen et al. 2001) and are associated with skin ulcers as well as reddening of fins and necrosis of the tail in eels (Esteve and Alcaide 2009). These bacteria have been linked to the tail and fin rod, haemorrhagic septicaemia and epizootic ulcerative syndrome (EUS) (Nielsen et al. 2001). The severity of the disease depends on the bacterial virulence as well as stress level and physiological condition of the host, but also on the degree of genetic resistance inherent within specific populations of fish (Cipriano et al. 2001). Several bacterial diseases causes inflammation at the site of infection and hence initiate a specific humoral response involving antibodies (Nielsen and Esteve-Gassent 2006).

Parasite infections

Several parasites including *Ichtyophthirius multifiliis, Pseudodactylogyrus anguillae* and *P. bini* (Mellergaard and Dalsgaard 1987) have been associated with immune reactions in eel. A well-described parasite in relation to immune reactions is *Anguillicola crassus* a nematode that invades the swim bladder of eels. It originates from East Asia with *Anguilla japonica* as its natural host. In the early 1980s it arrived with imported eels in Europe, where it successfully colonized most European countries. It is now known to occur in Asia, Europe, Africa and America (Kirk 2003) and is mainly found in freshwater but also in brackish water. In the natural host *A. japonica* the nematode normally does not cause serious pathological damage and parasitemia is low. However, the European host *A. anguilla* is much more susceptible and the parasite causes severe pathology (Nielsen 1997). This increased pathogenicity is probably due to a lack of adapted resistance, which is normally acquired after a long host-parasite co-evolution (Nielsen 1999). The main pathological effect that is observed during infestation of *A. crassus* is fibrosis in the swim bladder, which leads to a thickened and opaque appearance of this organ (Kirk 2003). A high parasite load can lead to an enlargement of the swim bladder, which leads to a swollen appearance of the abdomen of the eel. Additionally the pneumatic duct becomes dilated and inflamed if it is blocked by the nematodes. The lumen of the swim bladder is filled with adult nematodes, decaying worms, eggs and juveniles. In the most severe cases the swim bladder lumen collapses and forms a fibrotic complex with the other organs. Except for these extreme levels of infection parasites alone probably do not cause mortality however due to its negative effect on the condition of the fish it can lead to host death if other stress factors, such as high temperatures, decreased oxygen levels and handling, or secondary bacterial infections occur in addition to the infection (Kirk 2003).

In European eels infested with intestinal digenean parasites, rodlet cells and mucus cells accumulated at the site of infection (Dezfuli et al. 2009) and mucus secretion was enhanced and parasites covered with mucus. The same pattern has been observed in salmonids (Nielsen et al. 2004, Reite 1997). The main function of the rodlet cell has thus been speculated to be in the fight against parasites. Additionally, the rodlet cell shares features of the eosinophilic granulocyte, which in mammals, is anti-parasitic. The rodlet cell is a relatively poorly described fish-specific cell type. It seems absent in some fish, but on the other hand it appears to be a highly inducible cell type with only few rodlet cells present under sterile laboratory conditions (Reite 2005, Manera and Dezfuli 2004).

Dezfuli et al. (2009) also found granulomas consisting of large numbers of fibroblasts and mast cells around encysted *Contracaecum rudolphii* A larvae. In fish, as in mammals, the mast cell seems to be involved in anti-parasitic defence—either directly or indirectly through the recruitment and activation of neutrophils (Reite and Evensen 2006). However, granulocytes seem quite heterogenous across different vertebrate taxa and the terminology is not firmly established. In mammals mast cells are armed with immunoglobulin E, and upon cross-linking of several of these receptor-bound IgE molecules, the mast cells discharges granules of histamine and other mediators. Fish do not have IgE, although one of the other isotypes could be speculated to function similarly. In fact, channel catfish have been shown to have cells armed with IgD and IgM via Fc receptors (FcRs) and that these cells degranulate upon cross-linking of the Igs (Shen et al. 2004, Chen et al. 2009). Similarly, Chen et al. (2009) reported on IgD-armed basophils in humans, and that these release cytokines such as IL-4 and IL-13 which in turn can induce IgE secretion from B cells. Likewise histamine was only recently found in fish, and only in the evolutionarily advanced perciform fishes (Mulero et al. 2007), not the European eel, which was investigated in the same study. Nevertheless, parasitic defences in fish do not seem to be relying on antibodies to a very great extent (Alvarez-Pellitero 2008), and no vaccines against piscine parasitosis are available.

In summary, present knowledge on the eel immune system is surprisingly limited considering the economic importance of eels. Partly this is probably explained by the (as yet) dependence on wild-caught glass eels in anguilliculture. However, investigations have provided few surprises and the eel immune system probably functions much like other related teleosts.

Summary

The fish immune system comprises the innate and the adaptive immune system, whereby the innate system is the most important branch of the immune defence. This part of the immune system consists of humoral factors such as the complement system, lectins, lysozyme and anti-microbial peptides and proteins. In addition cells are involved in this first barrier against pathogens, these cells such as macrophages and neutrophils often have phagocytic activity. The recognise pathogen associated molecular patterns with specific receptors, phagocytose the pathogens and kills them.

The other arm of the immune system, the adaptive immune response, is less efficient compared to mammals but vaccination can improve resistance. Eels for example can be vaccinated against vibriosis. Humoral factors such as antibodies/immunoglobulins have been described in fish but besides IgM it is not known, which factors exist in eel. The cellular part of the specific immune response consist of antigen-presenting cells such as dendritic cells, macrophages, and B cells that present engulfed pathogen particles on MHC II peptides and present these to T cells. Signalling molecules, i.e. cytokines such as chemokines, interleukins and interferons, orchestrate the immune response and lead to the maturation, activation and recruitmend of leucocytes. The cytokines TGF-β1, -β3 and IL-1β have been identified in eels.

Various viral and bacterial diseases and parasites can infect eels and thus trigger the immune response.

Acknowledgement

The authors would like to thank senior scientist Dr. Thomas Lindenstrøm for stimulating suggestions, knowledge, and experience in reviewing of this chapter.

References

Acton, R.T., P.F. Weinheimer, S.J. Hall, W. Niedermeier, E. Shelton and J.C. Bennett. 1971. Tetrameric immune macroglobulins in three orders of bony fishes. PNAS 68(1): 107–11.

Alejo, A. and C. Tafalla. 2011. Chemokines in teleost fish species. Dev. Comp. Immunol. 35(12): 1215–22.

Alvarez-Pellitero, P. 2008. Fish immunity and parasite infections: from innate immunity to immunoprophylactic prospects. Vet. Immunol. Immunop. 126 (3-4): 171–198.

Aranichi, F., N. Mano, M. Nakane and H. Hirose. 1999. Effects of thermal stress on skin defence lysins of European eel, *Anguilla anguilla* L. J. Fish Dis. 22: 227–229.

Aranishi, F. 1999a. Lysis of pathogenic bacteria by epidermal cathepsins L and B in the Japanese eel. Fish Physiol. Biochem. 20: 37–41.

Aranishi, F. 1999b. Possible role for cathepsins B and L in bacteriolysis by Japanese eel skin. Fish Shellfish Immunol. 8: 61–64.

Aranishi, F., N. Mano, M. Nakane and H. Hirose. 1998. Epidermal response of the Japanese eel to environmental stress. Fish Physiol. Biochem. 19: 197–203.

Aranishi, F. 2000. High sensitivity of skin cathepsins L and B of European eel (*Anguilla anguilla*) to thermal stress. Aquaculture 182: 209–213.

Arnizaut, A.B. and L.A. Hanson. 2011. Antibody response of channel catfish after channel catfish virus infection and following dexamethasone treatment. Dis. Aquat. Org. 95: 189–201.

Baldus, S.E., J. Theiele, Y.-O. Park, F.-G. Hanisch, J. Bara and R. Fischer. 1996. Characterization of the binding specificity of *Anguilla anguilla* agglutinin (AAA) in comparison to Ulex europeus agglutinin I (UEA-I). *Glycoconjugate J.* 13: 585–590.

Bartos, J.M. and C.V. Sommer. 1981. *In vivo* cell mediated immune response to *M. tuberculosis* and *M. salmoniphilum* in rainbow trout (*Salmo gairdneri*), Dev. Comp. Immunol. 5(1): 75–83.

Bengtén, E., L.W. Clem, N.W. Miller, G.W. Warr and M. Wilson. 2006. Channel catfish immunoglobulins: repertoire and expression. Dev. Comp. Immunol. 30(1-2): 77–92.

Bird, S., J. Zou, T. Kono, M. Sakai, J.M. Dijkstra and C. Secombes. 2005. Characterisation and expression analysis of interleukin 2 (IL-2) and IL-21 homologues in the Japanese pufferfish, *Fugu rubripes*, following their discovery by synteny. Immunogenetics 56(12): 909–23.

Boltaña, S., N. Roher, F.W. Goetz and S.A. Mackenzie. 2011. PAMPs, PRRs and the genomics of gram negative bacterial recognition in fish. Dev. Comp. Immunol. 35(12): 1195–203.

Boshra, H., J. Li and J.O. Sunyer. 2006. Recent advances on the complement system of teleost fish. Fish Shellfish Immunol. 20(2): 239–62.

Buchmann, K., L. Ostergaard and J. Glamann. 1992. Affinity purification of antigen-specific serum immunoglobulin from the European eel (*Anguilla anguilla*). Scand. J. Immunol. 36(1): 89–97.

Carlson, E. and J.T. Zelikoff. 2008. The immune system of fish: A target organ of toxicity. *In*: R.T. Di Giulio and D.E. Hinton [eds]. The Toxicology of Fishes. CRC Press. United Kingdom. pp. 489–529.

Caruso, G., G. Maricchiolo, V. Micale, L. Genovese, R. Caruso and M.G. Denaro. 2010. Physiological responses to starvation in the European eel (*Anguilla anguilla*): effects on haematological, biochemical, non-specific immune parameters and skin structures. Fish Physiol. Biochem. 36(1): 71–83.

Castro, R., D. Bernard, M.P. Lefranc, A. Six, A. Benmansour and P. Boudinot. 2010. T cell diversity and TcR repertoires in teleost fish. Fish Shellfish Immunol. 31(5): 644–654.

Chang, P.H., Y.H. Pan, C.M. Wu, S.T. Kuo and H.Y. Chung. 2002. Isolation and molecular characterization of herpes-virus from cultured European eels *Anguilla anguilla* in Taiwan. Dis. Aquat. Org. 50(2): 111–118.

Chen, K., W. Xu, M. Wilson, B. He, N.W. Miller, E. Bengtén, E.-S. Edholm, P.A. Santini, P. Rath, A. Chiu, M. Cattalini, J. Litzman, J.B. Bussel, B. Huang, A. Meini, K. Riesbeck, C. Cunningham-Rundles, A. Plebani and A. Cerutti. 2009. Immunoglobulin D enhances immune surveillance by activating antimicrobial, proinflammatory and B cell-stimulating programs in basophils. Nat. Immunol. 10(8): 889–98.

Cho, J.H., I.Y. Park, H.S. Kim, W.T. Lee, M.S. Kim and S.C. Kim. 2002. Cathepsin D produces antimicrobial peptide parasin I from histone H2A in the skin mucosa of fish. FASEB Journal 16: 429–431.

Cipriano, R.C., G.L. Bullock and S.W. Pyle. 2001. Aeromonas hydrophila and motile aeromonad septicemias of fish. Fish disease leaflet 68.

Dacanay, A., B.E. Bentley, L.L. Brown, A.J. Roberts and S.C. Johnson. 2006. Unique multimeric immunoglobulin crosslinking in four species from the family Gadidiae. Fish Shellfish Immunol. 21(2): 215–9.

Danilova, N., J. Bussmann, K. Jekosch and L.A. Steiner. 2005. The immunoglobulin heavy-chain locus in zebrafish: identification and expression of a previously unknown isotype, immunoglobulin Z. Nat. Immunol. 6(3): 295–302.

Davidse, A., O.L.M. Haenen, S.G. Dijkstra, A.P. Van Nieuwstadt, T.J.K. Van der Vorst, F. Wagenaar and G.J. Wellenberg. 1999. First isolation of herpesvirus of eel (*Herpesvirus anguillae*) in diseased European eel (*Anguilla anguilla* L.) in Europe. Bull. Eur. Assn. Fish P. 19(4): 137–141.

Demers, N.E. and C.J. Bayne. 1997. The Immediate Effects of Stress on Hormones and Plasma Lysozyme in Rainbow Trout. Dev. Comp. Immunol. 21 (4): 363–373.

Dezfuli, B.S., C. Szekely, G. Giovinazzo, K. Hills and L. Giari. 2009. Inflammatory response to parasitic helminths in the digestive tract of *Anguilla anguilla* (L.). Aquaculture 296(1-2): 1–6.

Dunkelberger, Jason R. and Wen-Chao Song. 2010. Complement and its role in innate and adaptive immune responses. Cell research 20 (1) (January): 34–50. doi:10.1038/cr.2009.139.

Ebran, N., S. Julien, N. Orange, B. Auperin and G. Molle. 2000. Isolation and characterization of novel glycoproteins from fish epidermal mucus: correlation between their pore-forming properties and their antibacterial activities. Biochim. Biophys. Acta 1467(2): 271–80.

Ebran, N., S. Julien, N. Orange, P. Saglio, C. Lemaître and G. Molle. 1999. Pore-forming properties and antibacterial activity of proteins extracted from epidermal mucus of fish. Comparative biochemistry and physiology. Comp. Biochem. Physiol., Part A: Mol. Integr. Physiol. 122(2): 181–9.

Edholm, E.-S., E. Bengten and M. Wilson. 2011. Insights into the function of IgD. Dev. Comp. Immunol. 35(12): 1309–16.

Edholm, E.-S., E. Bengtén, J.L. Stafford, M. Sahoo, E.B. Taylor, N.W. Miller and M. Wilson. 2010. Identification of two IgD+ B cell populations in channel catfish, *Ictalurus punctatus*. J. immunol. 185(7): 4082–94.

Esteve, C. and E. Alcaide. 2009. Influence of diseases on the wildeel stock: the case of Albufera Lake. Aquaculture 289: 143–149.

Esteve-Gassent, M.D., B. Fouz and C. Amaro. 2004a. Efficacy of a bivalent vaccine against eel diseases caused by *Vibrio vulnificus* after its administration by four different routes. Fish Shellfish Immunol. 16(2): 93–105.

Esteve-Gassent, M.D., R. Barrera and C. Amaro. 2004b. Vaccination of market-size eels against vibriosis due to *Vibrio vulnificus* serovar E. Aquaculture 241(1-4): 9–19.

Esteve-Gassent, M.D., B. Fouz, R. Barrera and C. Amaro. 2004c. Efficacy of oral reimmunisation after immersion vaccination against *Vibrio vulnificus* in farmed European eels. Aquaculture 231(1-4): 9–22.

Esteve-Gassent, M.D., M.E. Nielsen and C. Amaro. 2003. The kinetics of antibody production in mucus and serum of European eel (*Anguilla anguilla* L.) after vaccination against *Vibrio vulnificus*: development of a new method for antibody quantification in skin mucus. Fish Shellfish Immunol. 15(1): 51–61.

Etlinger, H.M., H.O. Hodgins and J.M. Chiller. 1977. Evolution of the lymphoid system. II. Evidence for immunoglobulin determinants on all rainbow trout lymphocytes and demonstration of mixed leukocyte reaction. Eur. J. Immunol. 7(12): 881–887.

Feinstein, A. and E.A. Munn. 1969. Conformation of the free and antigen-bound IgM antibody molecules. Nature 224(5226): 1307–1309.

Feng, J., R. Guan, P. Lin and S. Guo. 2009. Molecular cloning and characterization analysis of immunoglobulin M heavy chain gene in European eel (*Anguilla anguilla*). Vet. Immunol. Immunop. 127(1-2): 144–7.

Fouz, B., M.D. Esteve-Gassent, R. Barrera, J.L. Larsen, M.E. Nielsen and C. Amaro 2001. Field testing of a vaccine against eel diseases caused by *Vibrio vulnificus*. Dis. Aquat. Org. 45(3): 183–9.

Gercken, J. and L. Renwrantz. 1994. A new mannan-binding lectin from the serum of the eel (*Anguilla anguilla* L.): isolation, characterization and comparison with the fucose-specific serum lectin. Comp.Biochem. Physiol., Part B: Biochem. Mol. Biol. 108(4): 449.

Gutierrez, M.A. and T. Miyazaki. 1994. Responses of Japanese eels to oral challenge with *Edwardsiella tarda* after vaccination with formalin-killed cells or lipopolysaccharide of the bacterium. J. Aquat. Anim. Health 6(2): 110–117.

Haenen, O., S. Dijkstra, P. Van Tulden, A. Davidse, A. Van Nieuwstadt, F. Wagenaar and G. Wellenberg. 2002. *Herpesvirus anguillae* (HVA) isolations from disease outbreaks in cultured European eel, *Anguilla anguilla* in The Netherlands since 1996. Bull. Eur. Assn. Fish. P22(4): 247–257.

Hangalapura, B.N., R. Zwart, M.Y. Engelsma and O.L.M. Haenen. 2007. Pathogenesis of *Herpesvirusanguillae* (HVA) in juvenile European eel *Anguilla anguilla* after infection by bath immersion. Dis. Aquat. Org. 78: 13–22.

Hansen, J.D, E.D. Landis and R.B. Phillips. 2005. Discovery of a unique Ig heavy-chain isotype (IgT) in rainbow trout: Implications for a distinctive B cell developmental pathway in teleost fish. PNAS 102(19): 6919–24.

Hildemann, W.H. and R. Haas. 1960. Comparative studies of homotransplantation in fishes. J. Cell Compar. Physl. 55: 227–33.

Honda, S., M. Kashiwagi, K. Miyamoto, Y. Takei and S. Hirose. 2000. Multiplicity, structures, and endocrine and exocrine natures of eel fucose-binding lectins. J. Biol. Chem. 275(42): 33151–33157.

Hossain, M.M. and K. Kawai. 2009. Stability of Effective *Edwardsiella tarda* vaccine developed for Japanese eel (*Anguilla japonica*). Journal of Fisheries and Aquatic Science 4(6): 295–305.

Hossain, M.M., K. Kawai and S. Oshima. 2011. Immunogenicity of pressure inactivated *Edwardsiella tarda* bacterin to *Anguilla japonica* (Japanese Eel). Pak. J. Biol. Sci. 14(15): 755–767.

Hung, H.-W., C.-F. Lo, C.-C. Tseng and G.-H. Kou. 1996. Humoral immune response of Japanese eel, *Anguilla japonica* Temminck & Schlegel, to *Pleistophora anguillarum* Hoshina, 1951 (Microspora). J. Fish Dis. 19(3): 243–250.

Iida, T. and H. Wakabayashi. 1988. Chemotactic and leucocytosis-induced activities of eel complement. Fish Pathol. 23(1): 55–58.

Jakob, E., H. Neuhaus, D. Steinhagen, B. Luckhardt and R. Hanel. 2009. Monitoring of *Herpesvirus anguillae* (HVA) infections in European eel, *Anguilla anguilla* (L.), in northern Germany. Journal of Fish Diseases 32(6): 557–561.

Jenssen, H., P. Hamill and R.E.W. Hancock. 2006. Peptide antimicrobial agents. Clin. Microbiol. Rev. 19(3): 491–511.

Kaattari, S.L., H.L. Zhang, I.W. Khor, I.M. Kaattari and D.A. Shapiro. 2002. Affinity maturation in trout: clonal dominance of high affinity antibodies late in the immune response. Dev. Comp. Immunol. 26(2): 191–200.

Kawabe, M., H. Suetake, K. Kikuchi and Y. Suzuki. 2012. Early T-cell and thymus development in Japanese eel *Anguilla japonica*. Fisheries Sci. 78(3): 539–547.

Kim, H., J.H. Yu, D.W. Kim, S. Kwon and S.W. Park. 2012. Molecular evidence of anguillid herpesvirus-1 (AngHV-1) in the farmed Japanese eel, *Anguilla japonica* Temminck & Schlegel, in Korea. J. Fish Dis. 35(4): 315–319.

Kirk, R.S. 2003. The impact of *Anguillicola crassus* on European eels. Fisheries Manag. Ecol. 10(6): 385–394.

Kusuda, R. and C. Xia. 1992. Lymphocyte Growth-Promoting factor(s) Produced by the Leukocytes of Eel *Anguilla japonica*. B. JPN Soc. Sci. Fish 58(9): 1667–1671.

Laing, K.J., L. Pilström, C. Cunningham and C.J. Secombes. 1999. TGF-beta3 exists in bony fish. Vet. Immunol. Immunop. 72(1-2): 45–53.

Lee, N.S., J. Kobayashi and T. Miyazaki. 1999. Gill filament necrosis in farmed Japanese eels, *Anguilla japonica* (Temminck & Schlegel), infected with *Herpesvirus anguillae*. J. Fish Dis. 22(6): 457–463.

Li, J., D.R. Barreda, Y.-A. Zhang, H. Boshra, A.E. Gelman, S. LaPatra, L. Tort and J.O. Sunyer. 2006. B lymphocytes from early vertebrates have potent phagocytic and microbicidal abilities. Nat. Immunol. 7(10): 1116–24.

Liang, Y., R. Guan, W. Huang and T. Xu. 2011. Isolation and identification of a novel inducible antibacterial peptide from the skin mucus of Japanese eel, *Anguilla japonica*. Protein J. 30(6): 413–21.

Litman, G.W., J.P. Rast and S.D. Fugmann. 2010. The origins of vertebrate adaptive immunity. Nat. Rev. Immunol. 10(8): 543–53.

Lorenzen, N. and S. La Patra. 2005. DNA vaccines for aquacultured fish. Rev. Sci. Tech. 24: 201–13.

Manera, M. and B.S. Dezfuli. 2004. Rodlet cells in teleosts: a new insight into their nature and functions. J. Fish Biol. 65: 597–619.

Małaczewska, J., M. Wójcik, R. Wójcik and A.K. Siwicki. 2010. The in vitro effect of bovine lactoferrin on the activity of organ leukocytes in rainbow trout (*Oncorhynchus mykiss*), European eel (*Anguilla anguilla*) and wels catfish (*Silurus glanis*). Pol. J. Vet. Sci. 13(1): 83–8.

Mellergaard, S. and I. Dalsgaard. 1987. Disease Problems in Danish Eel Farms. *Aquaculture* 67: 139–146.

Mizutani, T., Y. Sayama, A. Nakanishi, H. Ochiai, K. Sakai, K. Wakabayashi, N. Tanaka, E. Miura, M. Oba, I. Kurane, M. Saijo, S. Morikawa and S. Ono. 2011. Novel DNA virus isolated from samples showing endothelial cell necrosis in the Japanese eel, Anguilla japonica. Virology 412(1): 179–187.

Montgomery, B.C., H.D. Cortes, J. Mewes-Ares, K. Verheijen and J.L. Stafford. 2011. Teleost IgSF immunoregulatory receptors. Dev. Comp. Immunol. 35(12): 1223–37.

Mulero, I., M.P. Sepulcre, J. Meseguer, A. García-Ayala and V. Mulero. 2007. Histamine is stored in mast cells of most evolutionarily advanced fish and regulates the fish inflammatory response. Proceedings of the National Academy of Sciences of the United States of America 104(49): 19434–9.

Mulero, I., E.J. Noga, J. Meseguer, A. García-Ayala and V. Mulero. 2008. The antimicrobial peptides piscidins are stored in the granules of professional phagocytic granulocytes of fish and are delivered to the bacteria-containing phagosome upon phagocytosis. Dev. Comp. Immunol. 32(12): 1531–1538.

Nakanishi, T., H. Toda, Y. Shibasaki and T. Somamoto. 2011. Cytotoxic T cells in teleost fish. Dev. Comp. Immunol. 35(12): 1317–23.

Nakao, M., M. Tsujikura, S. Ichiki, T.K. Vo and T. Somamoto. 2011. The complement system in teleost fish: progress of post-homolog-hunting researches. Dev. Comp. Immunol. 35(12): 1296–308.

Nielsen, M.E. 1997. Glutathione-s-transferase is an important antigen in the eel nematode *Anguillicola*. J. Helminthol. 71: 319–324.

Nielsen, M.E. 1999. An enhanced humoral immune response against the swimbladder nematode, *Anguillicola crassus*, in the Japanese eel, *Anguilla japonica*, compared with the European eel, *A. anguilla*. J. Helminthol. 73: 227–232.

Nielsen, M.E., L. Hoi, A.S. Schmidt, D. Qian, T. Shimada, J.Y. Shen and J.L. Larsen. 2001. Is Aeromonas hydrophila the dominant motile Aeromonas species that causes disease outbreaks in aquaculture production in the Zhejiang Province of China? Dis. Aquat. Org. 46(1): 23–29.

Nielsen, M.E., T. Lindenstrøm, J. Sigh and K. Buchmann. 2004. Thionine-positive cells in relation to parasites. *In*: Host-parasite Interactions eds. Wendelaar Bonga, G.F. Wiegertjes and G. Flick [eds.]. Garland Science. pp. 45–66.

Nielsen, M.E. and M.D. Esteve-Gassent. 2006. Review The eel immune system: present knowledge and the need for research. J. Fish Dis. 29: 65–78.

Nieuwstadt, A.P.V., S.G. Dijkstra and O.L.M. Haenen. 2001. Persistence of herpesvirus of eel *Herpesvirus anguillae* in farmed European eel *Anguilla anguilla*. Dis. Aquat. Org. 45(2): 103–107.

Pilström, L., G.W. Warr and S. Strömberg. 2005. Why is the antibody response of Atlantic cod so poor? The search for a genetic explanation. Fish. Sci. 71(5): 961–971.

Qin, Q.W., M. Ototake, H. Nagoya and T. Nakanishi. 2002. Graft-versus-host reaction (GVHR) in clonal amago salmon, Oncorhynchus rhodurus. Vet. Immunol. Immunop. 89(1-2): 83–9.

Reite, O.B. 2005. The rodlet cells of teleostean fish: their potential role in host defence in relation to the role of mast cells/eosinophilic granule cells. Fish Shellfish Immunol. 19(3): 253–67.

Reite, O.B. and O. Evensen. 2006. Inflammatory cells of teleostean fish: a review focusing on mast cells/eosinophilic granule cells and rodlet cells. Fish Shellfish Immunol. 20(2): 192–208.

Reite, O.B. 1997. Mast cells/eosinophilic granule cells of salmonids: staining properties and responses to noxious agents. Fish Shellfish Immunol. 7: 567–584.

Ren, T., S. Koshio, M. Ishikawa, S. Yokoyama, F.R. Micheal, O. Uyan and H.T. Tung. 2007. Influence of dietary vitamin C and bovine lactoferrin on blood chemistry and non-specific immune responses of Japanese eel, Anguilla japonica. Aquaculture 267(1-4): 31–37.

Rijsewijk, F., S. Pritz-Verschuren, S. Kerkhoff, A. Botter, M. Willemsen, T. van Nieuwstadt and O. Haenen. 2005. Development of a polymerase chain reaction for the detection of *Anguillid herpesvirus* DNA in eels based on the herpesvirus DNA polymerase gene. J. Virol. Methods 124(1-2): 87–94.

Sakai, D.K. 1981. Heat inactivation of complement and immune hemolysis reaction in rainbow trout, masu salmon, coho salmon, goldfish and tilapia. B. JPN Soc. Sci. Fish. 47: 565–571.

Salati, F. and R. Kusuda. 1985. Vaccine preparations used for immunization of eel *Anguilla japonica* against *Edwardsiella tarda* infection. B. JPN Soc. Sci. Fish. 51(8): 1233–1237.

Salinas, I., Y.-A. Zhang and J.O. Sunyer. 2011. Mucosal immunoglobulins and B cells of teleost fish. Developmental and Comparative Immunology 35(12): 1346–65.

Saurabh, S. and P.K. Sahoo. 2008. Lysozyme: an important defence molecule of fish innate immune system. Aquaculture Research 39(3): 223–239.

Secombes, C.J., T. Wang and S. Bird. 2011. The interleukins of fish. Dev. Comp. Immunol. 35(12): 1336–45.

Shen, L., T.B. Stuge, E. Bengtén, M. Wilson, V.G. Chinchar, J.P. Naftel, J.M. Bernanke, L.W. Clem and N.W. Miller. 2004. Identification and characterization of clonal NK-like cells from channel catfish (Ictalurus punctatus). Dev. Comp. Immunol. 28(2): 139–152.

Shih, H.-H. 2004. A polymerase chain reaction for detecting Herpesvirus anguillae in asymptomatic eels. Taiwania 49(1): 1–6.

Siwicki, A.K. and S. Robak. 2011. The innate immunity of European eel (*Anguilla anguilla*) growing in natural conditions and intensive system of rearing. Cent. Eur. J. Immunol. 36(3): 130–134.

Star, B., A.J. Nederbragt, S. Jentoft, U. Grimholt, M. Malmstrøm, T.F. Gregers, T.B. Rounge, J. Paulsen, M.H. Solbakken, A. Sharma, O.F. Wetten, A. Lanzén, R. Winer, J. Knight, J.-H. Vogel, B. Aken, Ø. Andersen, K. Lagesen, A.T. Klunderud, R.B. Edvardsen, K.G. Tina, M. Espelund, C. Nepal, C. Previti, B.O. Karlsen, T. Moum, M. Skage, P.R. Berg, T. Gjøen, H. Kuhl, J. Thorsen, K. Malde, R. Reinhardt, L. Du, S.D. Johansen, S. Searle, S. Lien, F. Nilsen, I. Jonassen, S.W. Omholt, N.C. Stenseth and K.S. Jakobsen. 2011. The genome sequence of Atlantic cod reveals a unique immune system. Nature 477(7363): 207–10.

Tsutsui, S., K. Iwamoto, O. Nakamura and T. Watanabe. 2007a. Yeast-binding C-type lectin with opsonic activity from conger eel (*Conger myriaster*) skin mucus. Mol. Immunol. 44(5): 691–702.

Tsutsui, S., T. Iwamoto, O. Nakamura and T. Watanabe. 2007b. LPS induces gene expression of interleukin-1beta in conger eel (*Conger myriaster*) macrophages: first cytokine sequence within Anguilliformes. Fish Shellfish Immunol. 23(4): 911–6.

Uchida, D., H. Hirose, P.K. Chang, F. Aranishi, E. Hirayabu, N. Mano, T. Mitsuya, S.B. Prayitno and M. Natori. 2000. Characterization of Japanese eel immunoglobulin M and its level in serum. Comp. Biochem. Physiol., Part B: Biochem. Mol. Biol. 127(4): 525–32.

van Beurden, S.J., M.Y. Engelsma, I. Roozenburg, M.A. Voorbergen-Laarman, P.W. van Tulden, S. Kerkhoff, A.P. van Nieuwstadt, A. Davidse and O.L.M. Haenen. 2012. Viral diseases of wild and farmed European eel *Anguilla anguilla* with particular reference to the Netherlands. Dis. Aquat. Org. 101: 69–86.

van Beurden, S.J., M. Forlenza, A.H. Westphal, G.F. Wiegertjes, O.L.M. Haenen and M.Y. Engelsma. 2011. The alloherpesviral counterparts of interleukin 10 in European eel and common carp. Fish Shellfish Immunol. 31(6): 1211–7.

Van Beurden, S.J., A. Bossers, M.H.A. Voorbergen-Laarman, O.L.M. Haenen, S. Peters, M.H.C. Abma-Henkens, B.P.H. Peeters, P.J.M. Rottier and M.Y. Engelsma. 2010. Complete genome sequence and taxonomic position of anguillidherpesvirus 1. J. Gen. Virol. 91: 880–887.

Van Ginneken, V.J.T., B. Ballieux, R. Willemze, K. Coldenhoff, E. Lentjes, E. Antonissen, O. Haenen and G. van den Thillart. 2005. Hematology patterns of migrating European eels and the role of EVEX virus. Comp. Biochem. Physiol., Part C Toxicol. Pharmacol. 140: 97–102.

Verrier, E.R., C. Langevin, A. Benmansour and P. Boudinot. 2011. Early antiviral response and virus-induced genes in fish. Dev. Comp. Immunol. 35(12): 1204–1214.

Wang, W.-S., S.-W. Hung, Y.-H. Lin, C.-Y. Tu, M.-L. Wong, S.-H. Chiou and M.-T. Shieh. 2007. The Effects of Five Different Glycans on Innate Immune Responses by Phagocytes of Hybrid Tilapia and Japanese Eels *Anguilla japonica*. J. Aquat. Anim. Health 19(1): 49–59.

Wilson, M., E. Bengtén, N.W. Miller, L.W. Clem, L. Du Pasquier and G.W. Warr. 1997. A novel chimeric Ig heavy chain from a teleost fish shares similarities to IgD. Proceedings of the National Academy of Sciences of the United States of America 94(9): 4593–7.

Workenhe, S.T., M.L. Rise, M.J.T. Kibenge and F.S.B. Kibenge. 2010. Review: The fight between the teleost fish immune response and aquatic viruses. Mol. Immunol. 44: 2525–2536.

Yoder, J.A., M.G. Mueller, K.M. Nichols, S.S. Ristow, G.H. Thorgaard, T. Ota and G.W. Litman. 2002. Cloning novel immune-type inhibitory receptors from the rainbow trout, *Oncorhynchus mykiss*. Immunogenetics 54(9): 662–70.

Yoder, J.A. 2004. Review: Investigating the morphology, function and genetics of cytotoxic cells in bony fish. Comp. Biochem. Physiol., Part C Toxicol. Pharmacol. 138: 271–280.

Yoder, J.A. and G.W. Litman. 2011. The phylogenetic origins of natural killer receptors and recognition: relationships, possibilities, and realities. Immunogenetics 63(3): 123–41.

Zapata, A. 1979. Ultrastructural study of the teleost fish kidney. Dev. Comp. Immunol 5: 55–65.

Zhang, Y.-A., I. Salinas and J.O. Sunyer. 2011. Recent findings on the structure and function of teleost IgT. Fish Shellfish Immunol. 31(5): 627–634.

CHAPTER 10

Stress Physiology

Jonathan Mark Wilson

Introduction

Chrousos (1998) has defined stress "as a state of threatened homeostasis, which is reestablished by a complex repertoire of physiologic and behavioral adaptive responses of the organism". Thus stress is only a bad thing if the adaptive responses are inadequate for reestablishing homeostasis or are excessive and prolonged, and if a healthy steady state is not attained, in which case pathology may result. In fishes, stress can negatively impact growth, health, reproduction, and welfare and ultimately result in mortality, which is of relevance in aquaculture and in wild populations, where stress can also shape the composition of fish communities. Since there are already a number of comprehensive reviews on stress physiology in fishes that include general-whole animal (e.g., Sumpter 1997, Wendelaar-Bonga 1997, Barton and Iwama 1991, Barton 2002), ontogeny (Nesan and Vijayan 2012), specific aspects of physiology [metabolism (Mommsen et al. 1999, van der Boon et al. 1991), osmoregulation (McCormick 2001), immune system (Tort 2011, Weyts et al. 1999), reproduction (Leatherland et al. 2010, Milla et al. 2009, Schreck 2010), feeding and growth (Bernier and Peter 2001, van Weerd and Komen 1998)] and the cellular stress response (e.g., Iwama et al. 2004, Kultz 2005), the focus of this chapter will be to be eel-centric. There is a

Ecofisiologia, Centre for Interdisciplinar Marine and Environmental Research (CIIMAR) Porto, Portugal.
E-mail: wilson.jm.cimar@gmail.com

rich literature on the use of eels as an experimental animal that spans from pioneering work on elucidating basic neuroendocrine mechanisms (e.g., pituitary-interrenal axis, Maetz 1969) to the application of stress physiology in field studies to assess population health (Geeraerts and Belpaire 2010). Compared to other teleost fishes eels are generally very tolerant of various environmental stressors (temperature, hypoxia, hypercapnia, ammonia and pH; e.g., Cruz-Neto and Steffensen 1997, Sadler 1981, McKenzie et al. 2003), and can be found in environments that are polluted or have poor water quality; however, their complex and often mysterious life-history strategy makes them vulnerable at crucial stages in their life cycle (Robinet and Feunteun 2002, Geeraerts and Belpaire 2010).

This chapter will review the eels' complex life history, the organismal stress response and the key player cortisol, the cellular stress response, the effects of stress on growth and reproduction, and specifically the stress (cortisol) response to water physico-chemistry (temperature, hypoxia and anoxia, hypercapnia and pH disturbances, ammonia), handling and capture stress, pollutants and anesthetics, osmoregulation, social stress, disease and infection, and finally ends with a section on suggestions of future research directions for eel stress physiology (Fig. 1).

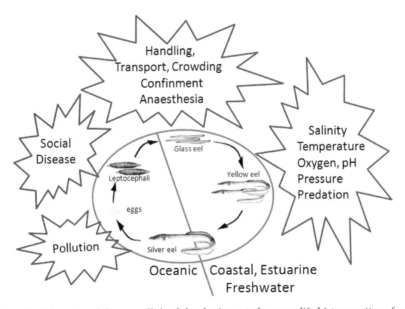

Figure 1. Schematic of the anguillid eel facultative-catadromous life-history pattern from oceanic spawning of the silver eel, migratory leptocephalus larvae, and post-metamorphic glass eel, to the feeding yellow eel. Different stressors encountered in the wild and under culture conditions by most if not all life history stages are listed.

Complex Life History

The Japanese (*Anguilla japonica*, Temminck and Schlegel), European (*Anguilla anguilla*, Linnaeus), and American (*Anguilla rostrata*, Le Sueur) anguillid species are the three most abundant, commercially important (fisheries and aquaculture) and studied of the 18 species/subspecies of this genus (Tesch 2003, Aoyama 2009). Aoyama (2009) has recently reviewed the life history and evolution of migration in anguillid eels which is characterized as being complex. Anguillid eels are facultatively catadromous, spending the majority of their juvenile growth phase in freshwater, estuarine and/or coastal habitats as yellow eels (Tsukamoto and Arai 2001). Once a sufficient size and energy (lipid) reserve are built up, animals will start to undergo a gradual morphological and physiological transition (metamorphosis) into migratory pre-pubescent silver eels (Durif et al. 2005). In European eels, males typical take 6–12 yr and females 9–20 yr to reach silvering size and condition. The oldest eel on record was 85 yr old (Fahay 1978). Eels are semelparous (terminal spawners) and since the spawning migration can be considerable (4000km in *A. japonica*, and *A. rostrata* and 5000–6000 km in *A. anguilla*) and sexual maturation is not completed until some point in the transoceanic migration (Tesch 2003, Palstra and Van den Thillart 2010), the quality of spawners is of paramount importance for successful reproduction [i.e., energy reserves and fitness need to be sufficient for both the long migration and reproductive success (viable larvae); see Robinet and Feunteun (2002)]. Adverse factors affecting spawner quality have become the topic of intense research over the past decade (Robinet and Feunteun 2002, Van Ginneken et al. 2005a, b, 2009). The spawning of eels remains a mystery with the exact location of the spawning grounds of European and American eels still being unknown although believed to be somewhere within the Sargasso sea (Schmidt 1923, Tesch 2003). The spawning grounds of the Japanese eel have been identified as a seamount near the Mariana Islands in the western North Pacific (Tsukamoto 1992, 2006, Chow et al. 2009). Following spawning, the transparent leaf-shaped leptocephalus larvae begin the long drift in the plankton for an average of 7–11 months or longer back to coastal waters. Little is known about this phase in the life cycle although recent work on artificially spawned Japanese eels is proving very enlightening (Tanaka et al. 2001, 2003). At the continental shelf the leptocephalus ceases feeding and undergoes a metamorphosis into the "glass eel" stage that has the typical anguillid body form although retaining its transparency. Upon entering freshwater or settling in coastal or estuarine waters, the glass eels gain pigmentation, transforming into elvers and starting to feed once more. After a year in this trophic life stage elvers then become known as yellow eels and continue feeding until they reach reproductive condition for the cycle to begin again.

The global decline of anguillid eel populations, notably in the North Atlantic, have been linked to oceanic modifications which can affect the larval trans-oceanic migration, and secondary productivity; freshwater habitat destruction from pollution and access loss from damming; overfishing of all freshwater life history stages and the introduction of the nematode parasite *Anguillicola crassus* (Robinet and Feunteun 2002, Knights 2003, Stone 2003). Reduction in fisheries, and the implementation of management and restoration programs for inland waters have not reversed the declining trends. *Anguilla anguilla* was added to CITES Appendix II in 2009 (Freyhof and Kottelat 2010) and *A. japonica* will be added in 2012. The Committee on the Status of Endangered Wildlife in Canada (COSEWIC) has listed *A. rostrata* as of species of special concern in 2006.

Stress Response

In the following section the stress response, which is highly conserved in vertebrates, is briefly reviewed. The general adaptation syndrome (GAS) described by Selye (1973) in response to the perception of a stressor [e.g., social, osmoregulatory, temperature, hypoxia, hypercapnia, handling/capture/transport, and chemical (pollution)] by the brain can be broken down into the following phases:

1. Neuroendocrine responses through the brain-sympathetic-chromaffin cell axis and the brain-pituitary-interrenal axis result in the release of catecholamines (adrenaline and nonadrenaline) and cortisol, respectively. Chromaffin cells and interrenal tissues are located in the head kidney of fishes.
2. Fight or flight. The released catecholamines peak within minutes and have rapid respiratory, hematological, muscular and cardiovascular effects, and increase energy through hyperglycemia to help the fish respond immediately. Cortisol released into circulation usually peaks within hours of the initial stressor and activates intracellular corticosteroid receptors that are ligand-bound transcription factors, modulating downstream gene expression in target tissues. In fishes, depending on the species one to two glucocorticoid receptors (GR) in addition to splice variants have been identified (Stolte et al. 2006, Prunet et al. 2006). In eels, two GRs have been identified (Marsigliante et al. 2000). Cortisol may also have more rapid non-genomic effects. For example Dindia et al. (2012) have shown in trout biophysical changes to plasma membrane properties, triggered by stressor-induced glucocorticoid elevation, that can act as a nonspecific stress response and may rapidly modulate acute stress-signaling pathways (phosphorylation status of putative PKA, PKC and AKT substrate

proteins). Cortisol notably affects the metabolism of carbohydrates, protein and lipid (see Mommsen et al. 1999). Generally cortisol is hyperglycaemic, primarily as a result of increases in hepatic gluconeogenesis initiated as a result of peripheral proteolysis. The increased plasma fatty acid levels during hypercortisolaemia may assist to fuel the enhanced metabolic rates noted for a number of fish species (Mommsen et al. 1999). In addition, cortisol has immediate effects on immune response (Tort 2011, Weyts et al. 1999), feeding/appetite (Bernier and Peter 2001) and behavior (Ellis et al. 2012). Cortisol is also the mineralocorticoid in fish acting through GR or a mineralocorticoid receptor (MR) (Prunet et al. 2006) and thus having effects on osmoregulation (McCormick 2001).

3. If the stress continues, changes at the whole animal level will be manifested as longer term affects on growth (van Weerd and Komen 1998), osmoregulation (McCormick 2001), the immune system (Tort 2011, Weyts et al. 1999), reproduction (Leatherland et al. 2010, Milla et al. 2009, Schreck 2010) and ultimately survival. Generally, the tertiary response reflects the adverse affects of chronic stress where reallocation of the energy budget will have consequences.

Cortisol

Since plasma catecholamines change very rapidly, they are generally not a reliable indicator of stress because of sampling related changes (Gamperl et al. 1994). Fish need to be fitted with an indwelling catheter for blood sampling for undisturbed catecholamine analysis [Table 1]. On the other hand, plasma cortisol levels increase less rapidly post handling so are commonly used as an indicator of the physiological stress response (e.g., Barton 2002, Pankhurst 2011, Ellis et al. 2012) [Table 2]. However, cortisol levels should not be assumed to be constant under apparent non-stress conditions. In eels, Li and Takei (2003) have shown that plasma cortisol concentration display a consistent diurnal fluctuation, with relatively low and stable levels (~30 ng/ml) during the day, and high but variable levels during the night (<80 ng/ml) which would correlated with their nocturnal activity (Mastorakos et al. 2005). This pattern is similar to that observed in some salmonid species (*Salmo trutta* and *Oncorhynchus mykiss*), where plasma cortisol levels were also low during the day and high during the night (Pickering and Pottinger 1983, Rance et al. 1982). Seasonally, plasma cortisol levels were not found to vary significantly in field collected eels although there was a tendency for higher levels in spring-summer months (Kelly et al. 2000). Holding conditions for eels can have an unforeseen effect on cortisol levels too. Eels kept in black tanks had higher plasma cortisol

Table 1. Plasma catecholamine levels (A adrenaline and NA nonadrenaline) in cannulated *Anguilla rostrata* (Ar) and *A. anguilla* (Aa) yellow eels collected from the wild (w) or cultured (c) following an acute hypoxia, acute and chronic hypercapnia or chronic cadmium exposure. Control values are shown in parentheses ().

Stressor	Spp., origin	A (nM)	NA (nM)	Reference
Acute hypoxia 20–35 torr PO_2 30min	*Ar* w	(0.84) ~4–5	(0.53) ~3	Perry and Reid 1992
Acute hypercapnia 1–5% CO_2 30min	*Ar* w	(0.18) 0.87	(0.31) 0.25	Hyde and Perry 1990
Chronic hypercapnia 6 wk (0.8), 15, 30, 45 mmHg PCO_2	*Aa* c	(3.11) 3.42, 3.11, 3.92	(5.31) 5.72, 7.32, 6.9	McKenzie et al. 2003
Chronic cadmium (75, 150 µg/l Cd) exposure 2–16 wk[§]	*Ar* w	(0.09–0.36) 0.03–0.22, 0.06–0.25	(0.01–0.11) 0.02–0.14, 0.00–0.23	Gill and Epple 1992

[§]In this experiment, eels were exposed to a secondary stressor [acute hypercapnia (1min CO_2 bubbling)] and serially sampled over 32. Changes in plasma NA and A were variable.

levels than those kept in white tanks under a mild noise stressor (Baker and Rance 1981, Gilham and Baker 1985).

Cellular Stress Response

The cellular stress response is a universal mechanism of physiological significance that represents the defense reaction at the molecular level (Kultz 2005). DNA, protein and lipid damage are not generally stressor specific and the response to damage is interconnected and common elements are shared. However, some responses directed at re-establishing homeostasis are stressor specific and activated in parallel to the cellular stress response. Stress proteins, notably heat shock protein 70 (hsp70), are universally conserved and have been well studied in fishes (see Iwama et al. 2004) although there are very few studies done in eels (Leitão et al. 2008, Fazio et al. 2008). Stress proteins perform essential biological roles as molecular chaperones that facilitate the synthesis and folding of proteins as well as participating in protein assembly, secretion, trafficking, and protein degradation. During cell stress they play a central role in maintaining cell homeostasis (Yamashita et al. 2010).

Growth

In Pickering's 1993 review on growth and stress in fish production, glucocorticoids are described as having a negative effect on growth and that elevated plasma cortisol levels promote a catabolic state with increased

Table 2. Response of Anguilid species to different stressors. Mean values are presented. Control values are shown in parentheses (). Asterisk indicate significant difference.

Stressor	Species (stage)+	Cortisol (ng/ml)	Glucose (mM)	Lactate (mM)	Comments	Reference
Capture + transport stress 6h post capture (water/in air at 4°C or ambient)	Aa (g) w	(10)⑧ 50–59*	(5.6) 10.2–12.7*	---	Glass eels were transported under 3 different conditions [in water or air at ambient temp. and air on ice (4°C)].	Leitao et al. 2010
Transport stress Capture, transport without water, anoxia)	Aa (y) w	(51)⑧ 175–353	(5.3) 11*	(1.4) 19.6*	Capture and transport conditions lack details. Fishing method. Time.	Santos and Pacheco 1996
Transport stress Capture, transport without water, anoxia)	Aa (y) c	(70) ⑧ 35	(4.4) 6.2	(2.8) 18.4*	(Same as previous).	Santos and Pacheco 1996
Handling (netting, air-exposure) 2–4h	Aa (y) w	(17) 56*	(4) 5.5–7.5*	---	Eels were netted and air-exposed for the duration of the experiment.	Gollock et al. 2004
+ Infected with *A. crassus*		(28) 65*	(4) 4.3–7.5*		(Same as previous). Naturally infected population of eels.	Gollock et al. 2004
Tank noise (moderate 5d)	Aa (y) w	(8) 46*	---	---	Fish were kept in black tanks.	Gilham and Baker 1985
WATER CONDITIONS						
Acute hypoxia Normoxia 158 torr PO$_2$ (23 torr PO$_2$) 4h, 8h	Aa (y) w	(15) 12, 5	(3.5) 6.4, 7.8	(3.0) 2.6, 2.7	No significant changes.	Gollock et al. 2005a
+ Infected with *A. crassus*	Aa (y) w	(12) 31*, 9	(5.2) 5.5, 4.4	(2.6) 2.1, 2.6	(Same as previous). Naturally infected population of eels.	Gollock et al. 2005a
Hypoxia (45 torr) 14 d	Aa (y) w	---	---	---	Decreased mRNA expression of cox1, atp6–8, cytc, nd5, sod2, cat. Mt mRNA expression increased§	Pierron et al. 2007b

Chronic hypercapnia (0.8), 15, 30, 45 mmHg	Aa (y) c	(22) 34, 28, 28	---	---	Fish cannulated. Not significantly different.	McKenzie et al. 2003
Anoxia 4h	Aa (y) c	(62) 143*	(4.7) 4.8	(2.8) 8.4*	Eels recovered 8 d from transport. Significant cortisol response	Santos and Pacheco 1996
Heat shock (no infection) 11°C, 22°C (2h), 28°C (6h)	Aa (y) w	(14) 40*, 28	(3.2) 3.8, 5.4*	---	Hct, Hb, and MCHC all increased	Gollock et al. 2005b DAO
+ Infected with *A. crassus*	Aa (y) w	(10) 24*, 24*	(2.8), 3.1, 5.5*	---	Hct, Hb, MCHC all increased	Gollock et al. 2005b DAO
TOXICANTS						
Cd (75, 150 µg/l Cd) 2–16 wk	Ar w	(17) 27–23*	Lower (37–67%)		75 µg/l Cd had no significant effect. Cortisol elevated from 2–16 wk. Only relative glucose levels given.	Gill et al. 1993
+1min hypercapnia	Ar w	(26) 36*	---	---	Cd fish give hypercapnic stress. Response measured within 32 min.	Gill et al. 1993
Cd (2, 10 µg/l) 14d	Aa (g) w	---	---	---	Decreased mRNA expression of cox1, atp6-8,cytc,nd5,sod2,cat Mt expression increased[§]	Pierron et al. 2007b
Cu 1, 2.5uM (24h)	Aa (y) w	(134) 180, 277*	(5.3) 7.8, 11.3*	(3.2) 7.9*, 5.7	Additional BNF treatment increase plasma glucose.	Teles et al. 2005b
Cu (0.2uM) 7d	Aa (y) w	(80) 30*	(4.7) 7.3			Oliveira et al. 2008
Cr 0.1+1 mM (24h)	Aa (y) w	(133) 215, 116	(5.7) 8.9, 8.1	(3.3) 3.7, 3.7	No significant differences. Additional BNF exposure increased plasma glucose.	Teles et al. 2005b
β-naphthaflavone (BNP) 2.7 µM 24h, 48h	Aa (y) w	(129, 149) 101, 158	(7.4,5.4) 17.8*, 24.4*	(2.4, 4.3) 2.4, 4.6	Massive increase in plasma glucose. 13.6 mM in fed eels (Cornish and Moon 1985)	Teles etal. 2005a

Table 2. contd....

Table 2. contd.

Stressor	Species (stage)[+]	Cortisol (ng/ml)	Glucose (mM)	Lactate (mM)	Comments	Reference
Bleached Kraft Mill Effluent. 4h BKME following 8d lab acclimation.	Aa (y) w	(67)⊛ 27	(5.3) 6	(1.4) 1.5	Lowered cortisol levels during post transport stress recovery. Interrenal cortisol levels high.	Santos and Pacheco 1996
BKME 4h	Aa (y) c	(70) ⊛ 1*	(4.4) 6.3	(2.8) 4.8	Eels recovered 8d before exposure.	Santos and Pacheco 1996
Gasoline water-soluble fraction 0.25% (3h–6d)	Aa (y) w	(29) 38	(15.4) 17.1	(3.6) 3.8		Pacheco and Santos 2001a
Diesel oil water-soluble fraction 2.5% (3h–6d)	Aa (y) w	(103–138) <5*	(10.6) 9.6	(3.6) 2.3*	7d acclimation to lab.	Pacheco and Santos 2001a
Diesel oil water-soluble fraction 2.5% (3h–4h)	Aa (y) w	(63) 121*	(3.3) 3.8	(3.4) 2.2*	1d acclimation to lab.	Pacheco and Santos 2001b
Naphthalene (0.1, 0.3, 0.9, 2.7 µM) Sampling 2, 4, 6, 8, 16, 24, 48, 72h	Aa (y) w	(106) 13*,2*,16*,68	---	---	Only time point with significant differences 6h.	Teles et al. 2003b
BKME resin acid: Retene 0.1–2.7µM 8–72h	Aa (y) w	(80–87) 26–29*	(4.6–7) 9–7.9	(0.6–4.8) 0.6–0.7	Dose and time responses variable. Only mean values of different groups shown.	Teles et al. 2003a
BKME resin acid: Dehydroabietic Acid (DHAA) 0.1–2.7 µM 8–72h	Aa (y) w	(167–191) 35–74*	(6.4–10) 19.8*	(8.1–28.8) 15.9–31.2	Response variable with time and dose. High control group cortisol levels and very high plasma lactate levels.	Teles et al. 2003a
BKME resin acid: Abietic Acid (AA) 0.1–2.7 µM 8–72h	Aa (y) w	(72–47) 168–200*	(2.1–4.5) 6.2–4.7	(1–1.5) 1.1–2.2	Dose and time responses variable. Not all values shown. Cortisol response.	Teles et al. 2003a
SALINITY						
FW-SW (35ppt) 2h	Aj (y) c	(38) 78*	---	---		Hirano 1969

SW (2mo)-FW 2h	Aj (y) c	(44) 60	---	Hirano 1969
ACTH	Aj (y) c	(38) 219*	---	Hirano 1969
FW-SW 35 ppt 2h	Aj (y) c	(38) 96	Transient increase by 4h then return back to fw level.	---	Hirano and Utida 1971
FW-SW (1–8 d)	Aa (y) w	(24)28–99*	---	Ball et al. 1971
SW-FW (1–13 d)	Aa (y) w	(24) 19–45	---	Ball et al. 1971
FW-DW (2–8d)	Aa (y) w	(25–32) 20–35	Plasma Na levels fell from 140 to 120 mM	---	Ball et al. 1971
FW-SW (1–8d)	Aa (s) w	(31) 62–135*	---	Ball et al. 1971
PARASITES				
Infection (20 L) *A. crassus*	*Aa* (y) w	(3) 9	---	Sures et al. 2001
ANASTHETICS				
MS-222 50–800 ppm	Aj (y) c	(3.6) 260–312*	One of the highest cortisol levels recorded in eel. Controls killed by electro-anesthesia	---	Chiba et al. 2006
2-Phenoxyethanol 0.1–0.8%	Aj (y) c	(3.5) 78–27	Controls killed by electro-anesthesia	---	Chiba et al. 2006
Iso-eugenol (300 mg/l) 3.8min, 7.6min	Aa (s) w	(67) 37, 47	Not significantly different.	---	Iversen et al. 2012
Eugenol (50 mg/l)	*Aa* (y) w	(162) 38*	Sampled during sedation. All fish killed by decapitation. Oxidative stress in brain sod, mt, and cat increase.§	---	Renault et al. 2011
Metomidate (40 mg/l) 2.6min, 6.5min	Aa (s) w	(76)148*, 179*	---	Iversen et al. 2012

Aa Anguilla anguilla Aj Anguilla japonica Ar Anguilla rostrata
*significantly different from control group in parentheses ().
®control group value from post-stress recovery group.
⁺stage: y yellow eel, e elver, g glass eel, s silver eel, w wild, c cultured
§Mitochondrial metabolism: cox1, cytochrome oxidase 1; atp6-8 ATP synthase subunit 6–8; cytc, cytochrome c; nd5, NADH dehydrogenase subunit 5. Oxidative stress: cat, catalase; sod2, mitochondrial superoxide dismutase. Detoxification: mt, metallothionein.

metabolic expenditures that determine growth suppression (Pickering 1993). In agreement, at the metabolic level cortisol treatment results in hyperglycemia and an increase in hepatic glycogen in American eel (*A. rostrata*) (Butler 1968), Japanese eel (*A. japonica*) (Inui and Yokote 1975, Chan and Woo 1978) and European eel (*A. anguilla*) (Lidman et al. 1979). Cortisol effects on lipolysis are also evident as increases in plasma fatty acids levels in eels (Butler 1973, Dave et al. 1979, Lidman et al. 1979). In a later appraisal by van Weerd and Komen (1998) they agreed that while the effects of stress on metabolism were not in dispute, extrapolation to growth was less than certain. They concluded that changes in plasma cortisol levels during chronic stress may not be unequivocally related to the ultimate growth response due to the involvement of regulatory mechanisms that can modulate the detrimental effects of cortisol-mediated responses.

In eels social stress has been shown to decrease growth rates in subordinate eels and to increase size variation (bimodal fast and slow growth fish groups; Wickins 1985). In subordinate eels (chronic social stress), degenerative changes in the stomach were observed which would reduce digestive efficiency and consequently growth (Peters 1982, Willemse et al. 1984). In the study by Wickins (1985) repeated handling stress from periodically weighing the fish decreased growth rate and resulted in a more uniform, less bimodal population. Chronic cadmium exposure decreased body mass and condition factor through increased lipolysis (Pierron et al. 2008). To the best of my knowledge the effects of chronic cortisol administration have not been assessed on eel growth rates. Single stressors such as the surgical implantation of passive integrated transponder (PIT) tags had no effect on growth rate in yellow shortfin eels *Anguilla australis* (Hirt-Chabbert and Young 2012). However, in yellow eels marked with floy tags (external plastic wire tags), negative growth was reported (Dekker 1989).

Reproduction

Silvering in yellow eels has been shown to be triggered by cortisol injection (Epstein et al. 1971). Huang et al. (1999) have also shown that cortisol in female silvering eels increases pituitary gonadotropin (GtH-II) production specifically through stimulation of GtH-II β subunit expression. Fish GtH-II, although more similar to mammalian luteinizing hormone (LH) at the amino acid level, behaves more like follicle stimulating hormone (FSH). Using a pharmalogical approach, cortisol was also found to be acting through a glucocorticoid receptor.

Apart from the pre-puberty stage of sexual maturation, the effects of stress on reproduction in eels is little studied because of the inability to naturally reproduce eels in captivity and the great difficulty (near impossibility; Chow et al. 2009) in collecting sexually mature silver eels from the field and also to directly assess spawning and reproductive success. In the case of the European and American eels, the exact location of the spawning grounds within the Sargasso Sea have not been identified (Tesch 2003). Reproduction in captivity is possible but requires exogenous hormone treatments and no studies have yet been undertaken on the effects of stress (Takana et al. 2001, 2003, Palstra and Van den Thillart 2009). However, in general in other fishes cortisol and stress have been found to decrease sex steroid levels and impair reproduction (Leatherland et al. 2010, Milla et al. 2009, Schreck 2010).

Since eels must endure a long non-feeding spawning migration, the prior deposition of energy stores as lipids for this migration is critical (Van den Thillart et al. 2009). Cortisol has been shown to have a lipolytic effect increasing plasma free fatty acid levels (Butler 1968, 1973, Chan and Woo 1978, Dave et al. 1979, Lidman et al. 1979) and is likely having a role in energy mobilization during migration where the stress of migration is predicted to elevate plasma cortisol levels (Huang et al. 1999, Sebert et al. 2009). However, swimming at less than 2 BL/s on its own has not been found to elevate cortisol levels (Van Ginneken et al. 2002) and eels likely swim at much lower speeds (optimal swimming speed 0.61–0.67 BL/s; Palstra and Van den Thillart 2010). High hydrostatic pressure experienced during the spawning migration might be a stressor (diurnal vertical migrations from 200 to 1000 m; Aarestrup et al. 2009, Sebert et al. 2009). The morphological and physiological transformation (Durif et al. 2005) from trophic yellow eels to migratory silver eels is also only triggered once sufficient lipid stores are attained (>20% body mass; Boetius and Boetius 1980). Thus stressors that can impair fat deposition will ultimately indirectly impact reproduction. In this respect, polycyclic aromatic hydrocarbon (PAH), polychlorinated biphenyls (PCBs; Van Ginneken et al. 2009) and cadmium (Pierron et al. 2007a), have been found to have profound effects. Also, as the lipid stores are consumed for the migration and gonad maturation, lipophillic contaminants will be released into circulation and deposited into the eggs, respectively. The consequently deleterious effects of accumulated contaminants on egg viability have been observed (PAH, Palstra et al. 2006). Pierron et al. (2008) also found that hormonally induced maturing female silver eels pre-exposed to Cd in a simulated swimming migration had a strongly stimulated pituitary-gonad-liver axis leading to early and enhanced vitellogenesis. This was followed by oocyte atresia and eel mortality. Oocyte maturation was not achieved and Cd was redistributed to the ovary.

Water Physico-chemistry

Changes in water physico-chemistry can be stressful to fish if the rate and magnitude of the change is sufficiently challenging. In addition, persistently poor water quality conditions can also be detrimental to survival and growth and thus are a concern in both aquaculture and with regards to the natural populations. In particular in intensive eel culture, with very high stocking densities (100–300 kg/m³) and recirculation (Steffensen and Lomholt 1990), water conditions can change very rapidly with lethal results. Common water physico-chemical parameters that have been studied include temperature, oxygen, carbon dioxide, pH and ammonia.

Temperature (Heat Shock)

The natural temperature range of *A. anguilla* was reported by Deelder (1985) to be from 0–30°C which reflects its wide geographical distribution. Fahay (1978) has reported that *A. rostrata* elvers have been collected at temperatures as low as –0.8°C. At the upper extreme, Sadler (1979) found an ultimate upper lethal temperature of 38°C in *A. anguilla* and Marcy (1973) reported that *A. rostrata* yellow eels were able to survive passage through the condenser cooling-water system of the Connecticut Yankee nuclear power plant (>28°C). Under culture conditions, the reported optimal temperature range is 24–28°C for *A. japonica* and 23–26°C for *A. anguilla* with glass eels/elvers preferring the higher side and larger eels low side of the temperature ranges (Tesch 2003). Below ~12°C Japanese eels are reported not to feed (Usui 1999) while in European and American species feeding ceases below 8°C and winter torpor is displayed (see Walsh et al. 1983). Torpor in European yellow eel has been reported at 1–3°C depending on the initial acclimation temperature (Sadler 1979). Migratory Japanese elvers were shown to have a temperature preference above their acclimation temperature 13–21°C which was not influenced by salinity (Chen and Chen 1990) while migrating *A. anguilla* glass eels prefer water of the same temperature as their acclimation water or lower temperature if only given the choice between higher or lower although not to very low temperatures (< 3°C) (Tongiorgi et al. 1986, Tosi et al. 1988). The preferred temperature range of artificially spawned Japanese eels was found to be 18–22°C (Dou et al. 2008) which corresponds to the temperature zone at 500 m in the Mariana Islands sea mount area the presumed spawning grounds of the Japanese eel (Tsukamoto 1992).

In 15°C acclimated glass eels, temperature shocks (acute transfer) greater than 27°C (+12°C) were found to be acutely lethal (Leitão et al. 2008). There did not appear to be any hepatic hsp70 response or cortisol response to acute temperature change after 24h in glass eels from the same study (JMW unpublished). In a gradual heat shock experiment with a temperature increase from 11 to 28°C over 6h, Gollock et al. (2005b) reported a significant increase in both plasma cortisol and glucose in *A. anguilla* yellow eels. Blood hematocrit, and hemoglobin concentration also increased. In the same study, *Anguillicola crassus* infection did not significantly alter the stress response. From these few studies it is thus clear that although eels are eurythermal, they may still experience a stress response in at least some life-history stages.

Hypoxia and Anoxia

In their natural environment, eels may experience hypoxia and/or anoxia when they burrow into the sediment during winter months when the surface is covered over in ice, as well as in summer when water bodies warm up and high plant and animal biomass can drive down oxygen levels at night, or when eels are out of water as sometimes occurs during overland transit (van Waarde et al. 1983, Hyde et al. 1987). Under culture conditions, fish biomass can also be very high (100–300 kg/m^3) and although oxygen supplementation is provided, mismatching of supply to demand can lead to periods of hypoxia (Cruz-Neto and Steffensen 1997). However, eels are well known to be able to withstand extended periods of extreme hypoxia and even anoxia (van Waarde et al. 1983, Hyde et al. 1987).

Van Waarde et al. (1983) reported an anoxia LT50 (time for 50% mortality) of 5.7 hr under low stress condition at 15°C which contrasts with mass mortality under culture conditions in 30 minutes (Van Ginneken et al. 2001). This apparent difference in hypoxia-anoxia tolerance may be due in large part to differences in temperature since eels are cultured intensively at higher temperatures (23–26°C; Tesch 2003). In addition, Sebert et al. (1995) have shown in yellow eel that Pcrit (lowest critical oxygen partial pressure that standard metabolic rate is maintained) increases with temperature (Pcrit 22 versus 28 mmHg at 17 and 21°C, respectively). Van Ginneken et al. (2001) have shown that anoxia was accompanied by a 70% reduction in metabolic rate without anaerobic ethanol production. Instead lactate is the end product of anaerobic metabolism. Interestingly, the group of Santos and Pacheco [(1996), Pacheco and Santos 2001a, Teles et al. 2003a, b] report using transport of eels out of water under anoxic conditions (30 min) in their experiments. In the study of Santos and Pacheco (1996), transport under these conditions resulted in high plasma cortisol levels at 4 hr recovery in wild eels (175–353 ng/ml). However, in farm collected eels the comparable cortisol levels were much lower (37 ng/ml). When farm eels were also exposed to 4 hr anoxia in the laboratory after a 20 hr recovery from the initial collection-anoxic-transport stress a cortisol response was observed (142 ng/ml). Thus anoxia alone is capable of inducing a cortisol response in eels. Notably, the cortisol response of 353 ng/ml in wild eels was the highest observed in the survey of the literature [Table 2].

From earlier studies it appeared as though eels were capable of efficiently air-breathing during emersion using their gills, swim bladder and/or skin (Berg and Steen 1965, 1966). However, more recent studies using indwelling catheters for undisturbed blood sampling have shown that during air-exposure, eels in fact become very hypoxic and instead are very tolerant of hypoxemia (25 mmHg arterial PO_2; Hyde et al. 1987). In contrast to other fishes, red blood cell intracelular pH was not defended and the resulting acidosis decreased blood oxygen carrying capacity through hemoglobin Root and Bohr shifts (corrected oxygen content of 0.78 volume percent from 5 vol. % under normoxia; Hyde et al. 1987).

In fishes, catecholamines increase oxygen uptake (via increased ventilation, branchial blood flow and oxygen diffusing capacity, and cardiac output), blood oxygen carrying capacity (increased hemoglobin-oxygen binding affinity, and erythrocyte release), and energy mobilization through glucose release from liver through glycogenolysis (Randall and Perry 1992, Wendelaar-Bonga 1997). The role of catecholamines in energy mobilization through free fatty acid release is less clear (Sheridan 1988, 1994). In a comparative study, Perry and Reid (1992) found that plasma catecholamine levels during hypoxia rise to a much greater extent in trout than in eels (100 nM versus 3–4 nM, respectively) and at different arterial PO_2 levels (23 versus 11 mmHg) but that release was trigger at around the P_{50} value (P_{50} is the arterial PO_2 at half-maximal Hb-O_2 saturation) in both species. The differences in catecholamine levels may be explained by differences in the responsiveness of the chromaffin cells to cholinergic stimulation but probably not by differences in levels of stored catecholamines (Reid and Perry 1994).

Hypercapnia and pH Disturbances

Intensive culture systems for European eel are closed cycle systems with recirculating water, and high fish stocking densities. In these systems, water O_2 levels are maintained at or above air saturation with the injection of pure O_2 at the tank inflow and pH maintained around neutral by the addition of base; however, under these circumstances metabolic CO_2 accumulation within the recirculation system may reach PCO_2 levels of 30–40 mmHg (Steffensen and Lomholt 1990). Although water oxygen levels are maintained, eel blood oxygen carrying capacity may be reduced by the respiratory acidosis caused by the high water PCO_2 levels through the Root shift (pH dependent change in the hemoglobin oxygen binding capacity) (Forster and Steen 1969).

Eels under these culture conditions have been shown to have high tolerance to acute and chronic hypercapnia (Gill and Epple 1992, Gill et al. 1993, McKenzie et al. 2002, 2003). Chronic hypercapnia (up to 45 mmHg PCO_2 for 6 weeks) resulted in an extracellular acidosis, high plasma HCO_3^- levels compensated for by lower Cl^- levels and lower blood oxygen content (up to 50% reduction) but heart and white muscle intracellular pH was defended. Eels showed no changes in hypoxia tolerance, aerobic metabolic scope or sustained swimming performance. The blood catecholamine (adrenaline and noradrenaline) and cortisol levels were not affected by acute or chronic hypercapnia indicating the absence of a chronic stress (Hyde and Perry 1990, McKenzie et al. 2003). The high tolerance of the eel to acute hypercapnia (up to 80 mmHg for 30 min) is probably a consequence of the tolerance of its heart to acidosis and ability to maintain cardiac output (Neilsen and Gesser 1984, Davie et al. 1992), and an apparent contribution from cutaneous respiration to maintain O_2 uptake rates at normocapnic levels (McKenzie et al. 2002). Eels will also become hypercapnic (PCO_2 8.7 mmHg) during air-exposure and experience a blood acidosis (pH 7.5 from 8.1) (Hyde et al. 1987).

Goss and Perry (1994) have shown that in response to a experimental metabolic alkalosis (HCO_3^- infusion), the rainbow trout (*Oncorhynchus mykiss*) relies on Cl^-/HCO_3^- exchange which results in an elevation of Cl^- influx, and a reduction of Cl^- efflux, in addition to intracellular buffering. In contrast, the American yellow eel relies almost exclusively on reduction of Cl^- efflux with no appreciable Cl^-/HCO_3^- exchange for HCO_3^- efflux. Manipulation of branchial ionocyte surface area to increase Cl^- uptake or the use of intracellular buffering appeared to be unimportant for regulating metabolic alkalosis in eel in contrast to trout.

Acid rain from industrial pollution has been a problem in Northern countries, and acidification of freshwater bodies has resulted (Beamish et al. 1975). In acid sensitive species, a clear stress response is observed and iono regulatory failure is a primary cause of mortality (McDonald and Wood 1981, Brown et al. 1989). Reynolds (2011) examined whether low water pH had any effect on glass eel development (pigmentation), survival or ion regulation. He concluded that *A. rostrata* elvers could tolerate low pH (4) without osmoreguatory difficulties and that low pH would not pose a significant mortality risk to elvers entering acidic freshwaters. However, effects of low pH on long term growth are unknown in eels although in acid-sensitive species low pH is known to be deleterious to growth and survival (Pickering 1993).

Ammonia

Ammonia is a unique toxicant in that it is toxic to fish and at the same time produced by the fish as a result of protein catabolism (Ip et al. 2001). Fish are also exposed to high environmental ammonia (HEA) as a result of the discharge from sewage treatment and industrial plants, runoff from applied fertilizers for agriculture and urban areas, atmospheric deposition, breakdown of vegetation and animal wastes, and excretion by other aquatic animals (US EPA 1999, Ip et al. 2001). The impact of ammonia on estuarine fishes, both migrant and resident, has been largely overlooked, although levels of ammonia in estuarine waters have been reported to be as high as 1.6 mM of total ammonia nitrogen (TAN) (Eddy 2005). In addition, HEA can emerge as a result of high-density fish transport (Portz et al. 2006) and in fact, glass eels transport to aquaculture or holding facilities that takes place in air exposed trays at high densities with only a film of freshwater can have TAN levels that reach concentrations of 3 mM (Moreira-Silva et al. 2009); a value close to the LC50 96h of 3.3 mM (Moreira-Silva et al. 2009, Yamagata and Niwa 1982). Although cortisol or other stress indicators were not measured in this study, Tomasso et al. (1981) have shown in channel catfish (*Ictalurus punctatus*) that ammonia exposure results in an acute stress response. Notably, the channel catfish and eels have as similar high ammonia tolerance (USEPA 1999, Moreira-Silva et al. 2009).

Handling and Capture Stress

Handling and capture stress should be of concern at all stages of our interaction with eels. This includes the glass eel fishery and transport to holding facilities, transport and stocking of elvers, as well as under daily conditions in aquaculture or for housing eels in the laboratory for experimental work. In field work, tagging or biometric data collection and release of yellow and silver eels are also situations where stress should be considered. Finally, during the slaughter of eels for consumption, humane non-stressful practices should be applied.

The glass eel fishery is the sole source of seed stock for both aquaculture and stocking programs (Ottolenghi et al. 2004). Fishing typically makes use of fyke nets (Portugal and Spain), small trawlers using 'wing nets' and 'trawls' (France), or hand nets (UK) in the winter to early spring when glass eels are migrating through estuaries. Generally, trawling produces the most damage to glass eels since they are compressed into the net with other fish and debris. Storage of glass eels on board during fishing is typically done under only damp conditions with exposure to air (i.e., the fish are not kept in a tank of water for holding). The fact that eels are apparently able to breath cutaneously and have a high tolerance of hypoxemia makes this form of storage possible and also very practical in the confines of small fishing boats. Fishing is done at night during the new moon phase during winter months when the glass eels are concentrated in estuaries making their migration into freshwater (Tesch 2003). Glass eels are transported "dry" to holding tanks before being sold onto larger distributors in the supply chain of glass eels for aquaculture and stocking.

Leitão et al. (2008) found that glass eels collected from the Minho River estuary (Portugal) using a fyke net and transported to the lab displayed a typical stress response (Fig. 2) characterized by a transient increase in plasma cortisol at 6 hr (post-fishing) and a delayed transient increase in plasma glucose at 24 hr. It would be expected that plasma cortisol levels would also have been elevated earlier during capture-holding-transport based on studies in other fishes in which cortisol levels peak with in hours of the initial stressor (Barton 2002). Leitão et al. (2008) also found that hepatic hsp70 levels increased gradually and were only significantly elevated after 96h (Fig. 3). This is also the only study in eels that I know of that has measured the heat shock protein (hsp) 70 changes at the protein level in response to a stressor.

Tagging studies of yellow and silver eels are important for monitoring resident populations and runs of fish on their spawning migration (Iversen et al. 2012). Although, this monitoring is necessary it can have unforeseen negative consequences on survival and performance of the released fish (Gale et al. 2011, Cooke and Schramm 2007). In wild yellow eels Gollock et al. (2004) found that a 2–4 h netting and air-exposure, handling stress resulted in a significant increase in plasma cortisol and glucose levels. In the same study, European yellow eels with a significant load of the nematode parasite *Anguillicola crassus* did not alter the stress response significantly. Santos and Pacheco (1996), also working with European yellow eels studied the effects of recovery conditions on stress induced from capture, transport "dry" and under anoxia of wild and farmed yellow eels. They found that the cortisol response to capture and transport was markedly higher in wild eels and was an order of magnitude lower in farmed eels [Table 2]. The difference may be down to differences in capture conditions and/or the resistance of cultured eels to handling. There appear to be no longer term studies looking at recovery in eels. However, Wickins (1985) suspected that initial handling stress contributed to growth variability in elvers.

Under culture conditions or when holding fish in the laboratory, factors such as tank colour can affect the cortisol response as mentioned earlier. Baker and Rance (1981) kept yellow eels in either black or white tanks and found that cortisol levels were higher in fish in black tanks. Gilham and Baker (1985), later found that tank colour affected the stress response to noise (vibration from an air pump), with a greater cortisol response in fish kept in black tanks. The higher cortisol levels in the black tank from the

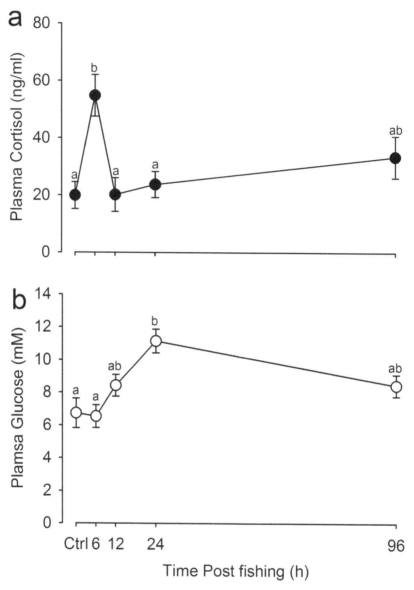

Figure 2. Time series of plasma cortisol (ng/ml) and glucose (mM) levels in glass eels after fishing. Fishing took place over 3h and the first sampling took place 6h later after transport to the lab. Fish were kept in freshwater aquaria at 15°C. The control group represents glass eels acclimated to laboratory conditions for 1 wk. Symbols with like letters are not statistically different. Data taken and replotted from Leitão et al. 2008.

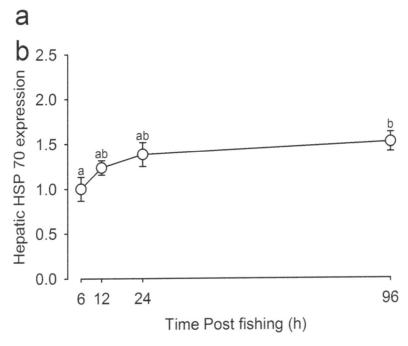

Figure 3. Time series of hepatic hsp70 levels in glass eels after fishing. Fishing took place over 3h and the first sampling took place 6h later after transport to the lab. Fish were kept in freshwater aquaria at 15°C to recover and subsequently sampled at 12, 24 and 96 h post-fishing. Representative immunoblots are shown in (a) and mean ± sem data normalized to the 6h group in (b). Symbols with like letters are not statistically different. Data taken and replotted from Leitão et al. 2008.

earlier study was accounted for by the higher background noise in the lab at that time. In white tanks fish have higher levels of the pituitary hormone melanin concentrating hormone (MCH), which has a direct effect on reducing skin pigmentation through melanophore contraction (Baker 1994). Although MCH does not have a direct suppressive effect on cortisol release by the interrenals, it has been shown to decrease ACTH levels indirectly by depressing corticotrophin releasing hormone (CRH) release. In this way the stress response is diminished in fish held in white tanks.

Two common commercial practices for killing eels at slaughter are decapitation and live brining. Verheijen and Flight (1997) found based on behavioral observation that these practices were inadequate with the decapitated head (from pectoral fins forward including heart) remaining active for up to 8 hr in air and that eels surviving brining (dry salt and wet slurry phases) could survive for up to 18 hr after being returned to water. Electrocephalograms (ECG) to measure brain activity were not made but results indicate that death was far from instantaneous. Electrical stunning has been proposed as a more humane manner of slaughtering eels (Algers et al. 2009, Lambooij et al. 2002a, b).

Chemical Stress (Pollution)

Yellow eels can be found in polluted (PCB, polycyclic aromatic hydrocarbons (PAH), pesticides and heavy metals) environments and are thus generally considered to be resistant at the individual level, yet when the anguillid eels' complex life history with its demanding spawning migration is taken into consideration, anguillid species are revealed to be very vulnerable to toxicant exposure. Given the importance of this topic to the survival of the anguillid species, a number of comprehensive reviews have been written specifically on the effects of contaminants in eels (e.g., Brusle 1991, Robinet and Feunteun 2002, Geeraerts and Belpaire 2010). Covered are laboratory studies on individual compounds or simple mixtures to field studies using either caged animals or sampling from native populations in waters with different pollution profiles. Monitoring data bases have also been set up to better gauge the effects and extent of pollutants on eels making eels a useful bioindicator of contaminants (e.g., INBO Eel Pollutant Monitoring Network database in Belgium; http://www.inbo.be/). Surprisingly given the extensive toxicology literature on eels, there are relatively few studies looking specifically at the cortisol response to toxicants that will be the focus of this section. Toxicant related stress effects on growth, and reproduction are found in those respective sections.

An elevation in plasma cortisol levels has been observed in fish after short-term exposure to pesticides, heavy metals, PAHs and crude oil, as well as bleach kraft mill effluent (BKME) and resin acids (RAs) (reviewed by Hontela 1997). In European yellow eels short term exposure to the diesel water-soluble fraction (3h DWSF 2.5% v:v) (Pacheco and Santos 2001b), and BKME resin acid component abietic acid (16–72h AA) (Teles et al. 2003b) resulted in a cortisol stress response. Exposure for 2 to 16 wk to copper also increased cortisol levels (Gill et al. 1993). However, in a number of other studies short to midterm exposures of *A. anguilla* have led to a decrease in plasma cortisol levels [BKME 4 h, Santos and Pacheco 1996; naphthalene at 6h, Teles et al. 2003b; BKME components AA, retene and dehydroabietic acid (DHAA) at 8 h, Teles et al., 2003a; copper 7 d, Oliveira et al. 2008] including a similar DWSF exposure study (Pacheco and Santos 2001a). The only apparent difference between the two DWSF studies being the acclimation time in the lab (1 d vs. 7 d). In the case of the DWSF, β-naphthoflavone (BNF), DHAA and naphthalene studies, cortisol levels in control fish were >100 ng/ml which would indicate a stressed state. Assuming the toxicant exposed fish are under similar housing and sampling conditions as the controls, this would indicate that their cortisol response is impaired. Ideally fish would be sampled with minimal disturbance, which would be possible using cannulated fish. However, only two studies were identified that used cannulated eels, and plasma cortisol levels in untreated fish were 10–22 ng/ml (Gill et al. 1993, McKenzie et al. 2003) which would be in the range of other non-stressed fishes (Barton 2002).

In terms of a mechanistic explanation for lower cortisol levels with stress, Hontela (1997) has found in a number of other species the endocrine disrupting effects of chronic exposure to sublethal levels of toxicants (BKME, PAHs, PCPs, and heavy metals) which exhaust the interrenals making them insensitive to ACTH stimulation resulting in a diminished cortisol response to a standardized capture stress. However, in the short term studies on eels it seems exhaustion of the interrenals would be unlikely and thus rearing conditions (tank colour, or social stress?) or collection site history (polluted? transport and handling conditions?) may play an important role in interpreting these results because of their effects of cortisol levels.

Alternatively, in trout, three day exposure to the PAH-like compound β-naphthoflavone (BNF) has been found to block the ACTH stimulated release of cortisol from the interrenals (Wilson et al. 1998). This effect of BNF has been found to be through stimulation of the aryl-hydrocarbon receptor and inhibition of steroidogenic acute regulatory protein (StAR) and cytochrome P450 cholesterol side chain cleavage (P450scc) enzymes, the rate-limiting steps in steroidogenesis (Alura and Vijayan 2007). In eels, laboratory exposure to either Cu or Cr with/without pre-exposure to BNF was without effect on plasma cortisol levels; however, control cortisol levels were high (>100ng/ml) perhaps masking a response to metals exposure (Teles et al. 2005b).

Changes in carbohydrate metabolism measured as plasma glucose and lactate have been used as general stress indicators in fishes (Hontela 1997, Mommsen et al. 1999), and their relation to cortisol function has been investigated in a number of toxicological studies in eels (Table 2). Generally, stress responses include increases in plasma glucose and lactate, however, sampling conditions and experimental design lead to inconsistent results.

In cage site studies where eels were placed in polluted sites, plasma cortisol and glucose levels were elevated in a BKME contaminated site (after 8 h but not 48 h of exposure) although not in the closest site to the pollution source and no response was observed at a contaminated harbour site (PAH, heavy metals, organometals) (Teles et al. 2004). The variability in cortisol levels in eels in response to contaminants makes its use as an indicator of chemical stress problematic, given that many factors can affect the stress status.

Anesthetics

Anesthetics, both chemical and physical (e.g., electro-anesthesis), have been used to reduce the stress response to during handling of fish (Barton and Iwama 1991, Small 2004, Harmon 2009, Ellis et al. 2012). However, anesthetics on their own are also capable of inducing a cortisol response (Barton and Peter 1982). Some commercially available chemical anesthetics used in fish culture, laboratory experiments, and field studies for catch and release of fish include MS-222 (3-aminobenzoic acidethyl ester methanesulfonate), benzocaine (p-aminobenzoic acid ethyl ester), 2-phenoxyethanol (1-hydroxy-2-phenoxyethane), metomidate [1-(1-phenylethyl)-1H-imidazole-5-carboxylic acid methyl ester] and more recently clove oil and its active ingredients eugenol (4-allyl-2-methoxyphenol) and iso-eugenol (4-propenyl-2-methoxyphenol) (see review by Iwama and Ackerman 1994). Electro-anesthesia includes electro-narcosis and electro-tetanus, and results in immobilization through inhibition of brain medullary motor paths and muscle excitation (tetanus), respectively (see Hudson et al. 2011). Electro-narcosis is achieved through the application of a continuous (non-pulsed) direct current (DC).

In cultured yellow eels, Chiba et al. (2006) found that anaesthetization with MS-222 at a dose of 50 to 800 ppm gave a high plasma cortisol response (260–312 ng/ml) in contrast to 2-phenoxyethanol (0.2–0.8%) and electro-anesthetization (240V AC, 500V–1000V DC). It should be noted that the MS-222 induced cortisol levels are amongst the highest found in this review. Chiba et al. (2006) noted that eels responded to the presence of anesthetic by cessation of water breathing and/or escaping by surfacing (head out of water) which may circumvent the effects of the anesthetic by limiting uptake from the water and contributing to the stress response. Similar cortisol responses to MS-222 although of a smaller magnitude have also been found in other species as well (e.g., *Ictalurus punctatus*; Small 2003)

In wild collected yellow eels, Renault et al. (2011) found that the anesthetic eugenol, a major component of clove oil, decreased plasma cortisol levels (38 ng/ml) compared to control fish that had cortisol levels greater than 150 ng/ml, which would be indicative of stressed fish (Barton 2002). They also observed that eugenol transiently increased the transcript expression of oxidative defense genes (superoxide dismutase, catalase, and metallothionein) in the brain, although not in the muscle. This response was related to the hypoxemia that occured during anesthesia. Transcripts of genes related to mitochondrial metabolism (ATP synthase subunit 6–8, cytochrome C oxidase subunit1, rRNA 12s, and NADH dehydrogenase subunit 5–6) were not affected by anesthesia with eugenol. In the same study, electro-anesthesia (DC) was found not to affected transcript expression of any of these genes in either brain or muscle. Unfortunately, plasma cortisol levels were not measured in this study.

In downstream migrating wild silver eels Iversen et al. (2012) compared the efficacy of iso-eugenol (Aqui-S vet.), and metomidate. Aqui-S vet. is a relatively new fish anaesthetic, containing among other things the active ingredient iso-eugenol (12% cis- and 88% trans-isomers) which is naturally found in clove oil. Metomidate is a rapid acting water-soluble non-barbiturate hypnotic and has also been shown to block cortisol synthesis and prevent handling-related glucose increase in a number of teleosts (see Iversen et al. 2012, Iwama et al. 1989). They found that iso-eugenol did not produce a cortisol response whereas metomidate did although the secondary indicator of stress, plasma chloride did not change with either anesthetic.

Similar anesthetic-induced increases in plasma cortisol concentrations have been reported in a number of other teleost including sea bream (*Sparus aurata*, Tort et al. 2002), red drum (*Sciaenops ocellatus*, Robertson et al. 1988), rainbow trout (*Oncorhynchus mykiss,* Barton and Peter 1982), Chinook salmon (*Oncorhynchus tshawytscha,* Strange and Schreck 1978) and striped bass (Davis et al. 1982). Although, in goldfish (*Carassius auratus*, Singley and Chavin 1975) and some of the same studies mentioned, rapid anesthesia (3–6 min) prior to handling reduces or completely blocks the stress response and in *O. mykiss* cortisol levels actually decrease (Iwama et al. 1989).

In eels at least, 2-phenoxyethanol, clove oil compounds (iso-eugenol and eugenol) and electrico-anesthesia would appear to be suitable anesthetics for avoid a direct cortisol response during fish handling such as during capture and tagging. Electro-anaesthesia has the additional advantage of not having direct effects on transcript level expression in at least the small set of seven genes studied by Renault et al. (2011). However, electro-anaesthesia is capable of causing mortality in some fish species as well as vertebral damage that requires longer term monitoring post-anesthesia to detect (Vandergoot et al. 2011).

Osmoregulatory Stress

Eels are predicted to experience osmotic stress in at least two stages in their life cycle. As catadromous fishes, the silver eel leaves freshwater to spawn at sea and it is the post metamorphic juvenile glass eel stage that makes the transition back into freshwater. In addition to these two predictable periods of environmental salinity challenge, some eels remain in the estuary or coastal zone representing a different ecotype and may experience more frequent changes in salinity (with the tide, high river flow) (Tsukamoto et al. 1998).

In teleost fishes, cortisol is also the main mineralocorticoid and has been coined the seawater adapting hormone (McCormick 2001). However, cortisol likely also has a role to play in freshwater adaptation as well (Mancera et al. 1994),[1] although in *A. anguilla* (Ball et al. 1971) and *A. japonica* (Hirano 1969, Hirano and Utida 1971) during transfer from seawater to freshwater or freshwater to distilled water, no increase in plasma cortisol was observed. Any increases observed with freshwater transfer were concluded to be due to handling stress because similar changes were seen in parallel controls. Epstein et al. (1971) reported that transfer of freshwater yellow American eels directly to seawater was lethal and required gradual acclimation since animals were incapable of adapting quickly to hypo-osmoregulation. However, in the other studies mentioned earlier, no such precaution appears to have been taken (Ball et al. 1971, Hirano 1969, Hirano and Utida 1971). This inconsistency may be due to species differences or more likely the salinity history of the eels used in these experiments. The eels used by Epstein et al. (1971) were clearly stated to be from fresh water and not from estuaries. Epstein et al. (1971) also found that injection of freshwater yellow eels with cortisol (40 µg/kg) improved survival and ion regulation after seawater transfer.

It is well established that the gills play a very important role in active ion regulation in fishes (Evans et al. 2005, Marshall and Grosell 2006) through specialized epithelial mitochondrion-rich ionocytes (Kaneko et al. 2008). In the current seawater fish model of the Cl⁻ secreting ionocyte (chloride cell), basolateral Na^+/K^+-ATPase drives the uptake of Cl⁻ against its electrochemical gradient via $Na^+:K^+:2Cl^-$ co-transporter (NKCC1; basolateral isoform). The intracellular Cl⁻ exits the cell via the apical cystic fibrosis transmembrane receptor (CFTR) Cl⁻ channel down its electrochemical gradient. Na^+ accumulates in the intercellular space, and exits across a leaky tight junction between neighbouring mitochondrion-rich cells following its electrochemical gradient (reviewed by Evans et al. 2005, Marshall and Grosell 2006). In freshwater fishes, the branchial mechanisms for Na^+ and Cl⁻ uptake are more variable, however, in the eel Na^+/H^+-exchange is most important with low uptake via Cl^-/HCO_3^- exchange and instead has very low overall Cl⁻ permeability (Goss and Perry 1994, Evans et al. 2005, Hwang et al. 2011). Tse and co-workers (Tse et al. 2011, Tse and Wong 2011) have recently shown that vacuolar H^+-ATPase, $Na^+: HCO_3^-$ cotransporter (NBC1) and Na^+/H^+-exchanger 3 (NHE3) are likely involved in Na^+ uptake in freshwater eels.

---

[1]includes reference to increase in plasma cortisol in freshwater challenged eels (Leloup-Hatey 1974) and other species. However, no salinity transfer data was found in Leloup-Hatey 1974 and a reference within reports no increase plasma cortsol with freshwater challenge (Ball et al. 1971).

In adult eels as in many other teleost fishes, cortisol has been shown to increase branchial Na$^+$/K$^+$-ATPase activity [e.g., *A. japonica* (Kamiya 1972); *A. rostrata* (Epstein et al. 1971, Doyle and Epstein 1972, Forrest et al. 1973)]. In freshwater adult eels (*A. rostrata*) cortisol treatment has been shown to stimulate Na$^+$ and Cl$^-$ uptake and ionocyte (chloride cell) number, apical exposure and development of the tubular membrane system that is associated with Na$^+$/K$^+$-ATPase (Doyle and Epstein 1972, Perry et al. 1992). In eels caught in Scottish estuaries and acclimated to either FW or SW for at least a month, those in SW had three times the Na$^+$/K$^+$-ATPase activity of those in FW (Sargent and Thomson 1974). Repeated cortisol injections have also been reported to result in silvering of yellow eels (Epstein et al. 1971). There are some data indicating that silvering of eel in fresh water is accompanied by branchial increases in Na$^+$/K$^+$-ATPase and the presence of seawater type chloride cells suggesting pre-adapted to seawater entry on the initiation of their spawning migration (Sasai et al. 1998).

The osmoregulatory challenge of glass eels crossing the estuary from the sea to enter freshwater has been addressed by the work of Wilson et al. (2004, 2007a, b). Freshwater challenge of estuarine glass eels resulted in a decline of gill Na$^+$/K$^+$-ATPase activity and whole body Na$^+$ and increase in whole body water (Wilson et al. 2004). Exogenous cortisol increased branchial Na$^+$/K$^+$-ATPase activity in a dose dependent manner in freshwater acclimated glass eels that was blocked by the glucocorticoid receptor blocker ru486 (mifepristone) (Wilson et al. 2004). In seawater acclimated glass eels Na$^+$/K$^+$-ATPase was already high and neither exogenous cortisol nor ru486 modulated activity. In glass eels challenged with ion poor water (distilled water), Na$^+$/K$^+$-ATPase was unresponsive even though osmoregulatory problems were evident (lower whole body Na$^+$ and high water content) (Wilson et al. 2007a) which would agree with findings in yellow eels (Ball et al. 1971). However, branchial NKCC expression, which is associated with hypoosmoregulation, decreased to its lowest level in distilled water acclimated eels.

The salinity responsive element osmotic stress transcription factor (OSTF1, TSC22-3b) was first identified in Mozambique tilapia *Oreochromis mossabicus* by Fiol and Kultz (2007). OSTF1 has since been identified in a number of different species including black porgy (*Acanthopagrus schlegeli*), Nile tilapia (*Oreochromis niloticus*), medaka (*Oryzias latipes*) and notably Japanese eel (*A. japonica*) (reviewed by Kultz 2011). Tse et al. (2007) found that *A. japonica* gill pavement cells respond to hypertonic shock *in vitro* with an increase in OSTF mRNA expression as well as NKA α and β subunits, NKCC and NHE1 protein expression. Protein expression also shows a positive response to (0.5–2 uM) dexamethasone (potent synthetic corticosteroid). Tse et al. (2008) went on to further characterize the expression of OSTF in eel. They found that basal expression levels were higher in gill ionocytes (chloride cells) than in pavement cells, however, in response to transfer to seawater transcriptional expression increased in both cell types although the magnitude of the changes was a reflection of the differences in basal levels of expression.

Kalujnaia et al. (2007) have developed an eel cDNA microarray (6,144 cDNAs) to identify a number of salinity-stress and salinity sensitive genes in European silver eels. In total, 229 unique differentially expressed genes of which 95 were identified and the remaining 59% unknown. Genes already known to be involved in osmoregulation were found to be differentially expressed and included myo-inositol monophosphatase 1, growth hormone, prolactin, angiotensin-converting enzyme, NKCC2, aquaporin 1 and the Na^+/K^+-ATPase (α subunit). Salinity induced changes in expression of unexpected genes included C-type lectin, myeloid protein 1, and the zymogen granule membrane protein 16 homolog, with their physiological roles and relationship to osmoregulation in the eel yet to be determined. Many of the genes identified as down regulated were related to the gut and likely reflect the cessation of feeding in migrating silver eels. In a more detailed follow up study Kalujnaia et al. (2010) established the importance of the enzyme inositol monophosphatase 1 and its product inositol as an osmolyte in a number of key osmoregulatory organs (gill, kidney, and esophagus) in seawater challenged eels.

Social Stress

Eels are generally solitary, territorial animals in the wild except when they are migrating. However, when kept communally in captivity for either culture or experimental work agonistic (fighting related) behaviors become prevalent and with escape not an option stunting, mortality, and cannibalism can result. At stocking densities <25 kg/m^3 growth decreases and greater size variability results (see Knights 1987). Conflict for social dominance has served as the stressor in a few studies with subordinate eels having a bimodal stress response (high and low responder groups) whereas dominant eels had cortisol levels lower than the population average (Peters et al. 1980). Peters et al. (1980) also observed higher blood glucose and lactate levels and lower liver glycogen and blood leukocyte numbers was also associated with subordinate eels. In these tournament experiments (one on one) threatening gestures, mock battles, biting matches and in some cases death of the loser was observed. As a consequence of dominance, Peters (1982) found that in subordinate yellow eels the stomach showed clear signs of morphological degeneration (loss of gastric glands and changes in epithelial cells) and in the gills ionocyte necrosis (Peters and Hong 1985). Willemse et al. (1984) also associated social stress from sharing of shelters with an increased occurrence of gallbladder enlargement and gut degeneration in elvers. It would appear from the high plasma cortisol levels of "control" fish kept in laboratory experiments (Table 2) that social stress is coming into play and that in future experiments greater attention should be paid to this aspect.

Since high variability of growth rates is a major problem for eel culture, some practical solutions to minimize size dependent dominance include increasing stocking densities, regular size grading, increase in shelter numbers, increase tank water flow rate, and photoperiod manipulation (see Knights 1987, Kushnirov and Degani 1991).

Disease (Parasite load)

Although there are numerous pathogens known to infect eels, it is notable that over the past 30 years the European eel has been most negatively impacted by the nematode parasite *Anguillicola crassus* (Kirk 2003). Thus in a number of studies the effects of infection on the stress response directly and in response to secondary stressors in eels have been examined. Importantly, the impact of infection on the critical spawning migration has also been investigated.

Anguillicola crassus was introduced into Europe in the 1980's with juvenile Japanese eels imported for experimental purposes and within a decade had spread across Europe (Kirk 2003). This parasite infects the swim bladder of the host resulting in a local inflammatory reaction, fibrosis and necrosis of the swim bladder impairing its function. Even though *A. crassus* is a blood feeder, infection is not associated with hematological effects (e.g., no decrease in hematocrit). The European and American eels possess ineffective defenses, unlike the native host, the Japanese eel, and consequently European populations have been ravaged and *A. crassus* infection has been acknowledged as a factor in the decline of this species (Kirk 2003, Knopf and Mahnke 2004). American eel populations have only been infected more recently but a similar fate is expected. *Anguillicola crassus* infections have been linked to mass mortality events in natural populations of European eels (Molnar et al. 1991) and been shown to decrease swimming performance in migrating silver eels (Palstra et al. 2007). Since stocks depend solely on natural spawning, successful spawning is critical to the survival on the species.

Experimental infection in the lab with *A. crassus* larvae (L3) itself causes a stress response in eels (Sures et al. 2001). However, after 60d of infection, cortisol levels returned to basal levels. This agrees with the finding that chronic infection in wild population of eel (Slapton Ley a coastal freshwater lake UK) does not elevate plasma cortisol levels in comparison with a population of eels with a low parasite loads (Otter River UK) (Kelly et al. 2000). Kelly et al. (2000) did not find any effects of parasite load on eel osmoregulation (plasma ion levels or gill Na^+/K^+-ATPase activity); in contrast to Fazio et al. (2008) who found significantly lower Na^+/K^+-ATPase α subunit mRNA expression with *A. crassus* infection. However, even though the eels from the study of Fazio were collected from the brackish water Salses-Leucate coastal lagoon the effects on Na^+/K^+-ATPase expression are not intuitive. It is also unknown what the relationship between transcript and pump protein or activity levels are in these fish although in glass eels and elvers Wilson et al. (2007b) found a positive correlation.

Parasite load has also been found not to alter the stress response to other secondary stressors [2–4 h netting and air-exposure (Gollock et al. 2004)] or temperature increase [11 to 28°C (Gollock et al. 2005b)]. However, Gollock et al. (2005a) did find that a stress response to hypoxia was only present in infected eels. In the case of hypoxia, there were no differences in hematocrit or hemoglobin concentration to explain the differences but impairment of swimbladder function and the possible increase in metabolic rate with parasite load are possibilities. In the study by Sures et al. (2006) in which the stress of infection was studied within the context of contaminant exposure [Cadmium and/or to 3,3k, 4,4k, 5-pentachlorobiphenyl (PCB 126)] a clear relationship does not emerge because of the confounding effects of excessive handling in the infection protocol. However, a negative correlation was found between cortisol levels and anti- *A. crassus* antibody titers. This latter finding is consistent with the immunosuppressive effects of cortisol on the immune system observed during stress (Pickering 1993).

Fazio et al. (2008) found that natural infection of eels with Digenea (flat worm) ecoparasites in the stomach correlated with lower hepatic hsp70 and metallothionein mRNA expression levels. The authors suggest that these infected fish may consequently be more susceptible to secondary stressors (temperature, pollution), although based on the work of *A. crassus* mentioned earlier this would require validation.

Van Ginneken et al. (2005b) examined the effects of viral infection of the rhabdovirus EVEX (Eel Virus European X) virus in eels on a simulated migration. EVEX causes the development of hemorrhage and anemia and has been shown to infect eels world-wide (perhaps spread by aquacultural practices). At the start of the simulated migration conducted in the lab with Blazka swim tunnels, no signs of infection were present but virus positive eels became anemic and died after swimming only 1000–1500 km whereas virus-negative eels were shown to be capable of completing the 5500 km swim. Van Ginneken et al. (2005b) have argued that the long migration in itself is stressful and given the well established immunosuppressive effects of stress and cortisol that has been demonstrated in other studies this seems a valid conclusion (Pickering 1993). However, swimming at these low speeds (0.5 BL/s) in itself is not stressful in the short term at least (Van Ginneken et al. 2002).

Future Directions

To date only a single glucocorticoid receptor has been cloned in the Japanese eel (*A. japonica*) (Genbank AB506765) despite there being evidence for at least two types in *A. anguilla* (Marsigliante et al. 2000) and a number of other teleosts (Stolte et al. 2006, Prunet et al. 2006). However, the sequencing of the Japanese (Henkel et al. 2012) and European eel genomes has now greatly simplified the prospects of identifying additional GR and MR isoforms (http://www.zfgenomics.org/sub/eel). Given the importance of receptor binding in the genomic effects of cortisol, additional work in this area is warranted.

Since eel aquaculture is still dependent on the capture of juvenile glass eels for stocking and will be for the foreseeable future (Ottolenghi et al. 2004), the consequences of capture and transport stress in glass eels on tertiary effects (growth, survival) are worth addressing (Wickins 1985). This is also prudent given that glass eel catches are declining despite the continuing high demand for aquaculture.

Epigenetic[2] is a burgeoning field of research, and studies of stress and cortisol on secondary and tertiary responses would be very enlightening on a mechanistic level. DNA methylation, which is a mechanism involved in epigenetic regulation, has been shown recently in mice to play a role during chronic administration of glucocorticoids to mimic chronic stress. Lee et al. (2010) were able to show that DNA methylation plays a role in mediating effects of glucocorticoid exposure on Fkbp5 function (a GR co-chaperone), with potential consequences on behavior. It may thus be possible to establish at least a role for stress related DNA methylation on tertiary effects of stress observed in eels.

Due to the difficulty in obtaining larval stages, very little is known about the development of the stress response in eels. In contrast in zebrafish (*Danio rerio*), which is an important developmental model species, it has been shown that maternally deposited cortisol has an important role in early developmental events while the development of the corticosteroid stress axis occurs only after hatch (Nesan and Vijayan 2012). The successful hatching and rearing of larval Japanese eels to metamorphosis into glass eels now makes it possible to address these events in eels (Tanaka et al. 2001, 2003). Conditions for improving larval culture are also being continuously improved (e.g., Okamura et al. 2009).

Genomic DNA studies using microsatellites have revealed that the European eel is not panmictic (single genetic population) and can be divided into three genetic subpopulations isolated by distance [Northern

---

[2]the study of heritable changes in gene expression or cellular phenotype caused by mechanisms other than changes in the underlying DNA sequence.

European subpopulation (consisting mainly of the Icelandic stocks); a Western European subpopulation (including the Baltic, the Mediterranean and Black Sea); a Southern sub-population (including stocks of Morocco) (Wirth and Bernatchez 2001, 2003). Given the latitudinal distribution of these populations (north to south), differences in temperature tolerance might be predicted. In contrast, *A. rostrata* does not have genetically isolated subpopulations and thus represents a panmictic species (Wirth and Bernatchez 2003).

In summary, access to new technologies and life history stages will enable us to have a more complete understanding of the role of stress in shaping the life of anguillid eels that have remained a mystery for so long.

Acknowledgments

Some of the work presented in this review was supported by the Portuguese Foundation for Science and Technology (FCT) grant Praxis XXI POCTI/BSE/34164/1999. JMW would like to thank C.M. Pereira-Wilson (U. Minho, Portugal) and P. Sebert (UBO, France) for comments on the manuscript.

References

Aarestrup, K., F. Økland, M.M. Hansen, D. Righton, P. Gargan, M. Castonguay, L. Bernatchez, P. Howey, H. Sparholt, M.I. Pedersen and R.S. McKinley. 2009. Oceanic Spawning Migration of the European Eel (*Anguilla anguilla*). Science 325: 1660.

Algers, B., H.J. Blokhuis, A. Botner, D.M. Broom, P. Costa, M. Domingo, M. Greiner, J. Hartung, F. Koenen, C. Muller-Graf, D.B. Morton, A. Osterhaus, D.U. Pfeiffer, M. Raj, R. Roberts, M. Sanaa, M. Salman, J.M. Sharp, P. Vannier and M. Wierup. 2009. Species-specific welfare aspects of the main systems of stunning and killing of farmed eels (*Anguilla anguilla*). Eur. Food Saf. Auth. J. 1014: 1–42.

Alura, N. and M.M. Vijayan. 2007. Hepatic transcriptome response to glucocorticoid receptor activation in rainbow trout. Physiol. Genomics 31: 483–491.

Aoyama, J. 2009. Life history and evolution of migration in catadromous eels (Genus *Anguilla*) Aqua-Bio. Sci. Mongr. 2: 1–42.

Baker, B.I. 1994. Melanin-Concentrating Hormone Updated Functional Considerations. Trends Endocrlnol. Metab. 5: 120–126.

Baker, B.I. and T.A. Rance. 1981. Differences in concentration of plasma cortisol in the trout and the eel following adaptation to black or white backgrounds. J. Endocrinol. 89: 135–140.

Ball, J.N., I. Chester Jones, M.E. Forster, G. Hargreaves, E.F. Hawkins and K.P. Milne. 1971. Measurement of plasma cortisol levels in the eel *Anguilla anguilla* in relation to osmotic adjustments. J. Endocrinol. 50: 75–96.

Barton, B.A. and R.E. Peter. 1982. Plasma cortisol stress response in fingerling rainbow trout, *Salmo gairdneri* Richardson, to various transport conditions, anaesthesia, and cold shock. J. Fish Biol. 20: 39–51.

Barton, B.A. 2002. Stress in Fishes: A Diversity of Responses with Particular Reference to Changes in Circulating Corticosteroids. Integ. Comp. Biol. 42: 517–525.

Barton, B.A. and G.K. Iwama. 1991. Physiological changes in fish from stress in aquaculture with emphasis on the response and effects of corticosteroids. Ann. Rev. Fish Dis. 1: 3–26.

Beamish, R.J., W.L. Lockhart, J.C. van Loon and H.H. Harvey. 1975. Long-term acidification of a lake and resulting effects on fishes. Ambio 4: 98–102.

Berg, T. and J.B. Steen. 1965. Physiological mechanisms for aerial respiration in the eel. Comp. Biochem. Physiol. 15: 469–484.

Berg, T. and J.B. Steen. 1966. Regulation of ventilation in eels exposed to air. Comp. Biochem. Physiol. 18: 511–516.

Bernier, N.J. and R.E. Peter. 2001. The hypothalamic–pituitary–interrenal axis and the control of food intake in teleost fish. Comp. Biochem. Physiol. B. 129: 639–644.

Boetius, I. and J. Boetius. 1980. Experimental maturation of female silver eels, *Anguilla anguilla*. Estimates of fecundity and energy reserves for migration and spawning. Dana 1: 1–28.

Brown, J.A., D. Edwards and C. Whitehead. 1989. Cortisol and thyroid hormone responses to acid stress in the brown trout, *Salmo trutta* L. J. Fish Biol. 35: 73–84.

Brusle, J. 1991. The eel (*Anguilla* sp.) and organic chemical pollutants. Sci. Total Environ. 102: 1–19.

Butler, D.G. 1968. Hormonal control of gluconeogenesis in the North American eel (*Anguilla rostrata*). Gen. Comp. Endocrinol. 10: 85–91.

Butler, D.G. 1973. Structure and function of the adrenal gland of fishes. Am. Zool. 13: 839–379.

Chan, D.K.O. and N.Y.S. Woo. 1978. Effect of cortisol on the metabolism of the eel, *Anguilla japonica*. Gen. Comp. Endocrinol. 35: 205–215.

Chen, Y.L. and H. Chen. 1990. Temperature selections of *Anguilla japonica* (L.) elvers, and their implications for migration Austral. J. Mar. Freshwater Res. 42: 743–750.

Chiba, H., T. Hattori, H. Yamada and M. Iwata. 2006. Comparison of the effects of chemical anesthesia and electroanesthesia on plasma cortisol levels in the Japanese eel *Anguilla japonica*. Fish. Sci. 72: 693–695.

Chow, S., H. Kurogi, N. Mochioka, S. Kaji, M. Okazaki and K. Tsukamoto. 2009. Discovery of mature freshwater eels in the open ocean. Fish. Sci. 75: 257–259.

Chrousos, G.P. 1998. Stressors, stress, and neuroendocrine integration of the adaptive response: the 1997 Hans Selye Memorial Lecture. Ann. New York Acad. Sci. 851: 311–335.

Cooke, S.J. and H.L. Schramm. 2007. Catch-and-release science and its application to conservation and management of recreational fisheries. Fish. Management Ecol. 14: 73–79.

Cruz-Neto, A.P. and J.F. Steffensen. 1997. The effects of acute hypoxia and hypercapnia on oxygen consumption of the freshwater European eel. J. Fish Biol. 50: 759–769.

Dave, G., M.-L. Johannsson-Sjöbeck, Å. Larsson, K. Lewander and U. Lidman. 1979. Effects of cortisol on the fatty acid composition of the total blood plasma lipids in the European eel, *Anguilla anguilla* L. Comp. Biochem. Physiol. 64A: 37–40.

Davie, P.S., A.P. Farrell and C.E. Franklin. 1992. Cardiac performance of an isolated eel heart: Effects of hypoxia and responses to coronary artery perfusion. J. Exp. Zool. 262: 113–121.

Davis, K.B., N.C. Parker and M.A. Suttle. 1982. Plasma corticosteroids and chlorides in striped bass exposed to tricaine methanesulfonate, quinaldine, etomidate and salt. Prog. Fish-Cult. 44: 205–207.

Deelder, C.L. 1985. Exposée synoptique des données biologiques sur l'anguille, *Anguilla anguilla* (Linnaeus, 1758). FAO Synop. Pêches. 80 Rev. 1: 1–71.

Dekker, W. 1989. Death rate, recapture frequency and changes in size of tagged eels. J. Fish Biol. 34: 769–777.

Dindia, L., J. Murray, E. Faught, T.L. Davis, Z. Leonenko and M.M. Vijayan. 2012. Novel nongenomic signaling by glucocorticoid may involve changes to liver membrane order in rainbow trout. PLoS ONE 7: e46859.

Dou, S.Z., Y. Yamada, A. Okamura, A. Shinoda, S.Tanaka and K. Tsukamoto. 2008. Temperature influence on the spawning performance of artificially-matured Japanese eel, *Anguilla japonica*, in captivity. Environ. Biol. Fish. 82: 151–164.

Doyle, W.L. and F.H. Epstein. 1972. Effects of cortisol treatment and osmotic adaptation on the chloride cells in the eel, *Anguilla rostrata*. Cytobiologie 6: 58–73.

Durif, C.M.F., S. Dufour and P. Elie. 2005. Impact of silvering stage, age, body size and condition on reproductive potential of the European eel. Mar. Ecol. Prog. Ser. 327: 171–181.

Eddy, F.B. 2005. Ammonia in estuaries and effects on fish. J. Fish Biol. 67: 1495–1513.

Ellis, T., H.Y. Yildiz, J. López-Olmeda, M.T. Spedicato, L. Tort, O. Overli and C.I.M. Martins. 2012. Cortisol and finfish welfare, Fish Physiol. Biochem. 38: 1–26.

Epstein, F.H., M. Cynamon and W. McKay. 1971. Endocine control of Na-K-ATPase and seawater adaptation in *Anguilla rostrata*. Gen. Comp. Endocrinol. 16: 323–328.

Evans, D.H., P.M. Piermarini and K.P. Choe. 2005. The multifunctional fish gill:dominant site of gas exchange, osmoregulation, acid base regulation and excretion of ntirogenous waste. Physiol. Rev. 85: 97–177.

Fahay, M.P. 1978. Biological and fisheries data on American eel *Anguilla rostrata* (Lesueur). NOAA Tech. Ser. Report 17: 1–96.

Fazio, G., H. Mone, R. Lecomte-Finiger and P. Sasal. 2008. Differential gene expression analysis in European eels (*Anguilla anguilla*, l. 1758) naturally infected by macroparasites. J. Parasitol. 94: 571–577.

Fiol, D.F. and D. Kultz. 2007. Osmotic stress sensing and signaling in fishes. FEBS J. 274: 5790–5798.

Forrest, Jr., J.N., W.C. MacKay, B. Gallagher and F.H. Epstein. 1973. Plasma cortisol response to saltwater adaptation in the American eel *Anguilla rostrata*. Am. J. Physiol. 224: 714–717.

Freyhof, J. and M. Kottelat. 2010. *Anguilla anguilla*. *In*: IUCN 2012. IUCN Red List of Threatened Species. Version 2012.2.

Forster, R.E. and J.B. Steen. 1969. The rate of the 'Root shift' in eel red cells and eel haemoglobin solutions. J. Physiol. 204: 259–282.

Gale, M.K., S.G. Hinch, E.J. Eliason, S.J. Cooke and D.A. Patterson. 2011. Physiological impairment of adult sockeye salmon in fresh water after simulated capture-and-release across a range of temperatures. Fish. Res. 112: 85–95.

Gamperl, A.K., M.M. Vijayan and R.G. Boutiler. 1994. Experimental control of stress hormone levels in fishes: techniques and applications. Rev. Fish Biol. Fish. 4: 215–255.

Geeraerts, C. and C. Belpaire. 2010. The effects of contaminants in European eel: a review. Ecotoxicol. 19: 239–266.

Gilham, I.D. and B.I. Baker. 1985. A black background facilitates the response to stress in teleosts. J. Endocrinol. 105: 99–105.

Gill, T.S., G. Leitner, S. Porta and A. Epple. 1993. Response of plasma-cortisol to environmental cadmium in the eel, *Anguilla-rostrata* Lesueur. Comp. Biochem. Physiol. C. 104: 489–495.

Gill, T.S. and A. Epple. 1992. Effects of cadmium on plasma catecholamines in the American eel *Anguilla rostrata*. Aqua. Toxicol. 23: 107–177.

Gollock, M.J., C.R. Kennedy and J.A. Brown. 2005a. European eels, *Anguilla anguilla* (L.), infected with *Anguillicola crassus* exhibit a more pronounced stress response to severe hypoxia than uninfected eels. J. Fish Dis. 28: 429–436.

Gollock, M.J., C.R. Kennedy and J.A. Brown. 2005b. Physiological responses to acute temperature increase in European eels (*Anguilla anguilla* (L.)) infected with *Anguillicola crassus*, compared to uninfected eels. Dis. Aquat. Org. 64: 223–228.

Gollock, M.J., C.R. Kennedy, E.S. Quabius and J.A. Brown. 2004. The effect of parasitism of European eels with the nematode, *Anguillicola crassus* on the impact of netting and aerial exposure. Aquaculture 233: 45–54.

Goss, G.G. and S.F. Perry. 1994. Different mechanisms of acid-base regulation in rainbow trout (*Oncorhynchus mykiss*) and American eel (*Anguilla rostrata*) during $NaHCO_3$ infusion. Physiol. Zool. 67: 381–406.

Harmon, T.S. 2009. Methods for reducing stressors and maintaining water quality associated with live fish transport in tanks: a review of the basics. Rev. Aquacult. 1: 58–66.

Henkel, C.V., R.P. Dirks, D.L. de Wijze, Y. Minegishi, J. Aoyama, H.J. Jansen, B. Turner, B. Knudsen, M. Bundgaard, K. Lyneborg Hvam, M. Boetzer, W. Pirovano, F.A. Weltzien, S. Dufour, K. Tsukamoto, H.P. Spaink, G.E.E.J.M. Van den Thillart. 2012. First draft genome sequence of the Japanese eel, *Anguilla japonica*. Gene 511: 195–201.

Hirano, T. 1969. Effects of hypophysectomy and salinity change on plasma cortisol concentration in Japanese eel, *Anguilla japonica*. Endocrinol. Jap. 16: 557–560.

Hirano, T. and S. Utida. 1971. Plasma cortisol concentration and the rate of intestinal water absorption in the eel, *Anguilla japonica*. Endocrinol. Jap. 17: 47–52.

Hirt-Chabbert, J.A. and O.A. Young. 2012. Effects of surgically implanted PIT tags on growth, survival and tag retention of yellow shortfin eels *Anguilla australis* under laboratory condition. J. Fish Biol. 81: 314–319.

Hontela, A. 1997. Endocrine and physiological responses of fish to xenobiotics: Role of glucocorticoid hormones. Rev. Toxicol. 1: 1–46.

Huang, Y.S., K. Rousseau, M. Sbaihi, N. Le Belle, M. Schmitz and S. Dufour. 1999. Cortisol Selectively Stimulates Pituitary Gonadotropin β-Subunit in a Primitive Teleost, *Anguilla anguilla*. Endocrinol. 140: 1228–1235.

Hudson, J.M., J.R. Johnson and B. Kynard. 2011. A portable electronarcosis system for anesthetizing salmonids and other fish. North Am. J. Fish. Manag. 31: 335–339.

Hwang, P.P., T.H. Lee and L.Y. Lin. 2011. Ion regulation in fish gills: recent progress in the cellular and molecular mechanisms. Am. J. Physiol. 11301: R28–R47.

Hyde, D.A. and S.F. Perry. 1990. Absence of red cell pH and oxygen content regulation in American eel (*Anguilla rostrata*) during hypercapnic acidosis *in vivo* and *in vitro*. J. Comp. Physiol. B. 159: 687–693.

Hyde, D.A., T.W. Moon and S.F. Perry. 1987. Physiological consequences of prolonged aerial exposure in the American eel, *Anguilla rostrata*: Blood respiratory and acid-base status. J. Comp. Physiol. B. 157: 635–642.

Inui, Y. and M. Yokote. 1975. Gluconeogenesis in the eel-IV. Gluconeogenesis in the hydrocortisone-administered eel. Bull. Jap. Soc. Sci. Fish. 41: 973–981.

Ip, Y.K., S.F. Chew and D.J. Randall. 2001. Ammonia toxicity, tolerance, and excretion. *In*: Wright P.A. and P.M. Anderson [eds]. Nitrogen Excretion. Academic Press, San Diego, California, USA. pp. 109–148.

Iversen, M.H., F. Økland, E.B. Thorstad and B. Finstad. 2012. The efficacy of Aqui-S vet. (iso-eugenol) and metomidate as anaesthetics in European eel (*Anguilla anguilla* L.), and their effects on animal welfare and primary and secondary stress responses. Aquacult. Res. 2012: 1–10.

Iwama, G.K. and P.A. Ackerman. 1994. Anesthetics. *In*: P. Hochachka and T.P. Mommsen [eds]. Biochemistry and molecular biology of fishes, vol. III. Analytical Techniques. Elsevier Science, Amsterdam, Netherlands. pp. 1–15.

Iwama, G.K., J.C. McGeer and M.P. Pawluk. 1989. The effects of five fish anaesthetics on acid–base balance, hematocrit, blood gases, cortisol, and adrenaline in rainbow trout. Can. J. Zool. 67: 2065–2073.

Iwama, G.K., L.O.B. Afonso, A. Todgham, P. Ackerman and K. Nakano. 2004. Are hsps suitable for indicating stressed states in fish? J. exp. Biol. 207: 15–19.

Kalujnaia, S., I.S. McWilliam, V.A. Zaguinaiko, A.L. Feilen, J. Nicholson, N. Hazon, C.P. Cutler and G. Cramb. 2007. A transcriptomic approach to the study of osmoregulation in the European eel (*Anguilla anguilla*). Physiol. Genomics 31: 385–401.

Kalujnaia, S., J. McVee, T. Kasciukovic, A.J. Stewart and G. Cramb. 2010. A role for inositol monophosphatase 1 (IMPA1) in salinity adaptation in the euryhaline eel (*Anguilla anguilla*). FASEB J. 24: 3981–3991.

Kamiya, M. 1972. Sodium potassium-activated adenosinetriphosphate in isolated chloride cells from eel gills. Comp. Biochem. Physiol. B. 43: 611–617.

Kaneko,T., S. Watanabe and K.M. Lee. 2008. Functional morphology of mitochondrion-rich cells in euryhaline and stenohaline teleosts. Aqua-BioSci. Monogr. 1: 1–62.

Kelly, C.E., C.R. Kennedy and J.A. Brown. 2000. Physiological status of wild European eels (*Anguilla anguilla*) infected with the parasitic nematode, *Anguillicola crassus*. Parasitol. 120: 195–202.

Kirk, R.S. 2003. The impact of *Anguillicola crassus* on European eels. Fish. Manag. Ecol. 10: 385–394.

Knights, B. 1987. Agonistic behavior and growth in the European eel, *Anguilla anguilla* L., in relation to warm-water aquaculture. J. Fish Biol. 31: 265–276.

Knights, B. 2003. A review of the possible impacts of long-term oceanic and climate changes and fishing mortality on recruitment of anguillid eels of the Northern Hemisphere. Sci. Total Environ. 310: 237–244.

Knopf, K. and M. Mahnke. 2004. Differences in susceptibility of the European eel (*Anguilla anguilla*) and the Japanese eel (*Anguilla japonica*) to the swim-bladder nematode *Anguillicola crassus*. Parasitol. 129: 491–496.

Kultz, D. 2005. Molecular and evolutionary basis of the cellular stress response. Annu. Rev. Physiol. 67: 225–57.

Kültz, D. 2011. Osmosensing. A.P. Farrell [ed]. *In:* Encyclopedia of Fish Physiology: From Genome to Environment, Vol. 2, Academic Press, San Diego, USA pp. 1373–1380.

Kushnirov, D. and G. Degani. 1991. Growth performance of European eel (*Anguilla anguilla*) under controlled photocycle and shelter availability. Aquacult. Eng. 10: 219–226.

Lambooij, E., J.W. van de Vis, H. Kuhlmann, W. Munkner, J. Oehlenschlager, R.J. Kloosterboer and C. Pieterse. 2002a. A feasible method for humane slaughter of eel (*Anguilla anguilla* L.): electrical stunning in fresh water prior to gutting. Aquacult. Res. 33: 643–652.

Lambooij, E., J.W. van de Vis, R.J. Kloosterboer and C. Pieterse. 2002b. Evaluation of head only and head-to-tail electrical stunning of farmed eel (*Anguilla anguilla* L.) for the development of a humane slaughter method. Aquacult. Res. 33: 323–331.

Leatherland, J.F., M. Li and S. Barkataki. 2010. Stressors, glucocorticoids and ovarian function in teleosts. J. Fish Biol. 76: 86–111.

Lee, R.S., K.L.K. Tamashiro, X. Yang, R.H. Purcell, A. Harvey, V.L. Willour, Y. Huo, M. Rongione, G.S. Wand and J.B. Potash. 2010. Chronic corticosterone exposure increases expression and decreases deoxyribonucleic acid methylation of Fkbp5 in mice. Endocrinol. 151: 4332–4343.

Leloup-Hatey, J. 1974. Influence de l'adaptation a l'eau de mer sur la function interrenalienne de l'Anguille (*Anguilla anguilla* L.). Gen. Comp. Endocrinol. 24: 28–37.

Leitão, A., A. Damasceno-Oliveira, C.M. Pereira, J.C. Coimbra and J.M. Wilson. 2008. Transport stress in glass eels. *In:* J.A. Muñoz-Cueto, J.M. Mancera and G. Martínez-Rodríguez [eds]. Avaços em Endocrinologia Comparativa Vol. 4. Universidad de Cadiz, Cadiz, Spain. pp. 63–68.

Li, Y. and Y. Takei. 2003. Ambient salinity-dependent effects of homologous natriuretic peptides (ANP, VNP, and CNP) on plasma cortisol level in the eel. Gen. Comp. Endocrinol. 130: 317–323.

Lidman, U., G. Dave, M.-L. Johansson-Sjöbeck, Å. Larsson and K. Lewander. 1979. Metabolic effects of cortisol in the European eel, *Anguilla anguilla* (LeSueur). Comp. Biochem. Physiol. 63B: 339–344.

Maetz, J. 1969. Observations on the role of the pituitary-interrenal axis in the ion regulation of the eel and other teleosts. Gen. Comp. Endocrinol. (supplement) 2: 299–316.

Mancera, J.M., J.M. Perezfigares and P. Fernandezllebrez. 1994. Effect of cortisol on brackish water adaptation in the euryhaline gilthead sea bream (*Sparus aurata* I). Comp. Biochem. Physiol. A. 107: 397–402.

Marcy, Jr., B.C. 1973. Vulnerability and survival of young Connecticut River fish entrained at a nuclear power plant. J. Fish. Res. Bd. Can. 30: 1195–1203.

Marshall, W.S. and M. Grosell. 2006. Ion transport, osmoregulation and acid-base balance. pp. 177–230. *In*: D.H. Evans and J.B. Claiborne [eds]. Physiology of Fishes, Vol. 3. CRC Press, Boca Raton, USA.

Mastorakos, G., M. Pavlatou, E. Diamanti-Kandarakis and G.P. Chrousos. 2005. Exercise and the Stress System. Hormones 4: 73-89.

Marsigliante, S., S. Barker, E. Jimenez and C. Storelli. 2000. Glucocorticoid receptors in the euryhaline teleost *Anguilla anguilla*. Mol. Cell. Endocrinol. 162: 193–201.

McCormick, S.D. 2001. Endocrine control of osmoregulation in teleost fish. Am. Zool. 41: 781–794.

McDonald, D.G. and C.M. Wood. 1981. Branchial and renal acid and ion fluxes in the rainbow-trout, *Salmo-gairdneri*, at low environmental pH. J. exp. Biol. 93: 101–118.

McKenzie, D.J., E.W. Taylor, A.Z. Dalla Valle and J.F. Steffensen. 2002. Tolerance of acute hypercapnic acidosis by the European eel (*Anguilla anguilla*). J. Comp. Physiol. B. 172: 339–346.

McKenzie, D.J., M. Piccolella, A.Z. Dalla Valle, E.W. Taylor, C.L. Bolis and J.F. Steffensen. 2003. Tolerance of chronic hypercapnia by the European eel *Anguilla anguilla*. J. exp. Biol. 206: 1717–1726.

Milla, S., N. Wang, S.N.M. Mandiki and P. Kestemont. 2009. Corticosteroids: friends or foes of teleost fish reproduction? Comp. Biochem. Physiol. A. 153: 242–882.

Molnar, K., C. Szekely and F. Baska. 1991. Mass mortality of eel in Lake Balaton due to *Anguillicola crassus* infection. Bull. Eur. Assoc. Fish Pathol. 11: 211–212.

Mommsen, T.P., M.M. Vijayan and T.W. Moon. 1999. Cortisol in teleosts: dynamics, mechanisms of action, and metabolic regulation. Rev. Fish Biol. Fish. 9: 211–268.

Moreira-Silva, J.C., J.C. Coimbra and J.M. Wilson. 2009. Ammonia sensitivity of the glass eel (*Anguilla anguilla* L.): Salinity dependence and the role of branchial $Na^+/K^+(NH_4^+)$-ATPase. Environ. Toxicol. Chem. 28: 141–147.

Neilsen, K.E. and H. Gesser. 1984. Eel and rainbow trout myocardium under anoxia and/or hypercapnic acidosis, with changes in (Ca^{2+}) and (Na^+). Mol. Physiol. 5:189–198.

Nesan, D. and M.M. Vijayan. 2012. Role of Glucocorticoid in Developmental Programming: Evidence from Zebrafish. Gen. Comp. Endocrinol. doi:pii: S0016-6480(12)00411-X. 10.1016/j.ygcen.2012.10.006.

Okamura, A., Y. Yamada, N. Mikawa, N. Horie, T. Utoh, T. Kaneko, S. Tanaka and K. Tsukamoto. 2009.Growth and survival of eel leptocephali (*Anguilla japonica*) in low-salinity water. Aquaculture 296: 367–372.

Oliveira, M., A. Serafim, M.J. Bebianno, M. Pacheco and M.A. Santos. 2008. European eel (*Anguilla anguilla* L.) metallothionein, endocrine, metabolic and genotoxic responses to copper exposure. Ecotoxicol. Environ. Saf. 70: 20–26.

Ottolenghi, F., C. Silvestri, P. Giordano, A. Lovatelli and M.B. New. 2004. Capture-Based Aquaculture: The Fattening of Eels, Groupers, Tunas and Yellowtails. Food and Agriculture Organization of the United Nations, Rome, Italy.

Pacheco, M. and M.A. Santos. 2001a. Biotransformation, endocrine, and genetic responses of *Anguilla anguilla* L. to petroleum distillate products and environmentally contaminated waters. Ecotoxicol. Environ. Saf. 49: 64–75.

Pacheco, M. and M.A. Santos. 2001b. Tissue distribution and temperature-dependence of *Anguilla anguilla* L. EROD activity following exposure to model inducers and relationship with plasma cortisol, lactate and glucose levels. Environ. Int. 26: 149–155.

Palstra, A. and G. Van den Thillart. 2009. Artifical maturation and reproduction of the European eel. *In*: G. Van den Thillart, S. Dufour and J.C. Rankin [eds]. Spawning Migration of the European Eel. Springer, New York, USA. pp. 309–331.

Palstra, A.P., D.F.M. Heppener, V.J.T. Van Ginneken, C. Szekely and G.E.E.J.M. Van den Thillart. 2007. Swimming performance of silver eels is severely impaired by the swim-bladder parasite *Anguillicola crassus*. J. Exp. Mar. Biol. Ecol. 352: 244–256.

Palstra, A.P. and G. Van den Thillart. 2010. Swimming physiology of European silver eels (*Anguilla anguilla* L.): energetic costs and effects on sexual maturation and reproduction. Fish Physiol. Biochem. 36: 297–322.

Palstra, A.P., V.J.T. Van Ginneken, A.J. Murk and G.E.E.J.M. Van den Thillart. 2006. Are dioxin-like contaminants responsible for the eel (*Anguilla anguilla*) drama? Naturwissenschaften 93:145–148.

Pankhurst, N.W. 2011. The endocrinology of stress in fish: An environmental perspective. Gen. Comp. Endocrinol. 170: 265–275.

Perry, S.F. and S.D. Reid. 1992. Relationship between blood O_2 content and catecholamine levels during hypoxia in rainbow trout and American eel. Am. J. Physiol. 263: R240–R249.

Perry, S.F., G.G. Goss and P. Laurent. 1992. The interrelationships between gill chloride cell morphology and ionic uptake in four freshwater teleosts. Can. J. Zool. 70: 1775–1786.

Peters, G. 1982. The Effect of Stress on the Stomach of the European eel, *Anguilla-anguilla* L. J. Fish Biol. 21: 497–512.

Peters, G. and L.Q. Hong. 1985. Gill structure and blood electrolyte levels of European eels under stress. *In*: A.E. Ellis [ed]. Fish and Shellfish Pathology, Academic Press, London, UK. pp. 183–196.

Peters, G., H. Delventhal and H. Klinger. 1980. Physiological and morphological effects of social stress in the eel, (*Anguilla-anguilla* L.). Arch. Fischereiwissenschaft 30: 157–180.

Pickering, A.D. 1993. Growth and stress in fish production. Aquaculture 111: 51–63.

Pickering, A.D. and T.G. Pottinger. 1983. Seasonal and diel changes in plasma cortisol levels of the brown trout, *Salmo trutta* L. Gen. Comp. Endocrinol. 49: 232–239.

Pierron, F., M. Baudrimont, A. Bossy, J.P. Bourdineaud, D. Brethes, P. Elie and J.C. Massabuau. 2007a. Impairment of lipid storage by cadmium in the European eel (*Anguilla anguilla*). Aqua. Toxicol. 81: 304–311.

Pierron, F., M. Beaudrimont, P. Gonzalez, J.P. Bourdinaud, P. Elie and J.C. Massabuau. 2007b. Common pattern of gene expression in response to hypoxia or cadmium in the gills of the European glass eel (*Anguilla anguilla*). Environ. Sci. Technol. 41: 3003–3005.

Pierron, F., M. Baudrimont, S. Dufour, P. Elie, A. Bossy, S. Baloche,N. Mesmer-Dudons, P. Gonzalez, J.P. Bourdineaud and J.C. Massabuau. 2008. How cadmium could compromise the completion of the European eel's reproductive migration? Environ. Sci. Technol. 42: 4607–4612.

Portz, D.E., C.M. Woodley and J.J. Cech Jr. 2006. Stress-associated impacts of short-term holding on fishes. Rev. Fish Biol. Fish. 16: 125–170.

Prunet, P., A. Sturm and S. Milla. 2006. Multiple corticosteroid receptors in fish: from old ideas to new concepts. Gen. Comp. Endocrinol. 147: 17–23.

Rance, T.A., B.I. Baker, G. Webley. 1982. Variations in plasma cortisol concentrations over a 24-hour period in the rainbow trout *Salmo gairdneri*. Gen. Comp. Endocrinol. 48: 269–274.

Randall, D.J. and S.F. Perry. 1992. Catecholamines. *In*: W.S. Hoar, D.J. Randall and A.P. Farrell [eds]. Fish Physiology, vol. 12b. Academic Press, San Diego, USA. pp. 255–300.

Reid, S.G. and S.F. Perry. 1994. Storage and differential release of catecholamines in rainbow trout (*Oncorhynchus mykiss*) and American eel (*Anguilla rostrata*) Physiol. Zool. 67: 216–237.

Renault, S., F. Daverat, F. Pierron, P. Gonzalez, S. Dufour, L. Lanceleur, J. Schafer and M. Baudrimont. 2011. The use of Eugenol and electro-narcosis as anaesthetics: Transcriptional impacts on the European eel (*Anguilla anguilla* L.). Ecotoxicol. Environ. Saf. 74: 1573–1577.

Reynolds, C. 2011. The Effect of Acidification on the Survival of American Eel (*Anguilla rostrata*). M.Sc. Thesis, Dalhousie University, Halifax, Nova Scotia, Canada.

Robertson, L., P. Thomas and C.R. Arnold. 1988. Plasma cortisol and secondary stress responses of cultured red drum (*Sciaenops ocellatus*) to several transportation procedures. Aquaculture 68: 115–130.

Robinet, T. and E. Feunteun. 2002. Sublethal effects of exposure to chemical compounds: a cause for the decline in Atlantic eels? Ecotoxicol. 11: 265–277.

Sadler, K. 1979. Effects of temperature on the growth and survival of the European eel, *Anguilla anguilla* L. J. Fish Biol. 15: 499–507.

Sadler, K. 1981. The toxicity of ammonia to the European eel (*Anguilla anguilla* L.). Aquaculture 26: 173–181.

Santos, M.A. and M. Pacheco. 1996. *Anguilla anguilla* L. stress biomarkers recovery in clean water and secondary-treated pulp mill effluent. Ecotoxicol. Environ. Saf. 35: 96–100.

Sargent, J.R. and A.J. Thomson. 1974. Nature and properties of inducible sodium-plus-potassium ion-dependent adenosine-triphosphatase in gills of eels (*Anguilla-anguilla*) adapted to fresh-water and sea-water Biochem. J. 144: 69–75.

Sasai, S., T. Kaneko, S. Hasegawa and K. Tsukamoto. 1998. Morphological alteration in two types of gill chloride cells in Japanese eels (*Anguilla japonica*) during catadromous migration. Can. J. Zool. 76: 1480–1487.

Schmidt, J. 1923. Breeding places and migration of the eel. Nature 111: 51–54.

Schreck, C.B. 2010. Stress and fish reproduction: the roles of allostasis and hormesis. Gen. Comp. Endocrinol. 165: 549–556.

Sebert, P., A. Pequeux, B. Simon and L. Barthelemy. 1995. Effects of hydrostatic pressure and temperature on the energy metabolism of the Chinese crab (*E. sinensis*) and the yellow eel (*A. anguilla*). Comp. Biochem. Physiol. A. 112: 131–136.

Sebert, P., A. Vettier, A. Amérand and C. Moisan. 2009. High pressure resistance and adaptation of European eels. *In*: G. Van den Thillart, S. Dufour and J.C. Rankin [eds]. Spawning Migration of the European Eel. Springer, New York, USA. pp. 99–127.

Selye, H. 1973. The evolution of the stress concept. Am. Sci. 61: 692–699.

Sheridan, M.A. 1988. Lipid dynamics of fish: aspects of absorption, transportation, deposition and mobilization. Comp. Biochem. Physiol. B. 90: 679–690.

Sheridan, M.A. 1994. Regulation of lipid metabolism in poikilothermic vertebrates. Comp. Biochem. Physiol. B. 107: 495–508.

Singley, J.A. and W. Chavin. 1975. Serum cortisol in normal goldfish (*Carassius auratus* L.). Comp. Biochem. Physiol. A. 50: 77–82.

Small, B.C. 2003. Anesthetic efficacy of metomidate and comparison of plasma cortisol responses to tricaine methanesulfonate, quinaldine and clove oil anesthetized channel catfish *Ictalurus punctatus*. Aquaculture 218: 177–185.

Small, B.C. 2004. Effect of isoeugenol sedation on plasma cortisol, glucose, and lactate dynamics in channel catfish *Ictalurus punctatus* exposed to three stressors. Aquaculture 238: 469–481.

Stefensen, J.F. and J.P. Lomholt. 1990. Accumulation of carbon dioxide in fish farms with recirculating water. *In*: R.C. Ryans [ed]. Fish Physiology, Fish Toxicology and Fisheries Management, Guangzhou, Peoples Republic of China. US-EPA.

Stolte, E.H., B.M.L. Verburg van Kemenade, H.F.J. Savelkoul and G. Flik. 2006. Evolution of glucocorticoid receptors with different glucocorticoid sensitivity. J. Endocrinol. 190: 17–28.

Stone, R. 2003. Freshwater eels are slip-sliding away. Science 302: 221–222.

Strange, R.J. and C.B. Schreck. 1978. Anesthetic and handling stress on survival and cortisol concentration in yearling Chinook salmon (*Oncorhynchus tshawytscha*). J. Fish. Res. Board Can. 35: 345–349.

Sumpter, J.P. 1997. The endocrinology of stress. *In*: G.K. Iwama, A.D. Pickering, J.P. Sumpter and C.B. Schreck [eds]. Fish Stress and Health in Aquaculture. Cambridge University Press, Cambridge, UK. pp. 95–118.

Sures, B., I. Lutz and W. Kloas. 2006. Effects of infection with *Anguillicola crassus* and simultaneous exposure with Cd and 3,3 ',4,4 ',5-pentachlorobiphenyl (PCB 126) on the levels of cortisol and glucose in European eel (*Anguilla anguilla*). Parasitol. 132: 281–288.

Sures, B., K. Knopf and W. Kloas. 2001. Induction of stress by the swimbladder nematode Anguillicola crassus in European eels, *Anguilla anguilla*, after repeated experimental infection. Parasitol. 123: 179–184.

Tanaka, H., H. Kagawa and H. Ohta. 2001. Production of leptocephali of Japanese eel (*Anguilla japonica*) in captivity. Aquaculture 201: 51–60.

Tanaka, H., H. Kagawa, H. Ohta, T. Unuma and K. Nomura. 2003. The first production of glass eel in captivity: fish reproductive physiology facilitates great progress in aquaculture. Fish Physiol. Biochem. 28: 493–497.

Teles, M., M.A. Santos and M. Pacheco. 2004. Responses of European eel (*Anguilla anguilla* L.) in two polluted environments: *in situ* experiments. Ecotoxicol. Environ. Saf. 58: 373–378.

Teles, M., M. Oliveira, M. Pacheco and M.A. Santos. 2005a. Endocrine and metabolic changes in *Anguilla anguilla* L. following exposure to beta-naphthoflavone - a microsomal enzyme inducer. Environ. Int. 31: 99–104.

Teles, M., M. Pacheco and M.A. Santos. 2005b. Physiological and genetic responses of European eel (*Anguilla anguilla* L.) to short-term chromium or copper exposure—Influence of preexposure to a PAH-Like compound. Environ. Toxicol. 20: 92–99.

Teles, M., V.L. Maria, M. Pacheco and M.A. Santos. 2003a. *Anguilla anguilla* L. plasma cortisol, lactate and glucose responses to abietic acid, dehydroabietic acid and retene. Environ. Int. 29: 995–1000.

Teles, M., M. Pacheco and M.A. Santos. 2003b. *Anguilla anguilla* L. liver ethoxyresorufin O-deethylation, glutathione S-tranferase, erythrocytic nuclear abnormalities, and endocrine responses to naphthalene and beta-naphthoflavone. Ecotoxicol. Environ. Saf. 55: 98–107.

Tesch, F.-W. 2003. The Eel. 5 ed. Blackwell Publishing Ltd., Oxford, UK.

Tomasso, J.R., K.B. Davis and B.A. Simco. 1981. Plasma corticosteroid dynamics in channel catfish (*Ictalurus punacatus*) exposed to ammonia and nitrite. Can. J. Fish. Aquat. Sci. 38: 1106–1112.

Tongiorgi, P., L. Tosi and M. Balsamo. 1986. Thermal preferences in upstream migrating glass-eels of *Anguilla anguilla* (L.). J. Fish Biol. 28: 501–510.

Tort, L. 2011. Stress and immune modulation in fish. Dev. Comp. Immunol. 35: 1366–1375.

Tort, L., M. Puigcerver, S. Crespo and F. Padros. 2002. Cortisol and haematological response in sea bream and trout subjected to the anaesthetics clove oil and 2-phenoxyethanol. Aquacult. Res. 33: 907–910.

Tosi, L., L. Sala, C. Sola, A. Spampanato and P. Tongiorgi. 1988. Experimental analysis of the thermal and salinity preferences of glass-eels, *Anguilla anguilla* (L.), before and during the upstream migration. J. Fish Biol. 33: 721–733.

Tse, W.K.F., S.C. Chow, K.P. Lai, D.W.T. Au and C.K.C. Wong. 2011. Modulation of ion transporter expression in gill mitochondrion-rich cells of eels acclimated to low-Na$^+$ or -Cl$^-$ freshwater. J. Exp. Zool. A. 315: 385–93.

Tse, W.K.F. and C.K.C. Wong. 2011. nbce1 and H$^+$-ATPase mRNA expression are stimulated in the mitochondria-rich cells of freshwater acclimating Japanese eels (*Anguilla japonica*). Can. J. Zool. 89: 348–55.

Tse, W.K.F., D.W. Au and C.K. Wong. 2007. Effect of osmotic shrinkage and hormones on the expression of Na$^+$/H$^+$exchanger-1, Na$^+$/K$^+$/2Cl$^-$ cotransporter and Na$^+$/K$^+$-ATPase in gill pavement cells of freshwater adapted Japanese eel, *Anguilla japonica*. J. exp. Biol. 210: 2113–20.

Tse, W.K.F., S.C. Chow and C.K.C. Wong. 2008. The cloning of eel osmotic stress transcription factor and the regulation of its expression in primary gill cell culture. J. exp. Biol. 211: 1964–8.

Tsukamoto, K. 1992. Discovery of spawning area for the Japanese eel. Nature 356: 789–791.

Tsukamoto, K. 2006. Spawning of eels near a seamount. Nature 439: 929.

Tsukamoto, K. and T. Arai. 2001. Facultative catadromy of the eel *Anguilla japonica* between freshwater and seawater habitats. Mar. Ecol. Prog. Ser. 220: 265–276.

Tsukamoto, K., I. Nakai and W.-V. Tesch. 1998. Do all freshwater eels migrate? Nature 396: 376 635–636.

[US EPA] U.S. Environmental Protection Agency. 1999. Update of ambient water quality criteria for ammonia. EPA 822/R-99/014. Technical Report. Office of Water, Washington, DC. 147 pp.

Usui, A. 1999. Eel Culture. Fishing News Books, Oxford, UK.

Van den Thillart, G., A. Palstra and V. van Ginnecken. 2009. Energy requirements of European eel for trans Atlantic spawning migration. *In*: G. Van den Thillart, S. Dufour and J.C. Rankin [eds]. Spawning Migration of the European Eel. Springer, New York, USA. pp. 179–199.

van der Boon, J., G. Van den Thillart and A. Addink. 1991. The effects of cortisol administration on intermediary metabolism in teleost fish, Comp. Biochem. Physiol. A. 100: 47–53.

Van Ginneken, V., A. Palstra, P. Leonards, M. Nieveen, H. van den Berg, G. Flik, T. Spanings, P. Niemantsverdriet, G. Van den Thillart and A. Murk. 2009. PCBs and the energy cost of migration in the European eel (*Anguilla anguilla* L.). Aqua. Toxicol. 92: 213–220.

Van Ginneken, V., B. Ballieux, R. Willemze, K. Coldenhoff, E. Lentjes, E. Antonissen, O. Haenen and G. Van den Thillart. 2005b. Hematology patterns of migrating European eels and the role of EVEX virus. Comp. Biochem. Physiol. C. 140: 97–102.

Van Ginneken, V., E. Antonissen, U.K. Müller, R. Booms, E. Eding, J. Verreth and G. Van den Thillart. 2005a. Eel migration to the Sargasso: remarkably high swimming efficiency and low energy costs. J. Exp. Biol. 208: 1329–1335.

Van Ginneken, V.J.T., M. Onderwater, O.L. Olivar and G.E.E.J.M. Van den Thillart. 2001. Metabolic depression and investigation of glucose/ethanol conversion in the European eel (*Anguilla anguilla* Linnaeus 1758) during anaerobiosis. Thermochim. Acta 373: 23–30.

Van Ginneken, V.J.T., P. Balm, V. Sommandas, M. Onderwater and G. Van den Thillart. 2002. Acute stress syndrome of the yellow European eel (*Anguilla anguilla* Linnaeus) when exposed to a graded swimming-load. Netherlands J. Zool. 52: 29–42.

van Waarde A., G. Van den Thillart and F. Kesbeke. 1983. Anaerobic energy metabolism of the European eel, *Anguilla anguilla* L. J. Comp. Physiol. B. 149: 469–475.

van Weerd, J.H. and J. Komen. 1998. The effects of chronic stress on growth in fish: a critical appraisal. Comp. Biochem. Physiol. A. 120: 107–112.

Vandergoot, C.S., K.J. Murchie, S.J. Cooke and J.M. Dettmers. 2011. Evaluation of two forms of electroanesthesia and carbon dioxide for short-term anesthesia in walleye. North Am. J. Fish. Manag. 31: 914–922.

Verheijen, F.J. and W.F.G. Flight. 1997. Decapitation and brining: experimental tests show that after these commercial methods for slaughtering of eel, *Anguilla anguilla* L., death is not instantaneous. Aquacult. Res. 28: 361–366.

Walsh, P.J., G.D. Foster and T.W. Moon. 1983. The effects of temperature on metabolism of the american eel *Anguilla rostrata* (LeSueur): Compensation in the summer and torpor in the winter. Physiol. Zool. 56: 532–540.

Wendelaar Bonga, S.E. 1997. The stress response in fish. Physiol. Rev. 77: 591–625.

Weyts, F., N. Cohen, G. Flik and B. Verburg-van Kemenade. 1999. Interactions between the immune system and the hypothalamo–pituitary–interrenal axis in fish. Fish Shellfish Immunol. 9: 1–20.

Wickins, J.F. 1985. Growth variability in individually confined elvers, *Anguilla anguilla* (L.). J. Fish Biol. 27: 469–478.

Willemse, J.J., L. Markussilvis and G.H. Ketting. 1984. Morphological effects of stress in cultured elvers, *Anguilla-anguilla* (L.). Aquaculture 36: 193–201.

Wilson, J.M., J.C. Antunes, P.D. Bouça and J. Coimbra. 2004. Osmoregulatory plasticity of the glass eel of *Anguilla anguilla*: freshwater entry and changes in branchial ion-transport protein expression. Can. J. Fish. Aquat. Sci. 61: 432–442.

Wilson, J.M., M.M. Vijayan , C. Kennedy, G.K. Iwama and T.W. Moon. 1998. β-naphthoflavone abolishes the interrenal sensitivity to ACTH in rainbow trout. J. Endocrinol. 157: 63–70.

Wilson, J.M., A. Leitão, A.F. Gonçalves, C. Ferreira, P. Reis-Santos, A.-V. Fonseca, J. Moreira da Silva, J.C. Antunes, C. Pereira-Wilson and J. Coimbra. 2007a. Modulation of branchial ion transport protein expression by salinity in glass eel (*Anguilla anguilla* L.). Mar. Biol. 151: 1633–1645.

Wilson, J.M., P. Reis-Santos, A.-V. Fonseca, J.C. Antunes, P.D. Bouça and J. Coimbra. 2007b. Seasonal changes in ionoregulatory variables of the glass eel *Anguilla anguilla* following estuarine entry: comparison with resident elvers. J. Fish Biol. 70: 1239–1253.

Wirth, T. and L. Bernatchez. 2001. Genetic evidence against panmixia in the European eel. Nature 409: 1037–1040.

Wirth, T. and L. Bernatchez. 2003. Decline of North Atlantic eels: a fatal synergy? Proc. R. Soc. Lond. B. 270: 681–688.

Yamagata, Y. and M. Niwa. 1982. Acute and chronic toxicity of ammonia to eel *Anguilla japonica*. Bull. Jpn. Soc. Sci. Fish. 48: 171–176.

Yamashita, M., T. Yabu and N. Ojima. 2010. Stress Protein HSP70 in Fish. Aqua-Bio. Sci. Monogr. 3: 111–141.

Index

Color Plate Section

Chapter 2

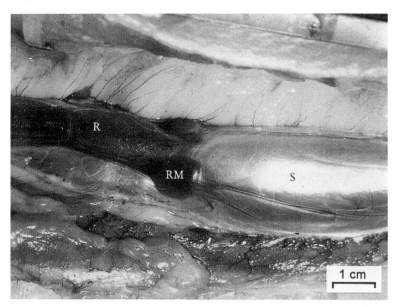

Figure 2. The swimbladder of a yellow stage European eel with a secretory section (S) and the pneumatic duct, which serves as a resorbing section (R) of the swimbladder. Two *retia mirabilia* (RM) are located next to the sphincter muscle separating the secretory and the resorbing section. The ductus pneumaticus is functionally closed, so that the eel cannot gulp air at the water surface to fill the swimbladder. Taken from Pelster (Pelster 2011) with permission.

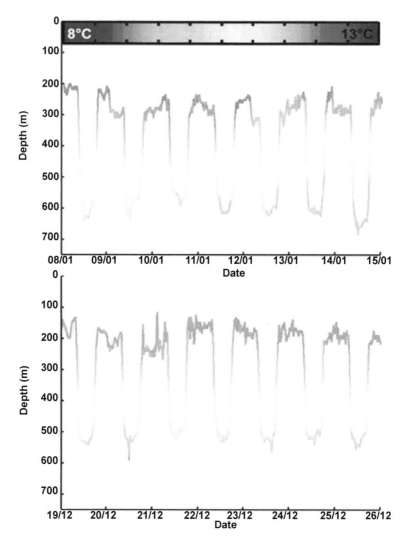

Figure 7. Oceanic daily vertical migration of two eels recorded by pop-up satellite archival transmitters attached to silver eels. Eels were released near the Irish west coast and followed for several days during their spawning migration to the Sargasso Sea. Depth values are colored to indicate the temperature encountered at the respective depth. Taken from Aarestrup et al. (Aarestrup et al. 2009), with permission.

Chapter 3

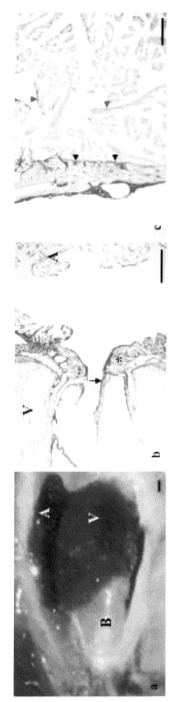

Figure 1. (a) Ventral view of *Anguilla anguilla* heart (A=atrium, B=bulbus, V=ventricle). Scale bars: 0,1 cm. (b) and (c): Sirius red. (b) shows the AV region; note the continuity between the AV and the atrial muscle; AV valves (arrows) anchor to a ring of compact, vascularized myocardium (asterisks). In (c) collagen at the *compacta/spongiosa* boundary (black arrowheads) and ventricular trabeculae are evident; note the collagen bundles localized at the subendocardial level (blue arrowheads). Scale bars: 100 μm. (Unpublished data).

Chapter 5

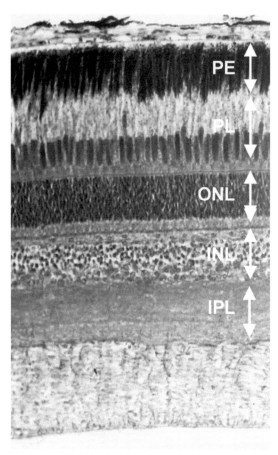

Figure 1. Transverse section of a typical teleost retina. PE, pigmented epithelium; PL, photoreceptor layer; ONL, outer nuclear layer, INL, inner nuclear layer; IPL, inner plexiform layer.

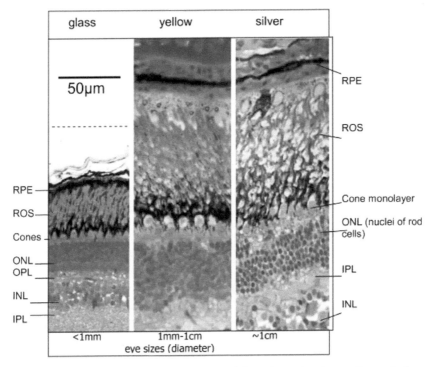

Figure 5. Changes in the thickness of the layers of the retina at the glass, yellow and silver stages of the life cycle. This is especially pronounced in the ONL of the glass eel compared to the later stages. The size of the eye also changes between the different stages, increasing tenfold in diameter. RPE, retinal pigment epithelium; ROS, rod outer segments; ONL, outer nuclear layer; INL, inner nuclear layer; OPL, outer plexiform layer; IPL, inner plexiform layer. From Cottrill et al. (2009).

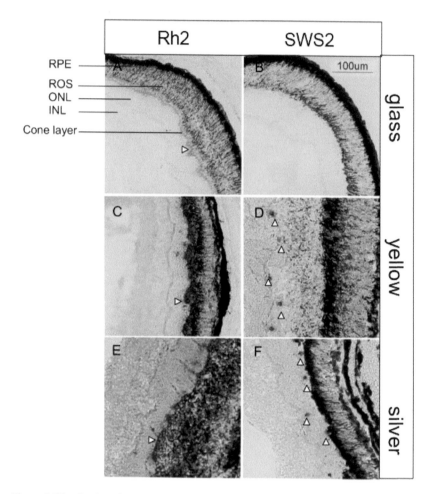

Figure 6. Distribution of opsin-expressing cone cells in eel retina. *In situ* hybridisation of *Rh2* or *SWS2* opsin probes of transverse sections of the eel retina from different developmental stages. Cone cell bodies in the eel are arranged in a single layer between the inner nuclear layer (INL) and the rod outer segments (ROS). Glass eel: (A) *Rh2* opsin expression (horizontal arrowhead), (B) no *SWS2* opsin expression. Yellow eel: (C) *Rh2* opsin (horizontal arrowhead), (D) occasional cone cells expressing *SWS2* opsin (vertical arrowheads). Silver eel: (E) monolayer of cone cells expressing *Rh2* opsin (horizontal arrowhead), (F) occasional cone cells expressing *SWS2* opsin (vertical arrowheads). RPE, retinal pigment epithelium; ROS, rod outer segments; ONL, outer nuclear layer; INL, inner nuclear layer. From Cottrill et al. (2009).

T - #0355 - 071024 - C380 - 234/156/17 - PB - 9780367379636 - Gloss Lamination